Fire Properties of Polymer Composite Materials

SOLID MECHANICS AND ITS APPLICATIONS
Volume 143

*Series Editor:*   G.M.L. GLADWELL
*Department of Civil Engineering*
*University of Waterloo*
*Waterloo, Ontario, Canada N2L 3GI*


*Aims and Scope of the Series*

The fundamental questions arising in mechanics are: *Why?, How?,* and *How much?*
The aim of this series is to provide lucid accounts written by authoritative researchers
giving vision and insight in answering these questions on the subject of mechanics as it
relates to solids.

The scope of the series covers the entire spectrum of solid mechanics. Thus it includes
the foundation of mechanics; variational formulations; computational mechanics;
statics, kinematics and dynamics of rigid and elastic bodies: vibrations of solids and
structures; dynamical systems and chaos; the theories of elasticity, plasticity and
viscoelasticity; composite materials; rods, beams, shells and membranes; structural
control and stability; soils, rocks and geomechanics; fracture; tribology; experimental
mechanics; biomechanics and machine design.

The median level of presentation is the first year graduate student. Some texts are
monographs defining the current state of the field; others are accessible to final year
undergraduates; but essentially the emphasis is on readability and clarity.

*For a list of related mechanics titles, see final pages.*

# Fire Properties of Polymer Composite Materials

by

A.P. MOURITZ
*RMIT University*
and
*CRC for Advanced Composite Structures*
*Melbourne, Victoria,*
*Australia*

and

A.G. GIBSON
*University of Newcastle-upon-Tyne*
*Centre for Composite Materials Engineering*
*England, UK*

 Springer

A C.I.P. Catalogue record for this book is available from the Library of Congress.

ISBN-10  1-4020-5355-X (HB)
ISBN-13  978-1-4020-5355-9 (HB)
ISBN-10  1-4020-5356-8 (e-book)
ISBN-13  978-1-4020-5356-6 (e-book)

---

Published by Springer,
P.O. Box 17, 3300 AA Dordrecht, The Netherlands.

*www.springer.com*

*Printed on acid-free paper*

# Table of Contents

# Preface

This book is the first to deal comprehensively with the important topic of the fire behaviour of polymer composite materials. Composites are used in a diverse range of applications, including land and marine transport, aerospace, the chemical industry, and most branches of civil engineering infrastructure. It is our belief that fire behaviour is the single most important factor limiting the wider use of composites in many of these areas. Our aim in producing this volume is therefore to stimulate the work that is needed to overcome this fundamental problem. The first step in such a journey is, of course, to summarise what is presently known, which is what we have attempted here. This book aims to be an authoritative reference source.

The book covers all of the key issues on the behaviour of polymer composites in fire. This includes a description of the thermal degradation and combustion mechanisms of composites, including the thermal decomposition reactions, reaction rates, flammable volatiles and toxic gases of organic polymers and fibres. The fire reaction properties that define the flammability and fire hazard of polymer composites are described, including time-to-ignition, heat release rate, flame spread, smoke and gaseous combustion products. The fire resistive properties of composites are also described, including the key properties of burn-through rate and mechanical integrity during and after fire. General principles and quantitative models are presented for predicting the development and spread of fire together with models for calculating the decomposition and thermal response of composites to fire. Low flammability and fire resistive materials for composites are described, including flame retardant organic and inorganic polymers, protective coatings, and polymer nanocomposites. Also covered are fire safety regulations, fire test methods, and the health hazards of burning composites. The book also identifies gaps and deficiencies in our current understanding of the fire performance of composites in order to determine the main issues that require further investigation.

Chapter 1 gives an introduction to the fire hazard of polymer composites. The chapter provides a general overview of fires, flames and the combustion of polymer composites. In addition, definitions to the key fire reaction and fire resistive properties that define the fire hazard of composites are given. Numerous examples and case studies of fire in aircraft and ships containing composites are provided to illustrate their potential hazard, and to demonstrate the need to develop more flame resistant materials.

Chapter 2 outlines the thermal degradation mechanisms that govern the fire reaction and fire resistive properties of polymer composites. The combustion process of composites is described, with special consideration given to those factors that control the process: heat flux, pyrolysis of the polymer matrix and organic fibres, evolution of flammable gases, and char formation. The pyrolysis reactions and combustion gases of organic polymers and fibres that are commonly used in composites are described, including polyesters, vinyl esters, epoxies, phenolics and several thermoplastics used for the matrix phase and organic fibres such as aramid and polyethylene for the reinforcement. The chapter also describes the microstructural damage suffered by laminates and sandwich composites in fire, such as charring, skin-from-core debonding, delamination and matrix cracking.

The fire reaction and fire resistive properties of polymer composites are outlined in Chapter 3. The fire reaction properties that influence the initiation, growth and spread of fire and determine the survival of humans exposed to fire are described. These properties are ignition time, heat release rate, limiting oxygen index, flame spread rate, smoke and gas emission. Many examples are provided of the fire reaction properties of thermoset laminates, thermoplastic laminates and sandwich composite materials. The fire resistive properties of composites are also discussed, with special attention given to burn-through resistance and heat transmission because of their influence on fire containment and flame spread. Later chapters are devoted to comprehensive descriptions of other fire resistive properties, namely degradation of mechanical properties in fire and post-fire structural properties.

The growth and spread of fire is described in Chapter 4 by Dr Brian Lattimer (Hughes Associates, Inc) and Ted Matthews (Materials Sciences Corporation). An overview of analytical models, finite element tools, computational fluid dynamic codes and other methods to theoretically analyse the development and spread of fire is described, with emphasis given to flame spread within enclosed spaces. Models are also presented for calculating several key fire reaction properties of polymers and polymer composites, including ignition time, heat release rate and flame spread. Included is a critique of the strengths and limitations of the various models for analysing fire behaviour. The chapter also introduces the many approaches towards modelling the structural response of composite to fire, which is a topic expanded upon in Chapter 6.

Chapter 5 outlines models for predicting the thermal and physical response of composites to fire. Models are described that range in complexity from simple analysis that only consider heat transmission in a thermally-stable material to complex models that consider a number of thermo-physical processes in a decomposing laminate, including heat conduction; radiation losses; internal heat evolution/absorption due to decomposition of the organic matrix; internal pressure and flow of gaseous reaction products; thermal swelling and strains. The reliability of the models to accurately predict the fire behaviour of composites is examined, and numerous examples of the application of the models to composites are given.

Chapters 6 and 7 describe the deterioration to the mechanical properties and structural performance of composites during and after fire, respectively. Models and experimental information on the degradation to the load-bearing properties of laminates and sandwich composites at elevated temperature, in fire, and after the fire has been extinguished are presented. The thermal damage responsible for the reduction to the mechanical properties is discussed as well as methods for improving the structural performance of composites in fire.

An overview of methods for reducing the fire risk of composites is given in Chapter 8. The efficacy of various types of flame retardant polymer systems are described, including polymers with flame retardant fillers, halogenated polymers, char forming polymers, and polymers that are chemically and structurally modified to increase flammability resistance. The fire behaviour properties of highly flame resistant composites are discussed, such as geopolymer laminates and other inorganic matrix materials. The fire protection provided by thermal barrier materials is also described, including flame retardant polymer, thermal barrier and intumescent coatings.

The fire properties of polymer nanocomposite materials are reviewed in Chapter 9 by Dr Dongyan Wang and Professor Charles Wilkie of Marquette University. The chemistry and molecular structure of nanocomposites are introduced, with emphasis given to polymers containing nano-clays. The techniques used to analyse the nanostructure of these polymers are briefly out-lined. The fire reaction properties of many varieties of thermoplastic and thermoset polymers loaded with nano-clays and other nanoparticles that impart improved fire resistance are described, and their flame retardant mechanisms are out-lined.

Fire safety standards are enforced on composites in many applications because of their potential fire risk and smoke toxicity. Chapter 10 provides a general review of the fire standards and regulations for composites when used in aircraft, ships, submarines, and other applications. An assessment of the scope and shortcomings of the different fire standards is given, and the impact of the regulations on the application of composites is discussed.

The experimental apparatus and test methods used to characterise the fire reaction and fire resistance properties of composites are reviewed in Chapter 11. A variety of methods that range in size from small bench-top techniques to full-scale fire tests are described. The small-scale fire tests include the cone calorimeter, Ohio State University calorimeter and limiting oxygen index test, and these are used for measuring properties such as flammability, heat release and smoke. Medium-scale and large-scale tests for assessing the fire behaviour of composite components, assemblies and structures are also reviewed. These include the intermediate-scale cone calorimeter, DTRC burn through test, furnace fire test, fire-under-load test, single burning item test, corner burn test, and room fire test. An assessment of the capabilities and limitations of the test methods is given, and the need for the further development of uniform, standardised fire tests is discussed.

Chapter 12 gives an overview of the health hazards to humans when exposed to toxic smoke from burning composite materials. The health effects of inhaling fibres, smoke and toxic fumes are described, and safe exposure limits to the main decomposition products are given. Empirical models and experimental apparatus for determining the toxic potency of the gases released from decomposing composites are also reviewed. A brief description of protective clothing and breathing apparatus that should be used when exposed a burning composite material is given.

The diverse scope, comprehensive nature and in-depth detail of this book ensures that it will be of great interest to professionals in the polymer and composites industries, fire safety engineers, researchers in the thermal and fire properties of polymeric materials, and the users of composites in the aerospace, marine, civil infrastructure, petrochemical, processing and automotive industries.

This book would not have been possible without the effort of many people. The authors are grateful to Dr Pat Potter of the United States Office of Naval Research for the provision of a grant (USONR Award No. N00014021022). AM also acknowledges the support of the Cooperative Research Centre for Advanced Composite Structures Ltd.

The authors thank Dr Brian Lattimer (Hughes Associates Inc.) and Ted Campbell (Materials Sciences Corporation) for writing the chapter entitled 'Fire Modelling of Composites' and Dr Dongyan Wang and Professor Charles A. Wilkie (Marquette University) for writing the chapter 'Fire Properties of Polymer Nanocomposites'. Dr Craig Gardiner was a co-author to the chapter 'Post-fire Properties of Composite Materials'. The many discussions with a large number of colleagues working on the fire properties of polymer composites have been invaluable, and special mention of those working on the ONR-funded program on the structural modelling of composites in fire led by Dr Luise Couchman: Zenka Mathys, Dr Stefanie Feih, Dr Yongshu Wu, Dr Jim Lua, Professor George A. Kardomateas, Professor John W. Holmes, Professor Victor Birman, Professor Jack Lesko, Professor Scott Case and Professor Paul DesJardin. The authors also thank the many publishers and colleagues who kindly allowed the reproduction of figures and tables used in this book. The advice and patience of Nathalie Jacombs and Anneke Pot from Springer is gratefully acknowledged, and Dr Henry Li for formatting the book is greatly appreciated. Lastly, we would like to acknowledge the support of our families; only they truly know the amount of time, effort and research given to the writing of this book.

Adrian P. Mouritz                                              Arthur G (Geoff) Gibson
*Melbourne, Australia*                                  *Newcastle-upon-Tyne, England*

# Chapter 1

# Introduction

## 1.1　Background

A sustained upsurge in the use of fibre reinforced polymer (FRP) composite materials has occurred over the last forty years. The use of polymer composites has grown at a phenomenal rate since the 1960s, and these materials now have an impressive and diverse range of applications in aircraft, spacecraft, boats, ships, automobiles, civil infrastructure, sporting goods and consumer products. The use of composites will continue to grow in coming years with emerging applications in large bridge structures, offshore platforms, engine machinery, computer hardware and biomedical devices. Figure 1.1 shows the growth in the use of composite materials by various industry sectors in the United States since 1960. Over this period consumption has increased about 30 times, and the growth rate is expected to continue. The greatest increases are occurring in the transport and construction markets, although the use of composites is also substantial in the corrosion protection (eg. piping), marine, and electrical/electronic markets as shown in Fig. 1.2.

Growth in the use of composites has reached a level where they are now challenging the use of traditional materials - most notably steels and aluminium alloys - in many markets; particularly the aircraft, boat-building and chemical processing industries. While composites will never replace steel as the most used engineering material, the value of the composite market is expected to remain strong. Sales of composites in the United States exceeded 1.5 million tons in 2001, and sales are expected to increase as these materials penetrate deeper into established markets such as construction and aerospace and infiltrate emerging markets such as rail. The drive to reduce the cost and increase the quality and structural performance of composites together with the emerging developments in polymer nanocomposites will be key factors supporting the increased use of FRP materials.

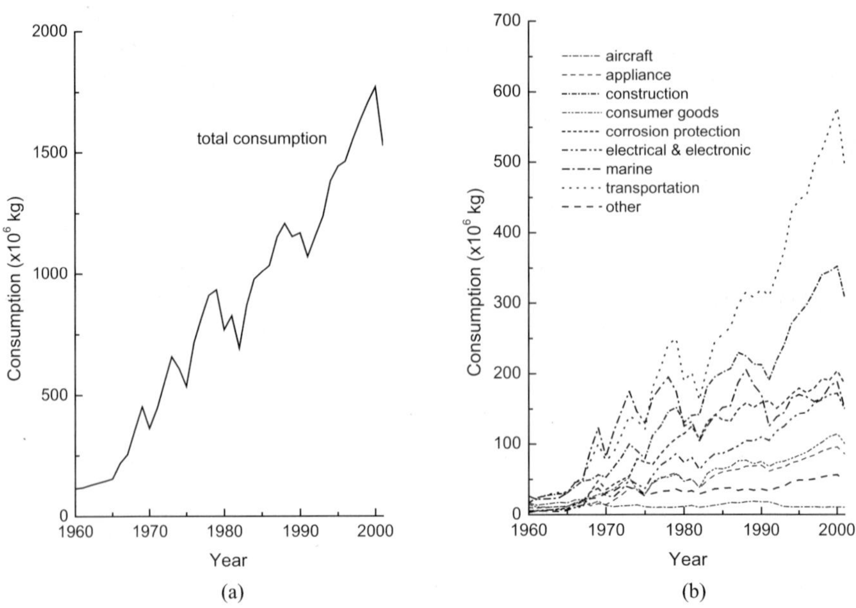



*Figure 1.1.  Growth in the (a) total use and (b) use by individual market sectors in the United States.  (Source: Composite Fabrication Association).*

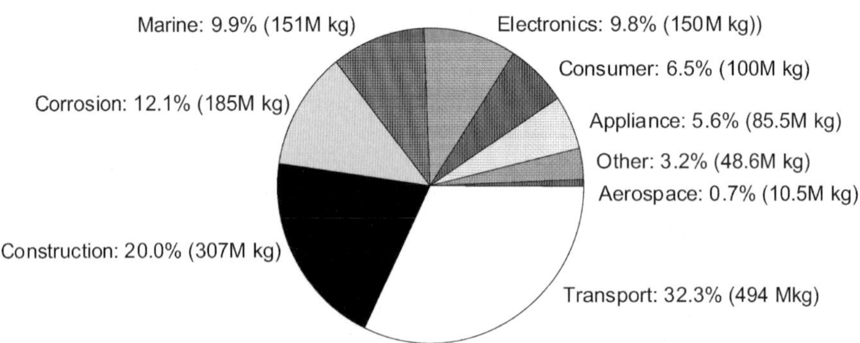



*Figure 1.2.  Use of composite materials by different market segments in the United States in 2001. (Source: Composites Fabrication Association).*

The use of composites in a wide variety of applications is due to their many outstanding physical, thermal, chemical and mechanical properties. Key advantages of composites over many metal alloys include low density, high specific stiffness and specific strength, good fatigue endurance, excellent corrosion resistance, outstanding thermal insulation and low thermal expansion. However, there are several disadvantages with composites that have impacted on their growth in some markets. Composites are plagued by problems such as low through-thickness mechanical properties, poor impact damage tolerance, and anisotropic properties.

A major disadvantage of many composite materials is poor performance in fire. When composites are exposed to high temperatures (typically above 300-400°C) the organic matrix decomposes with the release of heat, smoke, soot and toxic volatiles. Organic fibres used to reinforce composites, such as aramid and polyethylene, also decompose and contribute to the generation of heat, smoke and fumes. Composites also soften, creep and distort when heated to moderate temperature (>100-200°C), that can result in buckling and failure of load-bearing composite structures. The heat, smoke and gases released by a burning composite and the degradation in structural integrity can make fire-fighting extremely hazardous and increase the likelihood of serious injury and death. The susceptibility of composites to fire has been the key issue in curtailing their use in many infrastructure and public transportation applications.

Although many polymer composites are flammable, their resistance to pyrolysis can be improved. Furthermore, these materials possess some potentially useful properties in fire that are not inherent with metals. Composites have excellent thermal insulation properties and slow burn-through. The rate of heat conduction through composites is much slower than metals, and this is a significant benefit is slowing the spread of fire from room-to-room. Composites can provide an effective protective barrier against flame, heat, smoke and toxic fumes. For these reasons, composites are the material of choice in heat shields for re-entry spacecraft and rocket nozzle liners. Also, composites are being developed for heat protection in high fire risk applications such as offshore oil platforms.


## 1.2     Fire Reaction and Fire Resistance of Composites

The behaviour of polymer composite materials in fire has been a major concern for over thirty years, and much effort has been devoted to assessing and reducing their fire hazard. The fire properties of a diverse variety of composite materials have been analysed in terms of properties that provide a measure of their flammability, lethality or fire hazard. These properties include ignition time, heat release rate, heat of combustion, smoke and toxic potency of gas products. Degradation to the mechanical performance of structural composite materials in fire has also been a topic of intensive testing, analysis and development. Despite the knowledge gained since the 1970s into the fire behaviour of composites, significant gaps remain in our understanding of their fire properties. This book outlines the existing knowledge in the field, and identifies the critical gaps and deficiencies.

The fire hazard of composites is often defined by their *fire reaction* and *fire resistant* properties. *Fire reaction* is used to describe the flammability and combustion properties of a material that affect the early stages of fire, generally from ignition to flashover. Fire reaction also describes the smoke toxicity of a combustible material. Important fire reaction properties that affect fire growth are the heat release rate, time-to-ignition, flame spread rate and oxygen index. *Heat release rate* is considered the single most important fire reaction property because it is the best indicator of the fire hazard of a combustible material [1-3]. In fact, the importance of heat release rate as a measure of fire hazard out-weighs that of the other fire reaction properties including ignitability, flame spread and smoke toxicity. The heat release rate is a quantitative measure of the amount of thermal energy released by a material per unit area when exposed to a fire radiating a constant heat flux (or temperature). The unit for heat release rate is $kW/m^2$. The heat release rate value of a composite material is determined by the thermal energy liberated in several thermo-chemical decomposition processes, with the most important being exothermic combustion at the composite/flame boundary of flammable gas products released by the decomposing polymer matrix and (if present) organic fibres. The heat release rate value of a composite is not constant, but varies with exposure time to the fire as the material is progressively consumed and burnt-through. Therefore, the heat release rate is often described by two parameters: average heat release rate and peak heat release rate. The average heat release rate is the averaged value over a certain period of time (usually three or five minutes). The peak heat release rate is the maximum amount of heat liberated by a material during the combustion process, and it often occurs over a very short period of time (less than a few seconds). The peak heat release rate is considered a critical property controlling the maximum temperature and flame spread rate. Composite materials that have low values for peak and average release rates are often suitable in high fire risk applications to minimise the growth and spread of fire.

*Time-to-ignition* is the period that a combustible material can withstand exposure to a constant radiant heat flux before igniting and undergoing sustained flaming combustion. The ignition time can be used as a crude or approximate measure of the flammability resistance of a material. Obviously, it is desirable to use materials with long ignition times in high fire risk applications. *Flame spread rate*, as the term implies, describes the speed at which the flame front will travel over the surface of a combustible material. The flame spread rate is an experimentally measured value, and various experimental techniques with important differences in test configuration are used. Some tests are used to measure the rate of flame spread in a downward direction while other techniques measure it in a vertical or inclined direction. Consequently, the value for flame spread rate is test-dependent. *Oxygen index* is defined as the minimum oxygen content in the fire environment required to sustain flaming combustion of a material. Materials with high oxygen index values should be used in high fire risk applications, particularly for internal structures and components, because they offer the potential to self-extinguish when the fire becomes deprived of oxygen.

Two other important fire reaction properties are *smoke density* and *gas toxicity* because they have a major impact on the ability of humans to survive a fire. Most fatalities are not caused by heat and flame, but are due to thick smoke causing confusion and disorientation for people attempting to escape which increases their exposure time to toxic fumes that may lead to incapacitation and death. Smoke density is defined and measured in several ways, but basically it is the concentration of smoke particles (eg. soot, water vapour) within the plume of a fire. Gas toxicity is a generic term that describes the concentration and lethality of gas products within the smoke plume. A large number of toxic gases can be released from composites materials, including asphyxiants (eg. carbon monoxide), irritants (eg. hydrogen chloride) and carcinogens (eg. dioxins).

While many fire reaction properties are important in the development of fire up to the point of flashover, the fire resistant properties are critical when the fire has fully developed. *Fire resistance* defines the ability of a material or structure to impede the spread of fire and retain mechanical integrity. In other words, fire resistance describes the ability of a construction to prevent a fire from spreading from one room to neighbouring rooms. Fire resistance also describes the ability of a construction to retain structural integrity (ie. shape, load-bearing properties) in a fire. The main fire resistant properties are heat insulation, burn-through resistance, and structural integrity. *Heat insulation* is simply the resistive property that describes the rate of heat conduction through a material when exposed to fire. Obviously, materials that are good heat insulators are best suited for slowing the spread of fire from room-to-room, and this is one of the attributes of composites compared to metals. *Burn-through resistance* is the time taken for a flame to penetrate a material and emerge from the opposing side. Composites generally have better burn-through resistance than metals that melt at temperatures below the flame temperature, such as aluminium alloys. *Mechanical integrity* is another important fire resistant property, and this defines the ability of a material or structure to retain mechanical properties such as stiffness, creep resistance and strength when exposed to fire and after the fire has been extinguished.

## 1.3  Composites and Fire

It is not our intention to give a detailed account of fire and flame. The combustion mechanisms, thermochemistry, thermodynamics and airflow/smoke dynamics of fire is extraordinary complex and beyond the scope of this book. The topic of fire and flame is reviewed in several excellent books and articles (eg. [4-7]). Instead, a short and simple account of fire is given so the reader has a better understanding of the interactions that occur between a flame and composite material.

Fire is a complicated phenomenon that can develop in stages of increasing temperature and size before decaying. The fire event becomes more complex when a polymer composite material is involved because it can control the temperature, size and spread of the flame. A turbulent flame consists of three zones, and from the base to top can be divided into the solid flame region, intermittent flame region and thermal plume (Fig. 1.3). The solid flame region near the plume base is where the majority of the flammable vapours undergo exothermic chain-branching reactions that generate most of the heat. The temperature within this zone is reasonably constant at about 830-900°C for most types of solid fuels [8], although the maximum temperature in hydrocarbon pool fires and natural gas flames can reach 1150-1250°C [9]. Above the solid flame is the intermittent flame region, and the temperature drops continuously the higher up this zone. The average temperature at the visible tips of a flame is about 400°C [10], although it can vary over a wide range from 300 to 600°C [8]. The boundary between the solid flame and intermittent regions is not well defined in a turbulent flame, and some overlap or changes in the boundary location do occur. Above the flame tips is the thermal plume region, where no flames are visible and the temperature drops with height. The thermal plume consists of hot gases, vapours and soot particles that can carried upwards by convective heat.

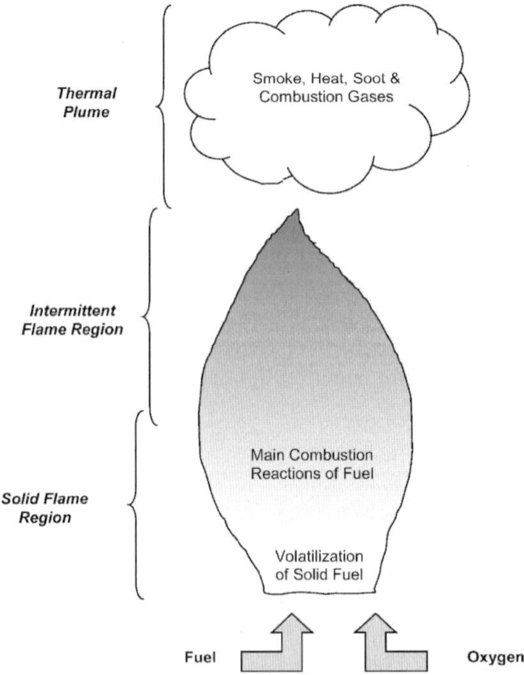


*Figure 1.3. Schematic showing the different zones within a turbulent plume.*

The initiation and growth of fire is determined by a multitude of factors including the type (caloric value) of fuel, fuel load, fuel size (area), oxygen content in the flame, wind speed, and whether the fire is within an open or enclosed space. In the case of polymer composites exposed to fire, the material itself can be a rich source of fuel that causes the temperature to rise and the flame to spread.

A serious concern with using composites in enclosed spaces, such as an aircraft cabin, ship compartment or rail carriage, is that the heat, smoke and toxic gases are trapped which seriously increases the fire hazard. It is therefore useful to examine the development of fire within a closed compartment. Figure 1.4 shows how the temperature can vary with time for a closed compartment fire. A fire can undergo several stages of growth, and in order these are:

- *Ignition*. This is the point when the fuel source ignites and undergoes sustained flaming combustion.
- *Growth*. The initial growth of a fire is dependent mainly on the fuel itself, with little or no influence from the combustible materials within the compartment. The fire will grow and the compartment temperature will continue to rise if sufficient fuel and oxygen are available. It is often in this stage that composite materials exposed to the flame will ignite when the temperature exceeds 350-500°C.
- *Flashover*. This occurs when the fire is fully developed and all combustible items in the compartment (including any composite materials) are involved in the fire. Flashover usually occurs when the average upper gas temperature in the room exceeds about 600°C.
- *Fully developed fire*. This stage occurs when the heat release rate and temperature of a fire are at their greatest. The peak temperature of a typical post-flashover room fire is 900-1000°C, although it can reach as high as 1200°C.
- *Decay*. The final decay stage occurs as the fuel and combustible materials become consumed, causing the compartment temperature to fall. Obviously, decay can also be caused by active fire suppression systems, such as sprinklers.

Polymer composite materials can provide a rich supply of hydrocarbon fuel that drives the growth of a fire, even after the fuel original source (eg. oil pool, gas jet) is depleted or extinguished. When a composite is heated to a sufficiently high temperature the polymer matrix and (if present) organic fibres will thermally decompose. Most polymer matrices and organic fibres decompose over the temperature range of about 350 to 600°C with the production of flammable gases. Decomposition occurs by a series of reactions that breaks down the polymer chains into low molecular weight volatiles that diffuse into the flame. Depending on the chemical composition and molecular structure of the polymer, the thermal degradation reactions may proceed by various paths. The majority of organic resins and fibres used in composites degrade thermally by a random chain scission process. This basically involves the break-down of the long organic chains at the lowest-energy bond sites into small fragments. Polymers can also decompose by other processes, including depolymerisation (that involves the break-down of the chain into monomers) and chain-end initiated scission (that involves the process starting from the chain ends and propagating along the chain length until it is completely degraded).  A full description of the thermal decomposition of organic polymers is provided in Chapter 2.

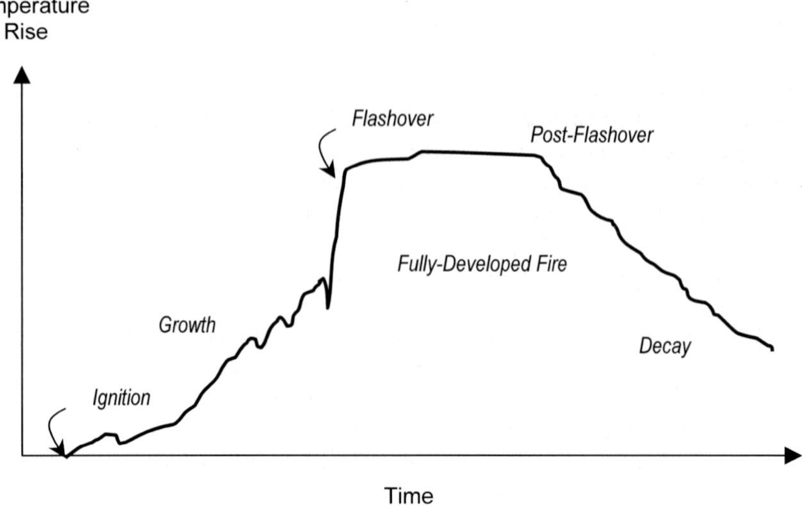


*Figure 1.4. Growth stages of a compartment fire.*

Regardless of the decomposition process, when the vapour pressure and molecular weight of the fragments from the polymer chain become sufficiently small they diffuse into the flame.  The majority of the evolved gases are hydrocarbon volatiles that are highly flammable and therefore become fuel to sustain the fire.  Depending on the chemical nature of the polymer, between about 30% and 100% of the organic matrix and fibres is volatised, and therefore large composite components can provide a plentiful supply of flammable gases.  Combustion of the gases occurs in the solid and (to a lesser extent) intermittent zones of the flame with the formation of highly active H· radicals.  This radical combines with oxygen in the flame to produce hydroxyl radicals (OH):

$$H^\cdot + O_2 \rightarrow OH^\cdot + O^\cdot \qquad (1.1)$$

$$O^\cdot + H_2 \rightarrow OH^\cdot + H^\cdot \qquad (1.2)$$

The main exothermic reaction that generates most of the heat in a flame is:

$$OH^\cdot + CO \rightarrow CO_2 + H^\cdot \qquad (1.3)$$

The H· radicals produced in reactions (1.2) and (1.3) feedback into reaction (1.1), and thereby the combustion process becomes a self-sustaining process when sufficient oxygen is available.  This is known as the combustion cycle of organic polymers, which is illustrated in Fig. 1.5.  The cycle stops only when the fuel source has been exhausted, which is usually when the organic components in a composite have been completely degraded.

It is common practice by fire scientists to quantify the intensity of a fire by the radiant heat flux rather than flame temperature.  Figure 1.6 shows the relationship between heat flux and temperature at the hot surface of a polymeric material.  There is an approximate relationship between fire type and heat flux, and examples are:

- Small smouldering fire: 2-10 $kW/m^2$
- Trash can fire: 10-50 $kW/m^2$
- Room fire: 50-100 $kW/m^2$
- Post-flashover fire: >100 $kW/m^2$
- Gas-jet fire: 150-300 $kW/m^2$.

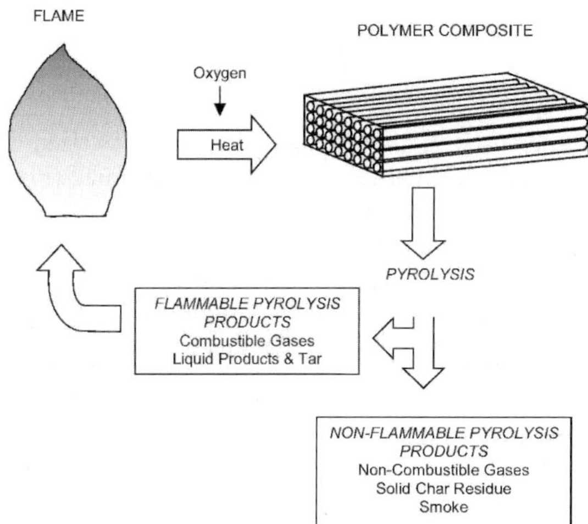


*Figure 1.5.  The combustion cycle of organic polymers.*

## 1.4    Case Studies of Composites in Fire

The diverse range of uses for composite materials means they can be exposed to a variety of fire threats, and their increasing use in high fire risk applications raises the likelihood of severe fire incidents.  Several case studies of fires in aircraft and ships are given to demonstrate the importance of understanding the fire behaviour of composites, and the need for more flame resistant polymeric materials.

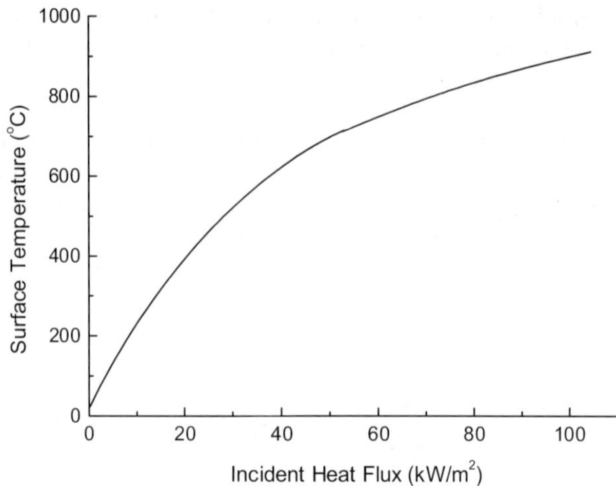



*Figure 1.6. Relationship between heat flux and surface temperature of a polymer material. Data from [11].*

The amount of composites used in aircraft and helicopters has risen dramatically since the 1970s, although the aerospace market remains relatively small (Figs. 1.1 & 1.2) with the consumption in the U.S. presently about ten thousand tons. Boeing and Airbus - the two largest aerospace companies - expect the amount of composites used in their aircraft to increase in the next 10 to 15 years. The percentage of the structure of large passenger aircraft made using composites is currently 5 to 10%, although this is projected to rise above 50% as carbon/epoxy laminates are being used increasingly in the airframe.

An important issue facing the growing use of composites in aircraft and helicopters is their high flammability. Stringent fire safety regulations are enforced by aviation authorities such as the FAA on materials used inside aircraft cabins. Most thermoset and thermoplastic composites fail to meet the requirements for low flammability and smoke toxicity. Fire resistant phenolic composites are the most commonly used laminates in cabins; accounting for 80%-90% of the interior furnishings of modern passenger aircraft. These composites are used in ceiling panels, interior wall panels, partitions, galley structures, large cabinet walls, structural flooring and overhead stowage bins. The most often used composite used in load-bearing aircraft structures is carbon/epoxy, which is flammable and readily decomposes when exposed to heat and fire. While flame retardant epoxies and other polymers with low flammability are being used increasingly in composite aircraft structures, these materials are often much more expensive and do not have the same mechanical performance as conventional aerospace-grade epoxies.

Figure 1.7 shows the causes of wide-body passenger aircraft crashes over a ten-year period. Over this time there were 180 crashes, but only six (or 3.5%) of these were caused by fire. In-flight fire ranks as the tenth most probable cause of aircraft accidents. Fire is a rare event because of the strict fire safety regulations and effective flame suppression systems on aircraft. Despite the relatively small number of crashes caused by fire, another interesting statistic is that fire is the fourth highest cause of fatalities (excluding unknown accidents). Figure 1.8 shows a breakdown of the causes of aircraft fatalities between 1992 and 2001, and 339 people were killed (4.9% of all fatalities) by in-flight fire. These statistics highlight the danger that in-flight fire poses to aircraft safety, and the tragically high death toll it can cause.

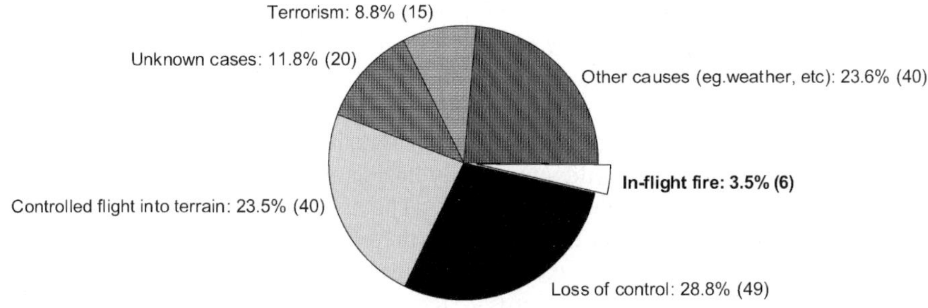


*Figure 1.7. Fatal aircraft passenger crashes between 1987 and 1996. The number of incidents is shown in brackets.*

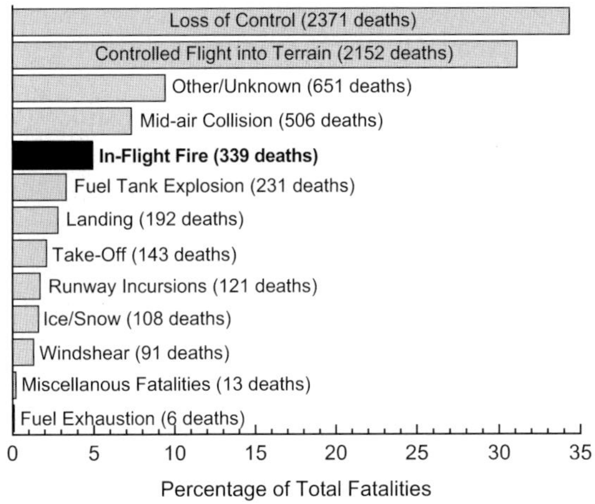


*Figure 1.8. Fatal aircraft passenger crashes between 1987 and 1996. The number for the different causes is shown in brackets.*

Aircraft fires are extremely hazardous because there is little time to combat and extinguish the fire before the crew and passengers are in serious danger. When a fire occurs in the cargo-hold the flight crew have about two minutes to extinguish the flames. Longer than this the fire will often grow too large to extinguish using on-board fire-suppression systems and the plane will most likely crash. If the aircraft has an extinguishable fire, then the pilots have about 14 minutes to land/ditch and evacuate before the risk of incapacitation from smoke and fumes.

The impact of fire on passenger survival becomes even more significant for impact-survivable aircraft accidents. The worst-case scenario for an impact-survivable crash is ignition of the aviation fuel. Such a fire can generate flame temperatures in excess of $1100^{\circ}C$ and radiant heat fluxes above 150 kW/m$^2$. Aircraft occupants have frequently survived the initial ground impact of a crash, only then to be quickly incapacitated from the heat and toxic smoke. Between 20% and 40% of fatalities in impact-survivable commercial aircraft crashes are due to fire. Figure 1.9 shows the temperature-time profile recorded in a passenger aircraft with a post-crash jet-fuel fire adjacent to an open cabin door [11]. The heat builds-up rapidly inside the cabin, and after only three minutes the temperature rises above 600-700$^{\circ}C$. The heat from such fires is the cause of some fatalities, but most deaths occur because the concentration of carbon monoxide, hydrogen cyanide and other toxic gases rapidly reach a lethal level. Full-scale aircraft cabin measurements of the fire hazards – temperature, smoke, oxygen deprivation, carbon monoxide, carbon dioxide, and irritant gases such as HCl and HF – indicate they all increase markedly at flashover [12,13]. The source of these toxic gases is burning aviation fuel and combustible materials in the aircraft. As mentioned, phenolic composites are used in about 80-90% of the interior furnishings in passenger aircraft, and while these materials are highly flame resistant they still release smoke and fumes. The anticipated growth in commercial airline traffic over the next ten years combined with a constant uniform accident rate is projected to lead to, on the average, one aircraft accident each week with about 20% of these accidents involving death due to fire.

The danger of a post-crash fire was demonstrated in August 1985 when a Boeing 737 operated by British Midlands experienced an engine explosion on take-off at Manchester Airport (Fig. 1.10). The explosion ignited over 4,000 litres of aviation fuel. The ensuing fire trapped many passengers inside the aircraft, and 55 people were killed by the smoke and toxic gases released from the burning fuel and cabin materials. The major source of the gases was the polymeric materials used in the seats, and not FRP laminates. The disaster was followed by calls for measures to prevent cabin materials giving off poisonous fumes, and the accident dramatically highlighted the importance of using materials with excellent fire reaction properties.

It is also essential that aircraft composites have excellent fire resistant properties, particularly when used in structural applications when it is essential that the load-bearing properties be retained during and after a fire. One of the worst accidents due to fire occurred in May 1996 when a ValuJet DC-9 crashed into the Florida everglades. Soon after take-off a fire developed in the forward cargo hold that caused the aircraft to crash, killing all 105 passengers and crew. The United States National Transport Safety Board investigated the accident, and concluded that the loss of control was most likely the result of failure of the flight control systems due to the extreme heat and structural collapse. While the accident was not caused by the combustion and subsequent structural failure of composites used in the aircraft, it does demonstrate the importance of excellent fire resistant properties to prevent structural failure.

The growing use of composites in aircraft high-lights the increasing importance of understanding their fire reaction and fire resistance properties to ensure passenger safety. To this end, the Federal Aviation Authority in the United States sponsored the ambitious 'Fire-Safe Materials' program to develop composites for use in aircraft that can maintain survivable aircraft cabin conditions for at least 10 to 15 minutes in post-crash fuel fires in order eliminate fatalities [14].

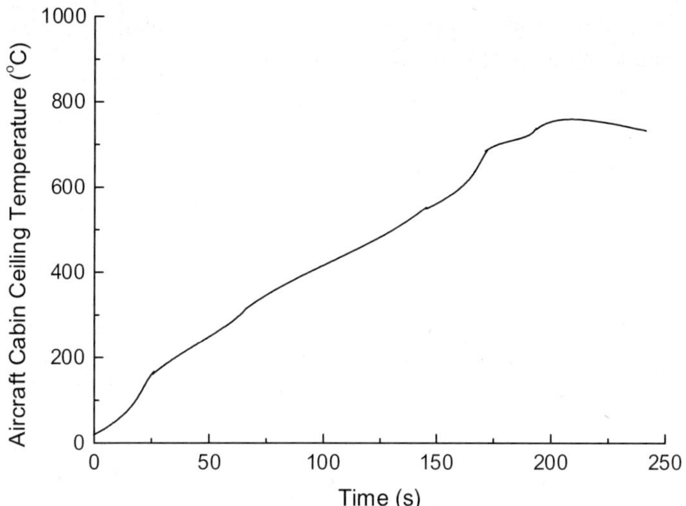

*Figure 1.9. Temperature-time profile above a aircraft cabin door during an external fuel fire. Data from [11].*

The application of composites to ships, submarines and other marine craft accounts for about 10% of current usage (Fig. 1.2). Composites are used in a variety of sea-craft, ranging from small yachts and powerboats through to large naval ships and passenger ferries. Different types of ship structures are made of composites, including the hull, superstructure, masts, bulkheads and piping systems. Composites are also used in small submersibles and in external structures on submarines, such as the non-pressure hull casing and sonar dome. However, composites are used sparingly inside submarines because of the fire hazard, particularly from smoke and toxic gases.

<table>
<tr><td>(a)</td><td>(b)</td></tr>
</table>

*Figure 1.10. Wreckage of the British Midlands aircraft that experienced a fire which killed 55 passengers and crew. (a) External and (b) internal cabin damage caused by the fire.*

Fire is unfortunately an all too frequent danger on boats (Fig. 1.11). Many boats are made using fibreglass composites that readily ignite when exposed to fire, which usually start from ignition of spilt fuel or electrical problems. The United States Coast Guard report that over two hundred boat fires occurred in 1998, with many involving the completion destruction of the composite hull. These fires resulted in 4 fatalities, 250 injuries – many from burns or smoke inhalation – and nearly \$4 million in property damage.

While fire is a concern for all types of marine craft, it is a greater hazard on large ships and submarines because it is difficult to easily escape from the flames, smoke and fumes [15-17]. Like aircraft, the inability of passengers and crew to quickly escape from a burning ship is a major safety concern. Fires on ships and submarines can be started by any number of causes, with the most common being electrical faults, open flame/welding operations, and ignition of flammable gases or liquids. During wartime there can obviously be other causes of fire, such as missile strikes.

Figure 1.12 gives the location of fires on ships, and most occur in enclosed machinery spaces such as the engine room [18]. This data is for fires on metal ships, although similar statistics can be expected for composite vessels because the source of the fire is not influenced by the material used in the ship construction. About 90% of fires that occur on the ships and submarines of the United States Navy are contained to the area where they started, and about three-quarters of fires are extinguished within thirty minutes [18]. Fire is often initiated by leaking fuel oil being sprayed under pressure onto hot engine parts following rupture of a fuel line. This is a reason for the high incidence of engine room fires. This type of fire is severe, with the radiant heat flux exceeding 75 kW/m$^2$ and the flame temperature above 1000°C. Ship fuel fires are notoriously dangerous because of dense smoke that reduces the visibility for crew, passengers and fire-fighters that may cause disorientation and confusion. Compounding the danger is that large ship fires often produce copious amounts of carbon monoxide and other toxic gases. For these reasons, strict fire standards are enforced on the use of composites in ships, such as the International Convention for the Safety of Life at Sea (SOLAS) regulations. The marine fire safety standards and regulations are outlined in Chapter 10.

*Figure 1.11. Fire on a composite boat. Photograph supplied courtesy of Boating.*

Despite the extensive use of composites in ships, the incidence of severe fire is rare. Large fires have occurred on two all-composite minehunter ships operated by the Royal Navy – HMS Ledbury and HMS Cattistock [19,20]. In both ships the fire started in an engine/machinery room, and in the case of HMS Cattistock the fire burned for over four hours before being extinguished. The fires extensively damaged the compartment to both ships, with the composite hull and bulkheads being heavily charred. However, the low thermal conductivity of the composite bulkheads and decks stopped the fire from spreading by heat conduction to surrounding compartments, which is more difficult to stop in steel ships.

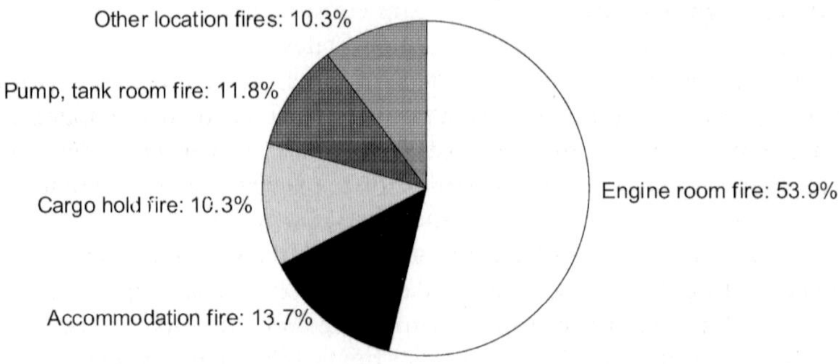


*Figure 1.12.  Locations of ship fires.  Data from [18].*

A more severe ship fire occurred in November 2002 on a Norwegian minesweeper that high-lights the potential hazard of composites.  A fire broke out in a propulsion system of KNM Orkla, which was a minesweeper built of sandwich composite material (Fig. 1.14).  The fire spread rapidly from the engine room, and the ship was soon engulfed by fire forcing the crew to abandon Orkla. The ship burned for more than 24 hours before capsizing, breaking up, and finally sinking.  This is the first reported case of the loss of a composite naval ship due to fire. Failure of the on-board fire suppression system was the principal cause for the rapid growth and spread of the fire.  However, it is believed the sandwich composite hull helped the spread of fire because of its high flammability. Furthermore, the burning composite produced large amounts of dense and toxic smoke, which forced the crew to abandon the ship.  This fire has concerned many navies that operate large vessels made using the same composite materials as the Orkla.

*Figure 1.14.  The composite minesweeper KNM Orkla sunk due to fire.  Photograph supplied courtesy of Summmørsposten.*

## 1.5      Concluding Remarks

The examples of fire on aircraft and ships serve to illustrate the importance of understanding the fire properties of composites and the need to use flame retardant polymers in composite materials.   A great deal of effort has been devoted to the development of flame resistant composites to use in high fire-risk applications, although many challenges remain.  One of the greatest impediments to the use of flame retardant composites is cost, with many advanced polymer systems having low flammability being too expense to use in all but the most exotic applications.  Figure 1.15 shows a plot of price against heat release capacity for a large number of polymers [21].  The heat release capacity is a parameter that correlates with the flammability of pure polymers, and a lower capacity value is indicative of better fire performance.  Also indicated in Fig. 1.15 are the approximate costs the aerospace, shipbuilding and civil infrastructure sectors are willing to pay for polymers used in composites, and it is apparent that many resin systems, despite their outstanding fire performance, are simply too expensive. Other problems with many of the highly flame retardant polymers are high viscosity that makes them difficult to process and moderate mechanical properties and environmental durability.

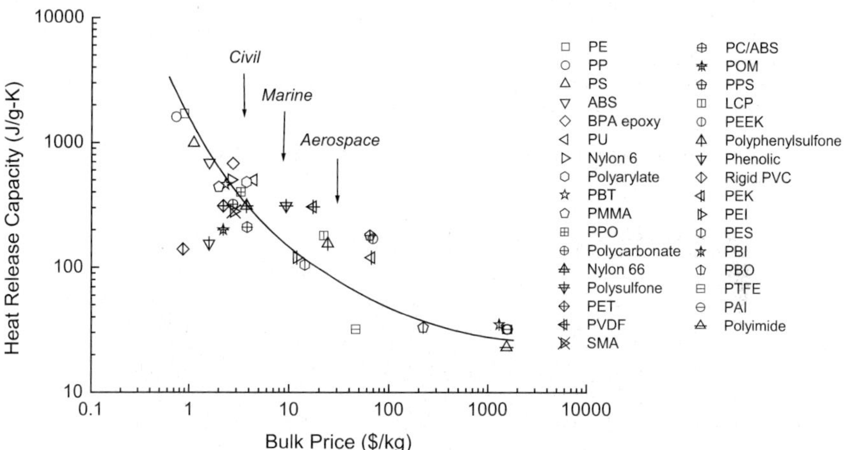

Figure 1.15.  Comparison of heat release capacity against price for polymers. The price is in USD for the year 2001.  Modified from [21].

This book covers all the key topics and issues concerning the fire behaviour of polymer composite materials.   This includes the combustion mechanisms, fire reaction properties, fire resistive properties, modelling, flame retardant materials, polymer nanocomposites, fire test methods, and fire safety standards. The fire behaviour of a wide variety of polymer laminates and sandwich composites are described, although most attention is given to the fire performance of those materials most commonly used in aerospace, marine, rail, automotive, civil infrastructure and chemical processing applications. This includes composites with a thermoset matrix of polyester, vinyl ester, epoxy or phenolic or a thermoplastic matrix of polyphenylene sulfide (PPS) or polyether etherketone (PEEK). The flammability of composites reinforced with non-combustible fibres (eg. carbon, glass) and combustible organic fibres (eg. aramid, polyethylene) are also considered. The book does not address the fire behaviour of other types of composites, such as metal- and ceramic-matrix composites or natural fibre composites. This is because the flammability of these composites is not a serious concern.

## References


1.  V. Babrauskas and R.D. Peacock. Heat release rate: the single most important variable in fire hazard. *Fire Safety Journal*, 1992; 18:255-272.
2.  V. Babrauskas. Why was the fire so big? HHR: The role of heat release rate in described fires. *Fire & Arson Investigator*, 1997; 47:54-57.
3.  M.M. Hirschler, 'Smoke and heat release and ignitability as measures of fire hazard from burning of carpet tiles', *Fire Safety Journal*, 1997; 18:305-324.
4.  D. Drysdale, An Introduction to Fire Dynamics, John Wiley & Sons, Chichester, 1985.
5.  W.D. Walton and P.H. Thomas, 'Estimating temperatures in compartment fires', in SFPE Handbook of Fire Protection Engineering, ed. P.J. DiNenno et al., 1995, pp. 3-134 to 3-147.
6.  B. Karlsson, and J.G. Quintiere, Enclosure Fire Dynamics, CRC Press, Washington, DC, 1999.
7.  J.G. Quintiere, Principles of Fire Behavior. Delmar Publishers, Albany NY, 1998.
8.  H. Ingason, Two dimensional rack storage fires', Proceedings of the Fourth International Symposium on Fire Safety Science, 1994, pp. 1209-1220.
9.  K.S. Mudan and P.A. Croce, 'Fire hazard calculations for large open hydrocarbon fires', in SFPE Handbook of Fire Protection Engineering, ed. P.J. DiNenno et al., 1995, pp. 2-13.
10. H. Inganson and J. de Ris, 'Flame heat transfer in storage geometries', *Journal of Fire Safety*, 1997.
11. R.E. Lyon, 'Fire-safe aircraft cabin materials', in *Fire & Polymers*, ACS Symposium Series No. 599, ed. G.L. Nelson, American Chemical Society, Washington DC, 1995, pp. 618-638.
12. C.P. Sarkos, R.G. Hill and W.D. Howell, 'The development and application of a full-scale wide body test article to study the behaviour of interior materials during a postcrash fuel fire', North Atlantic Treaty Organisation (NATO), Advisory Group for Aerospace Research and Development (AGARD) Lecture Series No 123 – Aircraft Fire Safety, Washington DC, 15-16 June 1982.
13. C.P. Sarkos and R.G. Hill, 'Effectiveness of seat cushion blocking layer materials against cabin fires', SAE Technical Paper 821484, Aerospace Congress and Exposition, Anahiem, CA, 25-28 October 1992.
14. R.E. Lyon, "Federal aviation administration research in fire safe materials for aircraft interiors", In: *Proceedings of the International SAMPE Symposium*, 41, (1996), Anaheim, CA, pp. 344-350.
15. D.R. Ventriglio, 'Fire safe materials for navy ships', *Naval Engineers Journal*, Oct 1982, 65-74.
16. K.J. Fisher, 'Is fire a barrier to shipboard composites?', *Advanced Composites*, May/June 1993, pp. 20-26.
17. U. Sorathia, R. Lyon, R. Gann and L. Gritzo, 'Materials and fire threat', *SAMPE Journal*, 1996; 32: 8-15.
18. E. Greene. *Marine Composites*. www.marinecomposites.com.
19. D.W. Chalmers, R.J. Osburn and A. Bunney, 'Hull construction of MCMVs in the United Kingdom', Proceedings of the International Symposium on Mine Warfare Vessels and Systems, 12-15 June 1984, London, Paper 2.
20. A.W.G. Daniel, R.S. Trask, D.M. Elliot and P.W. Lay, 'Repair of HMS Cattistock's GRO structure using resin infusion', Proceedings of Royal Institute of Naval Architects Conference on 'Lightweight Construction: Latest Developments, 24 February 2000, London.
21. R. Lyon, Presentation titled 'Solid State Thermochemistry of Flaming Combustion'.

# Chapter 2

# Thermal Decomposition of Composites in Fire

## 2.1    Introduction

The behaviour of composite materials in fire is governed largely by the chemical processes involved in the thermal decomposition of the polymer matrix and, if present, the organic fibres.    This chapter provides a description of these decomposition mechanisms.  The description is kept at a general level, and the reader can refer to the many excellent textbooks on polymer decomposition for more information [1-5]. Following this, the decomposition behaviour of polymer systems used in composites is described. A great number of different polymers can be used in composites, and it would be too exhaustive to describe the decomposition of each type.  Instead, the chemical nature and decomposition behaviour of the thermoset polymers and thermoplastics most commonly used in composites are reviewed.  This will include the thermosets: polyesters, vinyl esters, epoxies and phenolics, and the thermoplastics: polypropylene (PP), poly ether ether ketone (PEEK) and polyphenylene sulphide (PPS). The thermal decomposition of the organic fibres most often used in composites; namely aramid and UHMW polyethylene fibres, are also discussed.  The decomposition of other types of organic fibres that are presently used in niche applications are not reviewed, such as nylon 6,6 or PBO (Zylon$^{\circledR}$), or fibres that are still under development, such    as    M5    (poly(2,6-diimidazo[4,5-*b*:4',5'-*e*]pyridinylene-1,4-(2,5-dihydroxy) phenylene)).    Finally, the physical aspects of degradation of composites in fire are described, including char formation, delamination damage and matrix cracking.

## 2.2      Thermal Decomposition Mechanisms of Organic Polymers

The events involved in the decomposition of a composite material in fire are summarised in Fig. 2.1. When the material is exposed to a sufficiently large heat flux radiated from a fire, the polymer matrix and organic fibres will thermally decompose to yield volatile gases, solid carbonaceous char and airborne soot particles (smoke). The volatiles consist of a variety of vapours and gases, both flammable (eg. carbon monoxide, methane, low molecular organics) and non-flammable (carbon dioxide, water). These diffuse from the decomposing composite into the flame zone, where the flammable volatiles react with oxygen in the fire atmosphere leading to the formation of the final combustion products (usually carbon dioxide, water, smoke particles and a small amount of carbon monoxide) accompanied by the liberation of heat. In order for the process to be self-sustaining, it is necessary for sufficient heat to be fed-back into the composite to continue the production of flammable decomposition gases.

The overall process is a complex one, as shown in Fig. 2.1. It depends to a large extent on the fire scenario, and especially on the quantity of composite material present compared to other possible fuel sources. The 'fire' drives the initial decomposition and ignition of the composite. This may be an ignition source, for instance an electrical fault, overheated machinery, burning debris or the heat from a welding torch. Alternatively, when the composite is the 'minor' component in another type of structure, the fire may be an established conflagration resulting from a completely separate sequence of events, with a fuel source, such as oil, gas or the cellulosic content of a building. The on-going decomposition process may be driven by heat from the main fire, with additional feedback into the laminate of some heat generated by the local burning of decomposition products. Alternatively, when the composite material itself represents a significant fuel source compared to the other materials present, the burning decomposition products may feedback into the main fire, increasing its intensity.

Polymers decompose via a series of chemical reaction mechanisms when heated to a sufficiently high temperature. The main mechanisms that reduce molecular weight are random chain scission, chain-end scission ('unzipping') and chain stripping (removal of side groups). Two other thermally induced processes, cross-linking and condensation, have the opposite effect of increasing molecular weight. Although decomposition often involves more than one of the scission mechanisms, the dominant reaction in most polymer systems is random chain scission. This commences with the weakest bonds in the chain, which is usually where a 'irregularity' occurs in the molecular structure due for instance to the presence of a 'tertiary' carbon atom, as in polypropylene, or other relatively unstable linkages with low dissociation energies. Bond energies are discussed in reference [6]. Scission usually proceeds randomly throughout the length of the chain. With increasing temperature other chemical bonds with higher dissociation energies rupture causing the resulting segments to break-down further into monomers, oligomers (ie. polymer units with ten or fewer monomer units) and other low molecular weight species. It is noteworthy that while random chain scission can decompose long polymer chains into an extremely large number of fragments, in general only a few percent of the bonds need to rupture to drastically degrade the mechanical properties. A bond rupture level of about 10% is generally sufficient to generate organic compounds that are volatile in a fire. For the fragments to be small enough to diffuse through the polymer char into the fire their molecular weight must be lower than about 400, although with many volatile species the molecular weight is much less (for example, styrene MW = 96). It is these volatiles that decompose at the fire/composite interface that produce heat that sustains the decomposition process.

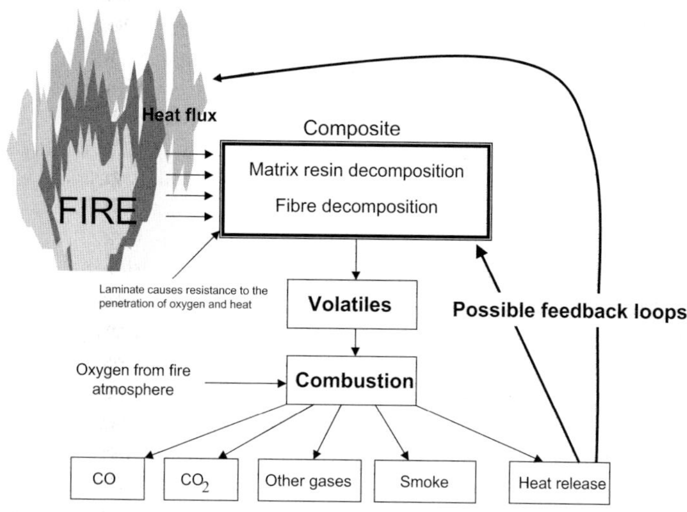


*Figure 2.1. Mechanisms involved in the thermal decomposition of polymer composites, showing feedback loops involving heat flux.*

Chain-end scission (unzipping or depolymerisation) is another important decomposition reaction that can compete with random scission in some polymer systems. Here, individual monomer units or volatile chain fragments are successively removed at the chain end until the polymer molecule has completely depolymerised. Chain stripping is a further decomposition reaction that involves the removal of side groups.

Cross-linking results, often temporarily, in an increase in molecular weight, in competition with the processes mentioned above. In most thermosets, for instance, it is well-known that 'post-cure' or further cross-linking occurs at elevated temperature (say above 100-150°C) and precedes the decomposition processes that occur at higher temperatures (typically above 250-400°C). Likewise in some thermoplastics (eg. polyethylene) a degree of cross-linking precedes chain scission.

The thermal decomposition reactions of polymers may proceed by oxidative processes or simply by the action of heat. The decomposition process is often accelerated by oxygen, but in thick composite sections it is generally only the surface region that decomposes in the presence of oxygen. The out-gassing of volatiles from the decomposition zone impedes the ability of oxygen to diffuse much beyond the surface layers of the composite. Therefore, atmospheric oxygen does not have a major influence as the decomposition process moves deep into thick section composites, where decomposition tends to be driven mainly by heat.

Finally, condensation is of prime importance in residual char formation. The formation, composition and structure of char is reviewed by Levchik and Wilkie [7]. Char is richer in carbon than the original polymer, although it is rarely pure carbon. Chars consist mostly of carbon (~85-98%) with trace amounts of aromatic-aliphatic compounds often with hetroatoms (O, N, P, S). Char is a highly porous material that can consist of crystalline (ie. graphitic) and/or amorphous regions, with the relative amounts of these phases determined by the original chemical composition of the polymer as well as the temperature and atmosphere of the fire.

The amount of char formed in a composite material is dependent on the chemical nature of the polymer matrix and (if present) organic fibres. Levchik and Wilkie [7] propose that polymers can be categorised into one of three classes depending on the chemical processes governing the thermal decomposition process and formation of char. The thermal degradation process of the first class is characterised by random chain scission reactions in which almost all of the molecular structure becomes fragmented into volatile gases, resulting in a negligible amount of char. This applies to some thermosets of the 'solvent monomer' type, as well as to polyolefin thermoplastics (eg. polypropylene, polyethylene) and thermoplastics such as polystyrene (PS) and poly methyl methacrylate (PMMA).

The second group undergoes random chain scission, end-chain scission and chain stripping reactions, which leads to the loss of hydrogen atoms, pendant groups and other low molecular weight organic groups from the main chain. These polymers yield a small amount of char - typically 5-20% of the original mass - and they include polyesters, vinyl esters, epoxies and polyvinyl chlorides (PVC).

The third group of polymers is characterised by a high aromatic ring content that decomposes into aromatic fragments that fuse via condensation reactions to produce moderate to high amounts of char. Aromatic rings are the basic building blocks from which char is formed, and therefore the higher the aromatic content of the polymer the higher the char yield. Parker and Kourtides [8] have shown that the char yield increases linearly with the concentration of multiple-bonded aromatic ring groups in the polymer system, as shown in Fig. 2.2. These aromatic groups are transformed at high temperature into pitch-like entities that eventually combine into char. The best-known polymer from the viewpoint of char formation is phenolic, in which 40-60% of the resin mass is converted to char. Several other polymer systems yield high amounts of char, and these include highly aromatic thermosets (eg. polyimides, phthalonitriles, epoxy novolacs, cyanate esters) and certain thermoplastics (eg. PPS, PPO, PEEK).

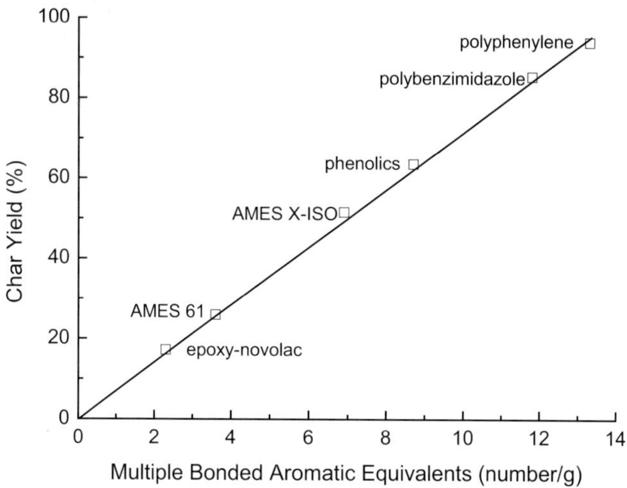


*Figure 2.2. Relationship between aromatic content and char yield of polymers. Parker, J.A. and Kourtides D.A., J. Fire Sci., 1, 1983, 432-458. Reference 49, Reproduced with permission Sage Publications.*

## 2.3     Rate Processes and Characterisation of Decomposition

Another important factor in the decomposition of polymers is the manner in which the heating rate controls the various decomposition reactions. When a composite laminate is exposed to one-sided heating from a fire, the heating rate is non-uniform through the material; being highest at the hot surface and decreasing rapidly towards the cold face. For example, Fig. 2.3 shows a log-linear plot of heating rate against normalised distance below the hot surface of a thick E-glass/polyester composite exposed to an incident heat flux of 50 kW/m$^2$. The normalised depth (x/L) is the distance below the hot surface (x) divided by the total thickness of the composite (L), which in this example is 12.5 mm. The heating rate at the hot surface initially reaches about 1000°C/min, although this usually lasts for less than one minute and then the rate slows considerably as the surface temperature approaches the temperature of the fire. The heating rate drops rapidly with increasing distance from the hot surface, due mainly to the 'thermal lag' that results from the relatively low thermal diffusivity of the material. Other factors also contribute to the reduced heating rate, including the convective cooling from the outward flow of combustion gases and the absorption of heat by the decomposition reactions of the polymer matrix, which are mainly endothermic. The heating rate in the example given drops to about 15°C/min at the back surface (x/L = 1). Even slower heating rates are encountered in thicker laminates and composites with surface fire protection.  This also applies in sandwich panels, where the cold face is insulated to a very significant extent by the effect of the low thermal conductivity of the core material.

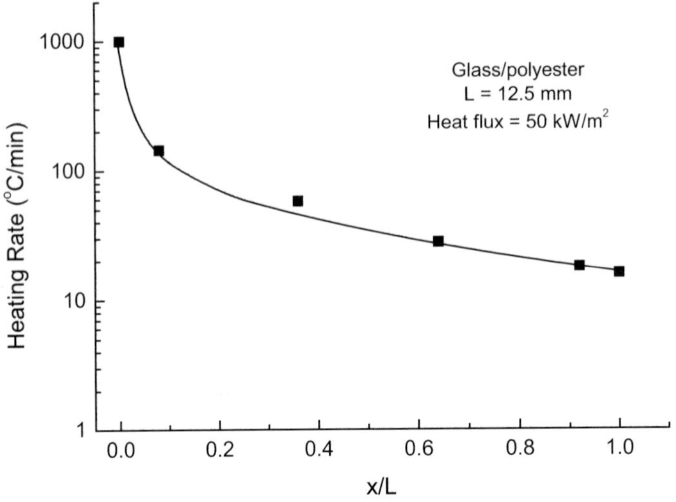


*Figure 2.3.  Variation in heating rate through of a glass/polyester composite exposed to a heat flux of 50 kW/m$^2$.*

The temperature range over which polymers decompose increases with the heating rate. For example, Fig. 2.4 shows the retained mass versus temperature curves for a phenolic resin measured using thermogravimetric analysis (TGA) at heating rates between 5 and 50$^{\circ}$C/min [9].  The temperature range over which the phenolic resin decomposes shifts progressively to higher temperature when the heating rate is increased.  The decomposition temperature of the polymer matrix and any organic fibres in a composite exposed to fire will not therefore be uniform, but instead will decrease in temperature from the hot to cold surface.

Various analytical techniques can be used to characterise the decomposition reactions and the chemical nature of the reaction volatiles of polymers.  The three most important methods are TGA, in which the weight loss is measured with increasing temperature or time; differential scanning calorimetry (DSC), in which heat absorption or evolution due to chemical changes of the polymer are measured; and gas chromatography/mass spectrometry (GC/MS), in which the chemical composition of the volatile gases are determined.  Other methods may also be used, including thermal volatilisation analysis (TVA) and differential thermal analysis (DTA). It is often necessary to use a combination of methods to obtain a complete understanding of the decomposition reaction rate, reaction mechanisms, volatile gases and char yield.

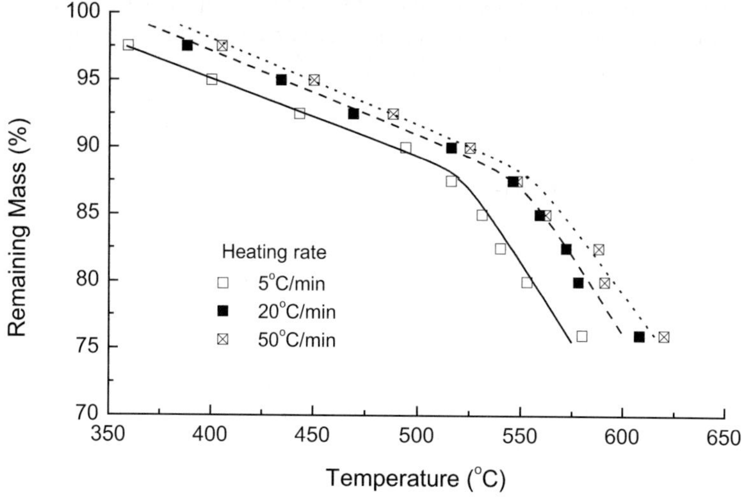


*Figure 2.4.  TGA curves for a phenolic at different heating rates.  Pektas, I. J. App. Poly.Sci., 68, 1998, 1337-1342, reference 7, John Wiley & Sons, Inc. Reproduced with permission.*

## 2.4      Polymers and Their Decomposition Processes

### 2.4.1    INTRODUCTION

This section provides a description of several polymer systems used as the matrix phase to composites, along with a brief summary of what is known of their thermal decomposition mechanisms. Included here are thermosetting polyesters, vinyl esters, epoxies and phenolics, as well as the thermoplastics polypropylene (PP), poly ether ether ketone (PEEK) and polyphenylene sulphide (PPS).  When possible, information is included on the combustion gases and char yield resulting from the decomposition process, which have a major influence on the fire hazard of composite materials.

### 2.4.2    THERMOSETTING RESINS FOR COMPOSITES

*Unsaturated Polyester Resins*
This class accounts for the largest tonnage of resins involved in the manufacture of composite materials, with approximately 400,000 tons used annually. Polyesters are used in many composite products because of their moderate cost, good mechanical properties, reasonable environmental durability, low viscosity at room temperature, and

low-temperature cure properties. Polyesters are used in all the 'wet resin' manufacturing processes for composites (hand lay-up, spray lay-up, resin transfer moulding, resin infusion) as well as in hot press moulding and injecton moulding of sheet moulding compounds and bulk moulding compounds. The so-called polyesters (the name being something of a misnomer, as explained below) are the most widespread example of 'solvent monomer' resin systems, which are cured by free radical polymerisation. The general principle is shown schematically in Fig. 2.5.

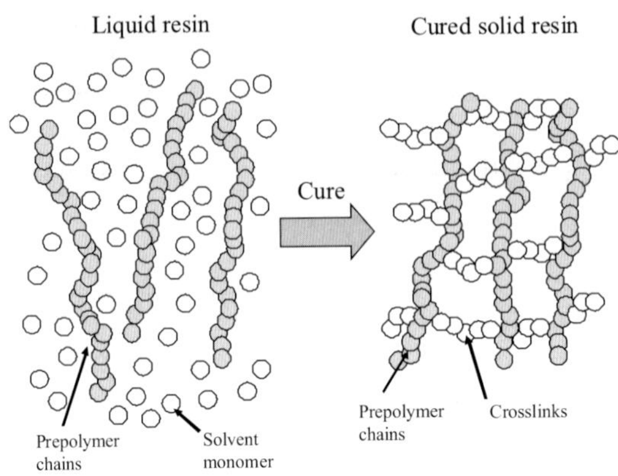


*Figure 2.5.  Principle of cross-linking 'solvent monomer' systems such as thermosetting polyester.*

The two components of a solvent monomer system are (i) a pre-polymer of relatively low molecular weight containing carbon-carbon double bonds (or unsaturation) in the back-bone chain and (ii) an unsaturated monomer, such as styrene, in which it is dissolved. Examples of solvent monomers are (Fig. 2.6):

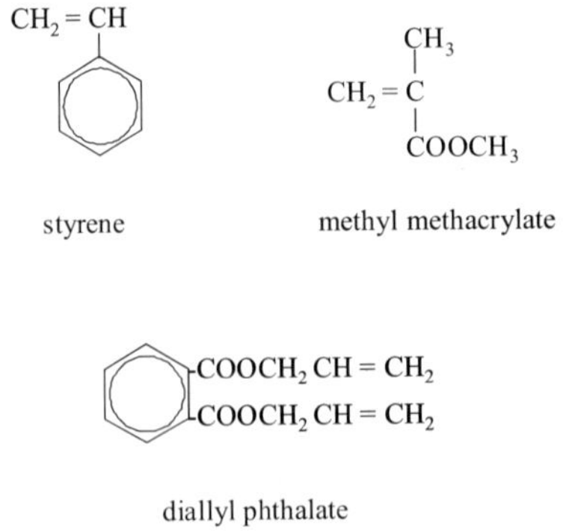

$$CH_2 = CH$$

styrene

$$CH_2 = C \begin{array}{c} CH_3 \\ | \\ | \\ COOCH_3 \end{array}$$

methyl methacrylate

$$COOCH_2\,CH = CH_2$$
$$COOCH_2\,CH = CH_2$$

diallyl phthalate

*Figure 2.6.  Commonly employed solvent monomers.*

Cure is caused to take place by adding an initiator, which is a source of free radicals. The free radical cure reaction involves the addition polymerisation of the monomer species. The double bonds in the pre-polymer are also involved, with the result being a polymeric network in which the pre-polymer chains are cross-linked by chains of polymerised monomer. In polyester resins the pre-polymer is an unsaturated polyester and the solvent monomer is usually styrene, so the final cross-linked product can be regarded as containing polyester chains with polystyrene cross-links. Indeed, a typical polyester resin may contain up to 35% by weight of styrene.

Styrene is by far the most widely used solvent monomer for cost reasons. Although, methyl methacrylate is sometimes used as a total or partial replacement of styrene in resins requiring improved optical properties or lower smoke production in fire. Diallyl phthalate is used as the monomer in polyester systems that are required to be solid at room temperature and in 'alkyd' moulding compounds. Other solvent monomer systems that use the general principle shown in Fig. 2.5 include vinyl esters and modified acrylics.

Polyesters are the result of condensation polymerisation between a di-carboxylic acid (or di-acid) and a di-ol (or glycol) (Fig. 2.7):

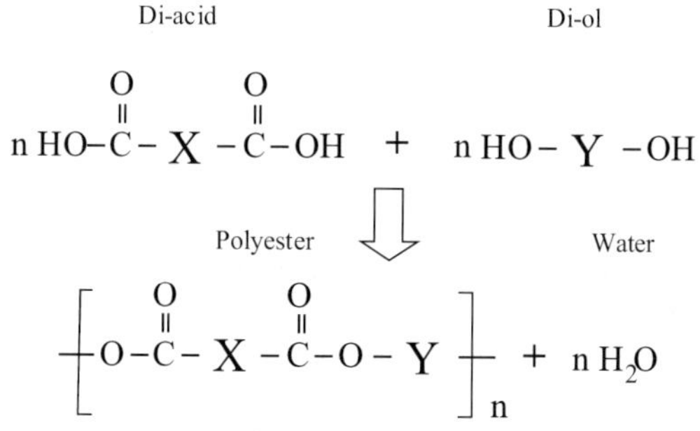

$$\text{n HO-}\underset{\text{Di-acid}}{\underset{\displaystyle \overset{O}{\overset{\|}{C}}-X-\overset{O}{\overset{\|}{C}}}{}}\text{-OH} \quad + \quad \text{n HO-}\underset{\text{Di-ol}}{Y}\text{-OH}$$

$$\underset{\text{Polyester}}{\left[\overset{O}{\overset{\|}{O-C}}-X-\overset{O}{\overset{\|}{C}}-O-Y\right]_n} \quad + \quad \text{n H}_2\text{O} \quad (\text{Water})$$

*Figure 2.7. Condensation of a di-carboxylic acid with a di-ol to form polyester resin.*

The final structure is determined largely by the nature of the components, X and Y. It is worth noting that this reaction, which is carried out at elevated temperature, results in the evolution of water as a by-product and that the length of the polyester chains is determined not only by the time and temperature in the process, but by the extent of water removal. The reverse reaction (hydrolysis) can occur when polyester is in the presence of water. Polyester-based polymers are fairly stable and environmentally resistant at ambient temperature, but the possibility for hydrolysis always exists when water is present, especially at higher temperatures or in the presence of acids or alkalis.

The unsaturated polyesters used in thermosetting resins are co-polyesters containing a mixture of saturated and unsaturated acids.  The 'reactivity' of the polyester is determined by the ratio of the unsaturated to saturated acids used.  This determines the concentration of double bonds in the chain, which is the factor that controls the crosslink density in the final product.  Polyesters with a high reactivity have a high $T_g$ and low permeability to water and other diffusing liquids, so they also have higher chemical and hydrolysis resistance.  The converse is also true: low reactivity can produce polymers with a lower $T_g$. These may be tougher or even flexible at room temperature, but they also tend to be less resistant to hydrolysis.

The most commonly used saturated acids are the three phthalic acid isomers (Fig. 2.8):

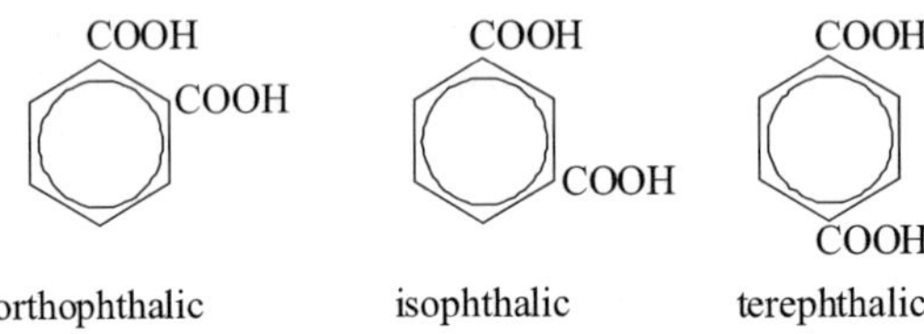


*Figure 2.8. Saturated acids commonly used in polyester resins.*

Terephthalic acid is sometimes used in unsaturated polyesters, but it is better known for its use in thermoplastics. When combined with ethylene glycol, it produces the important semi-crystalline thermoplastic polyethylene terephthalate (PET).  PET, already widely used in fibres, films and drinks bottles, is beginning to be employed in composites, but it should not be confused with the unsaturated polyesters (the so-called "polyester resins") that are thermosetting resins.

Of the other phthalic acid isomers, both orthophthalic and isophthalic are widely used in unsaturated polyesters.  Orthophthalic is the least expensive isomer and is thus employed in so-called 'general purpose' resins. It is also favoured because it can be used in the 'anhydride' form, phthalic anhydride (Fig. 2.9).

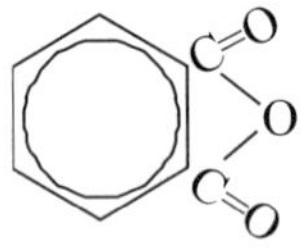

*Figure 2.9. Phthalic anhydride.*

Here, one molecule of water has already been abstracted prior to the polymerisation reaction. Employing isophthalic acid is more expensive, but isophthalic polyesters offer a significantly better combination of mechanical strength and hydrolysis resistance than orthophthalic.

Of the unsaturated acids, both the fumaric and maleic isomers are used, although maleic is often preferred, again because it is available in the anhydride form (Fig. 2.10).

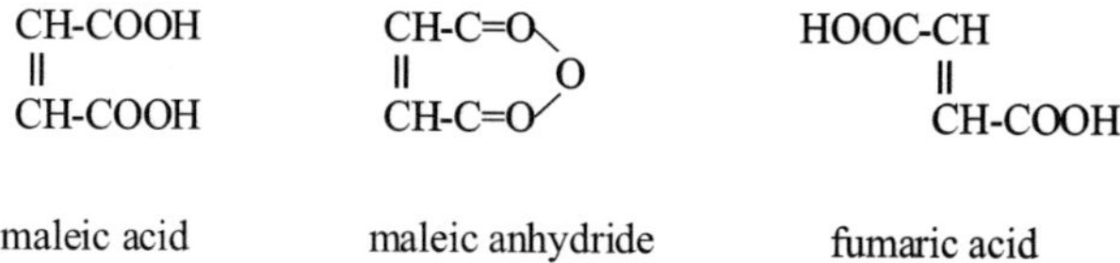

maleic acid          maleic anhydride          fumaric acid

*Figure 2.10.   Unsaturated di-carboxylic acids commonly used in polyester resins.*

A range of other acid types may be employed to achieve particular property modifications.  For instance 'nadic anhydride' may be used to improve thermal stability (Fig. 2.11).

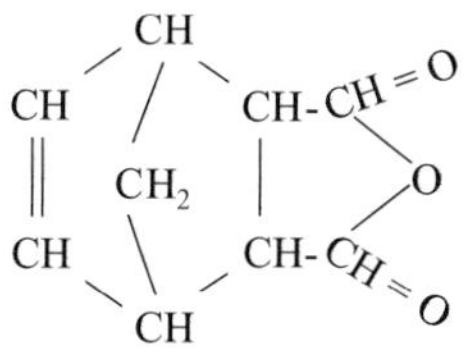

*Figure 2.11.  Nadic anhydride.*

Examples of commonly used glycols are (Fig. 2.12):

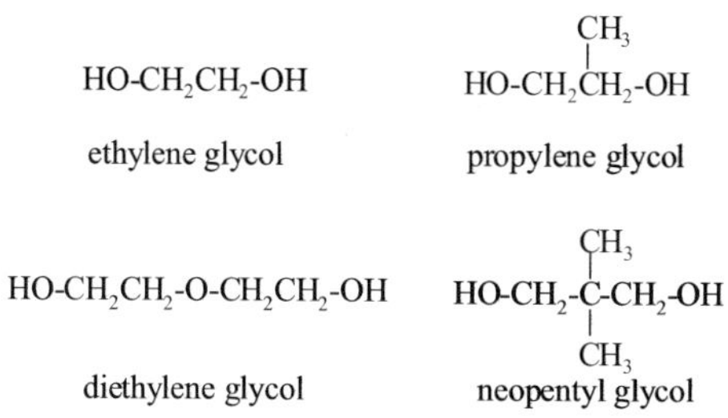

$$HO\text{-}CH_2CH_2\text{-}OH \qquad\qquad HO\text{-}CH_2\underset{|}{\overset{CH_3}{C}}H_2\text{-}OH$$

ethylene glycol          propylene glycol

$$HO\text{-}CH_2CH_2\text{-}O\text{-}CH_2CH_2\text{-}OH \qquad HO\text{-}CH_2\text{-}\underset{CH_3}{\overset{CH_3}{C}}\text{-}CH_2\text{-}OH$$

diethylene glycol          neopentyl glycol

*Figure 2.12.  Glycols commonly used in polyester resins.*

For unsaturated polyesters, glycols with larger numbers of carbon atoms, most notably propylene glycol, are preferred because the resulting polymers dissolve well in styrene and do not have a tendency to crystallise.  Neo-pentyl glycol (NPG) is used when greater toughness is required without compromising hydrolysis resistance.  NPG resins are often used in marine gel-coats.

Bisphenol A is also used in the role of a glycol in cases where improved chemical resistance is required, for instance in composite items for chemical plant (Fig. 2.13):

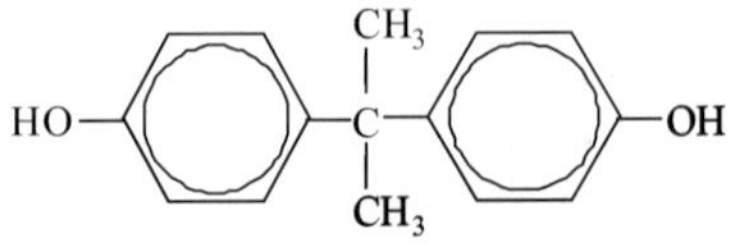

*Figure 2.13.  Bisphenol A.*

Following polymerisation, chips of solid unsaturated polyester are dissolved in styrene solvent monomer and shipped as polyester resin.  The cross-linking process in the manufacture of a composite part is started by the addition of the free radical initiator (universally and incorrectly referred to as the 'catalyst'), usually at a level of 0.5-1.5%. These are organic peroxide derivatives, and common initiators are methyl ethyl ketone peroxide (MEKP) and tertiary butyl perbenzoate (TBP).  These compounds are unstable and decompose to give free radicals. In the case of room temperature cure the decomposition of the initiator is catalysed by a small amount of 'accelerator', previously added to the resin.  The accelerator, often cobalt naphthenate, is a true catalyst for the decomposition of the initiator.  The cure rate of the resin is determined by the quantity and type of initiator added and the temperature. There are many different initiators, usually chosen for their tendency to decompose in a particular temperature range.

Inhibitors are additives that remove free radicals, and are added to resins to prolong their shelf life.  Additional inhibitor may be added at the time of processing in order to produce an induction time or delay period before cure commences.  This is advantageous in certain manufacturing processes when a prolonged working life of the resin is required before gelation, such as for instance in large area mouldings.

In the case of high temperature cure processes, as in pultrusion, hot press moulding or thermoset injection moulding, no accelerator is needed: decomposition of the peroxide is induced by temperature alone.  For these processes, several different initiators may be added to ensure that the cure reaction takes place over a wide temperature range as the composite is heated.

Regardless of the method used to cure the resin, full cure is seldom if ever achieved in the fabrication process itself. This is because the cure reaction slows down to a very low rate as cure progresses and the reacting species become less mobile after the system has gelled. The solution to this problem is usually a prolonged high temperature oven 'post-cure' to complete the process after the component has been de-moulded. Post-curing is desirable, where feasible, because it improves the glass transition temperature, mechanical properties and chemical resistance of the resin. Unfortunately, with large composite structures, such as large marine craft, high temperature post-cure may not be possible. In the marine industry, therefore, it is not uncommon for post-cure to occur under ambient conditions over the first few months (or years) of life of the craft.

The most common method of improving the short-term fire reaction performance of polyester resins is by the use of halogenated versions of the polyester constituents described above and by the use of halogenated additives. The effect of halogens on "mopping up" free radicals in the flame process is discussed in chapter 8. There are many ways of adding halogen to polyester, with one of the most common being the use of chlorendic acid (also known as HET acid), as shown in Fig. 2.14.

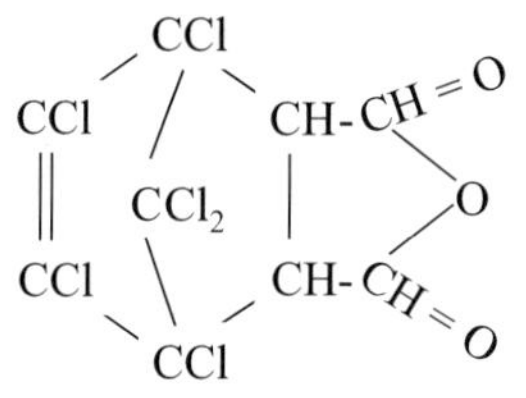

*Figure 2.14. Chlorendic acid used in low flammability resins.*

Addition of halogens is highly effective in reducing flammability and, despite environmental concerns, this is still the main method by which low flammability general-purpose resins are achieved. The main benefit is in cases where the flammability of the composite material itself is the principal threat. Halogenated resins are of little benefit in established fires where the polyester component is not the source of the fire. In this case the resin will burn alongside the other flammable components and release heat and toxic products. Halogenated resins are also known to have slightly poorer mechanical properties compared to conventional polyesters.

Resin suppliers are striving to develop non-halogen methods of modifying the flammability of polyesters. There is interest, for instance, in low toxicity polyesters in which part or all of the styrene has been replaced by methyl methacrylate, and the resin is highly filled with alumina trihydrate (ATH) [10]. The flame retardant properties of these polyester systems are similar to the modified acrylics [11] described later in this chapter.

A number of workers, including Bansal et al. [12] and Gibson and Hume [13], have reported on the decomposition of polyesters or the fire behaviour of polyester composites. The thermal decomposition process of all unsaturated polyesters is probably governed in the initial stages by scission of highly strained portions of the polystyrene cross-links, with the formation of free radicals that then go on to promote further decomposition, including some accompanying scission of the polyester backbone. This results in a variety of low molecular weight volatiles, including CO, $CO_2$, methane, ethylene, propylene, butadiene, naphthalene, benzene and toluene [14]. TGA curves for various polyester-type resins are presented in Fig. 2.15, and it can be seen that 90-95% of the original mass is decomposed into these volatiles, rather than char. This is the main reason for the relatively high flammability and heat release of polyester composites. The fire reaction properties and flammability of polyester composites is described in greater detail in the next chapter.

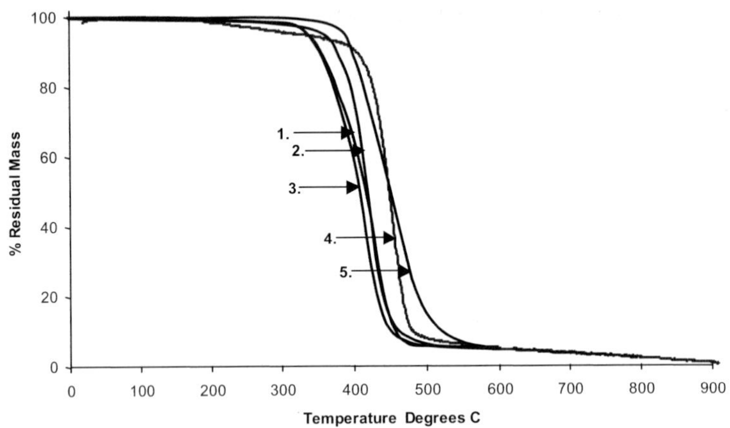


*Figure 2.15. Thermogravimetric analysis traces for solvent monomer resins at 25°C/min in nitrogen . 1. orthophthalic polyester; 2. isophthalic polyester; 3. halogenated (HET acid) polyester; 4. vinyl ester and 5. Bisphenol A polyester.*

Although polyester decomposition is arguably a two-stage process, single-stage Arrhenius kinetics is often sufficiently accurate to model the process [15]. While unsaturated polyesters containing different components will undoubtedly produce different molecular fragments, the overall decomposition behaviour is probably as heavily influenced by the degree of cure of the resin as by the polyester components themselves. The TGA comparison in Fig. 2.15 shows that the qualitative differences in thermal stability of the resins are not large.

In addition to flammability and heat release, another disadvantage of styrene-based solvent monomer systems in fire is that the styrene component itself tends to produce smoke. Furthermore these resins tend, during decomposition, to pass through a 'liquid' or low viscosity stage that can result in the formation of flaming droplets. These tendencies can be alleviated to a certain extent through the use of composites with high inorganic content, by employing, where possible, a high glass or filler content. Char promoting additives, such as those based on phosphorus are also beneficial [16].

*Epoxy Vinyl Ester Resins*
These are often referred to simply as vinyl ester resins. Like polyesters, they are solvent monomer systems, as described in Fig. 2.5, with typically up to 44% of styrene monomer. The structure of the most common vinyl ester pre-polymer is shown in Fig. 2.16. This is based largely on the structure of epoxy resin chains (see below). Vinyl esters are often said to combine the ease of processing of polyester-type resins (ie. free radical cure) with some of the chemical resistance and improved mechanical properties of epoxy resins. They generally have a higher $T_g$ than polyesters, often in the range of 110°-125°C, depending on the measurement technique.

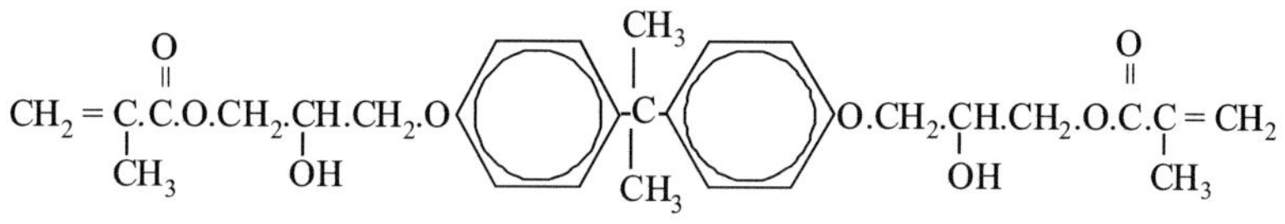

*Figure 2.16. Structure of epoxy vinyl ester pre-polymer.*

As with the polyesters, the thermal decomposition of vinyl esters is governed, at least initially, by decomposition of the styrene component. The similarity to the polyesters can be seen in the TGA curves in Fig. 2.15. The fire behaviour overall is very similar to the polyesters, although, due to the slightly higher styrene content, the time-to-ignition, heat release rate and smoke generation may often be slightly higher (see Chapter 3 for further information).

Regnier and Mortaigne [17] studied the thermal decomposition of glass/vinyl ester composites in air or nitrogen using TGA. As expected, decomposition occurs over a lower temperature range in air due to the action of oxygen. Pyrolysis commences with the elimination of small molecules at the chain ends, and this is followed by cleavage reactions involving the side chains and random chain scission of the main polymer chain. These reactions yield a large mass fraction of flammable volatiles including styrene, toluene and phenyl propane that provide fuel to a fire. Vinyl esters also yield a variety of other combustion gases, including CO and $CO_2$. As with unsaturated polyesters, most of the polymer is decomposed into volatiles and only 5-10% of the original mass is converted into char.

Because vinyl esters are derivatives of epoxy resin, the main chain can be modified in the same way by the addition of other polymers, such as novolacs. Epoxy novolac vinyl esters provide a means of further increasing the $T_g$, often up to about 150°C. The glass transition temperature and thermal stability of vinyl esters is ultimately governed and limited by the styrene content needed to facilitate processing by the free radical route.

*Modified Acrylic Resins (MODAR)*
Modar is also based on the solvent monomer principle (Fig. 2.5) where the pre-polymer is a polyurethane and the solvent monomer is methyl methacrylate. The distinguishing features of modar resins are their low viscosity in the uncured state and the very fast nature of the cure reaction. Although rapid cure is desirable in some economic respects it limits the applicability mainly to 'closed mould' processes, such as resin transfer moulding and pultrusion.

The low resin viscosity enables relatively high levels of fillers, usually alumina trihydrate (ATH), to be incorporated, while retaining processability. It is mainly this characteristic that accounts for the very good fire performance reported for composites based on the modar-ATH system [11]. It should also be re-iterated that the methyl methacrylate monomer has a lower tendency to produce smoke than styrene. ATH-modified modar has been claimed, therefore, to show an excellent combination of fire reaction properties. These properties are achieved mainly through the low polymer content and through the relatively benign decomposition products of methyl methacrylate, compared with styrene. It should be noted, however, that the high levels of ATH can limit the structural performance of this type of resin.

*Epoxy Resins*
This class of resins contains a very large number of chemical types and is widely used in the manufacture of high performance composites as well as in adhesives and surface coatings. All such resins rely on the reactivity of the three-membered epoxy (or oxirane) group at either end of the molecule. The most widely used epoxy resin is diglycidyl ether of Bisphenol A (DGEBA), which is prepared by reacting epichlorohydrin with Bisphenol A under alkaline conditions (Fig. 2.17).

Resins may be supplied partly polymerised, usually with the average value of $n$ in the region 0-3. Other more complex types of epoxy resin are employed when a higher level of functionality is needed, ie. when a greater number of cross-linking points per epoxy group is required, as for instance in some of the higher performance resins used in aerospace. The main cure process involves reactions with the epoxy end group, but further cross-linking reactions may take place involving the pendant hydroxyl groups attached to the main chain, more of which develop as a result of cure. Traditionally, epoxy resins are cured by reacting them with amines or anhydrides; these materials being known as 'hardeners'. There are also catalysts that will polymerise the resin via the end groups and hydroxyls without the need for a hardener.

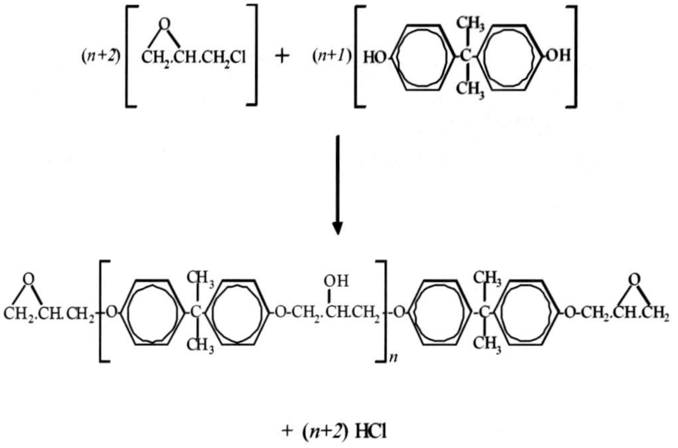

Figure 2.17.  *Manufacture of epoxy resin (diglycidyl ether of Bisphenol A).*

The reaction with amine hardener is fairly straightforward: a hydrogen atom in the amide group reacts with the epoxy end group from the resin to form a link as shown in Fig. 2.18.

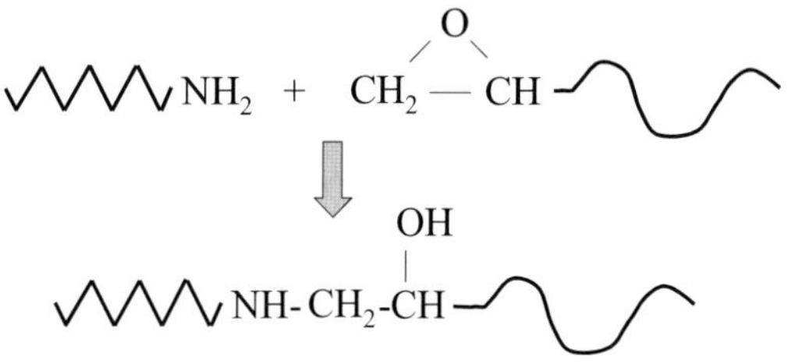

Figure 2.18.  *Epoxy resin: first stage of amine cure.*

It can be seen that one hydrogen on the amine group remains.  This is able to react with a further epoxy group to form an additional link, so the functionality of the amine group is 2 (Fig. 2.19).

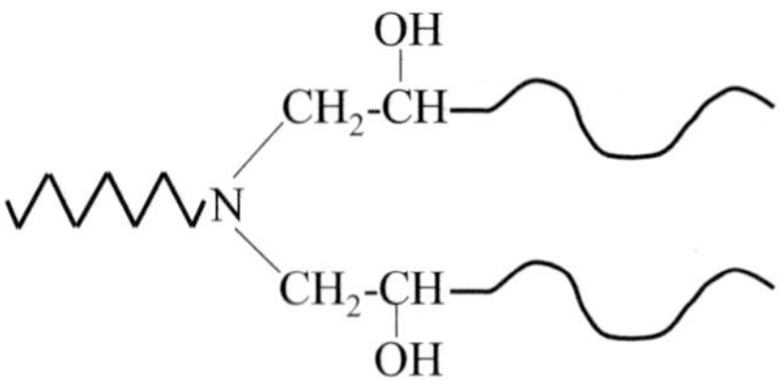

*Figure 2.19. Epoxy resin: amine group linking two epoxy resin chains.*

Multi-functional amines are usually employed as hardeners. Many different types are used depending on the needs of the application, and common examples include di-ethylene triamine, tri-ethylene tetramine, methylene dianiline and isophorone diamine. Figure 2.19 shows that the hardener forms a significant part of the cured resin, and therefore large amounts need to be added to the epoxy prior to processing in order to achieve the correct stoichiometric ratio to ensure equivalence between the reactive groups on the resin and hardener. It can be seen that further cross-linking side reactions are possible, most notably via the hydroxyl groups that form as a result of the opening of the epoxy ring.

The curing of epoxy resins using anhydrides is more complex than amine cure, and special catalysts are usually needed to achieve the necessary cure rate. Simplifying the reaction path, the overall reaction can be represented by (Fig. 2.20):

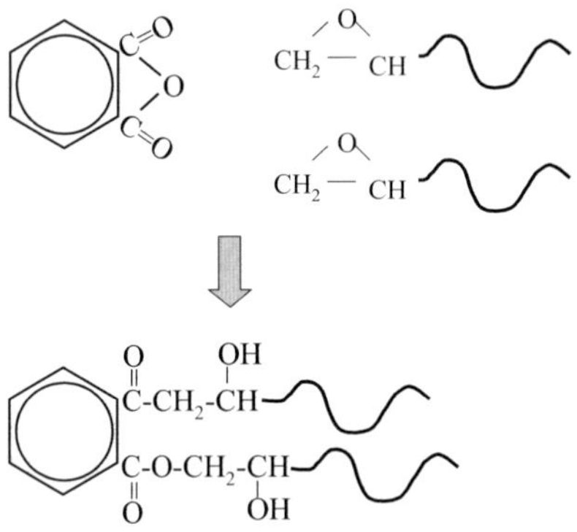

*Figure 2.20. Epoxy resins: anhydride cure- simplified reaction.*

It can be seen that the acid anhydride has an effective functionality of 2.  Although phthalic anhydride is shown here, several different types of anhydride are used.

In addition to the curing methods briefly outlined here, there are several other reagents commonly used in the manufacture of epoxy resins, including catalytic reagents that can act to cross-link the resin using its own reactivity, without the addition of hardeners.

The reactivity and adaptability of the epoxy group enable many variants of the basic epoxy system to be produced.  In relation to fire performance, for instance, the char-forming characteristics of the resin may be enhanced by incorporating novolac sequences into the main chain, as shown in Fig. 2.21.

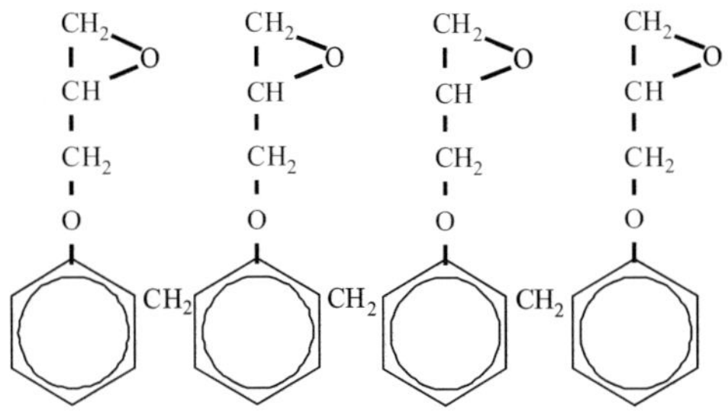

Figure 2.21.  Representative structure of epoxy novolac resin.

Processing of epoxy resins is less straightforward than unsaturated polyesters and vinyl esters. This is because the cure reactions are slower than with free radical cure.  Many epoxy resins and hardeners are viscous or even solid at room temperature and may require preheating or dilution to lower their viscosity prior to processing into composites.  Further heating, sometimes to temperatures up to 150-180°C, is often needed to promote resin gelation and curing.  Nevertheless, epoxies are used in all the traditional composite processes, including wet laminating, resin transfer moulding and filament winding.

Some of the highest performance epoxy composites are processed via the prepreg method.  Here the reinforcement is continuously impregnated with a precise quantity of resin in a highly controlled process.  Following this the resin is heated to commence the cure process.  Once cure has advanced to a point just short of gelation, the prepreg is cooled to below room temperature (usually –5 to –15°C), stored and shipped.  This partially cured resin in the prepreg is referred to as being in the 'B' stage. Full cure is then induced by heating in an oven or autoclave following fabrication of the part.

The TGA curve for a DGEBA-based epoxy tested in nitrogen is shown in Fig. 2.22. Decomposition of most epoxies occurs via random chain scission reactions over the temperature range of about 380 to 450°C. In amine- or amide-cured epoxies, the nitrogen linkages have lower bond dissociation energies than the ether or ester linkages, and therefore chain scission occurs at the C-N bonds. The hydroxyl groups are also vulnerable to degradation at high temperature. The scission reactions decompose 80-90% of the original polymer weight into almost 100 different volatile compounds, which are mainly various types of substituted alkylated phenols, aromatic ether derivations and other flammable organic species [18]. These compounds provide a fuel source for the decomposition reaction to continue until the epoxy is completely degraded. Between 10% to 20% of the original polymer weight is transformed into a highly porous char, and in the presence of air this will start to oxidise above 550°C. As with polyester composites, the high yield of flammable volatiles produced in the decomposition reaction is the main reason for the relatively poor fire performance of epoxy matrix composites. The fire reaction properties of epoxy composites are outlined in Chapter 3.

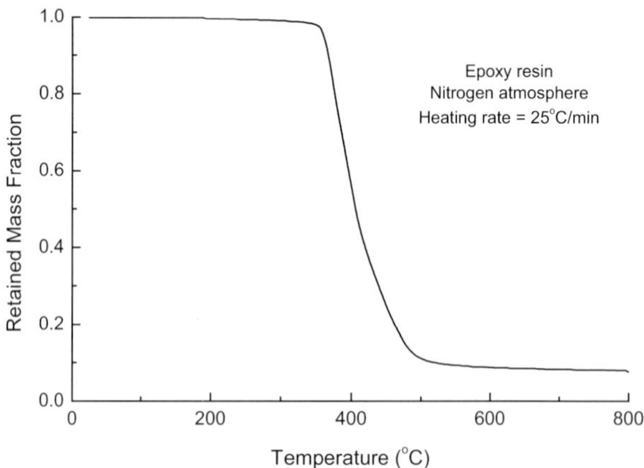


*Figure 2.22. TGA curve for an epoxy in nitrogen atmosphere.*

*Phenolic Resins*
Phenolics are the oldest form of fully synthetic polymer, and are still widely used, often in heat or fire-related applications. The overall polymerisation reaction involves the condensation of phenol and formaldehyde, which takes place with the elimination of water. Phenol has a functionality of 3 (ie. sites where cross-linking can occur) and formaldehyde has 2. Phenolic resins are never manufactured by direct combination of the two precursors because of the difficulty of dealing with the large quantity of heat and water evolved. Instead, pre-polymers of two different types - resoles and novolacs - are used. Resole pre-polymers contain a small excess of formaldehyde, over the

stoichiometric requirement, and are polymerised with an alkaline catalyst (Fig. 2.23). This results in a pre-polymer in which methylol groups are present in addition to methylene cross-links.  The excess formaldehyde is stored in the methylol groups.

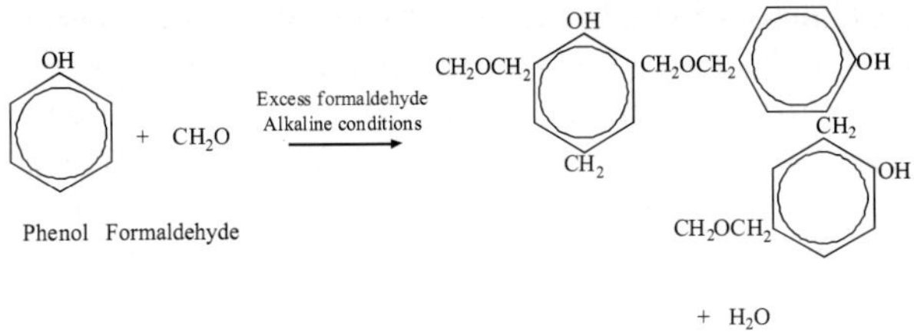

Figure 2.23.  *Phenolic resole-type resin preparation, showing an indicative structure.*

The majority of structural phenolic composites processed by hand laminating, filament winding and pultrusion involve liquid resoles as the pre-polymer.  These resins contain 7-15% of water resulting from the initial polymerisation, and further water is evolved during cross-linking and cure.  The final cured structure contains a random three-dimensional network of aromatic rings, cross-linked at some of the '2' and '4' carbon atoms by methyl groups. Although some steps may be taken to permit the release of water from the laminate during cure, this still leads to a high degree of porosity in the final product which affects both mechanical properties and subsequent water ingress. Nevertheless, resole-based phenolics may be used in structural applications, where they compete with polyesters in terms of properties.  The cure process for resoles usually involves strong Lewis acid catalysts, such as sulphonic acid, which can result in corrosion problems when metal dies or tooling are used.

Novolacs, on the other hand, contain a small excess of phenol and employ an acid catalyst to prepare the pre-polymer (Fig. 2.24).  Novolac phenolics are used in hot cure processes, where hexamethylene tetramine (HMT) is used to provide the additional methylene cross-links.  Cure takes place in mildly alkaline conditions.

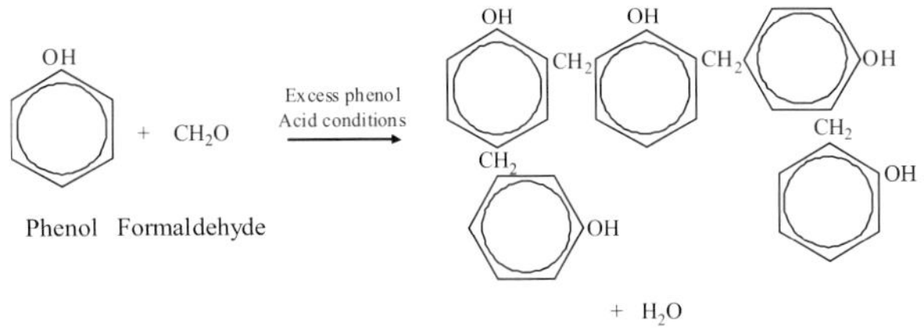

Figure 2.24.  *Phenolic Novolac-type resin preparation, showing an indicative structure.*

The thermal decomposition behaviour of phenolic resins has been studied extensively because of the usage of phenolic matrix composites in high temperature applications (eg. rocket nozzles, ablative heat shields) and fire resistant components [eg. 19-24]. A TGA curve for a glass/resole phenolic composite measured in nitrogen atmosphere is presented in Fig. 2.25. Unlike polyester, vinyl ester, epoxy and many other polymer systems, the retained mass of resole phenolic decreases with increasing temperature in several stages that is indicative of a multiple-order decomposition process. The first stage occurs between 100 and 300°C, and involves a small loss in mass due to the vaporisation of water. The volatilisation of unreacted monomers, phenols, formadehyde and free species from the catalyst also contribute to the mass loss in the first stage. During this stage the polymer network remains largely intact. Above ~300°C the second stage commences, which mainly involves scission reactions between the dihydroxydiphenylmethane units along the chain, with elimination of some volatile by-products. These reactions are partially oxidative in nature, and the resin itself can act as an oxygen source. The reaction rate reaches a maximum in the second stage, and a variety of volatile gases are produced including CO, methane, phenol, cresols and xylenols. In contrast to many other thermosetting systems, much of the higher molecular weight aromatic material remaining after the scission reactions is able to condense to form a solid material.

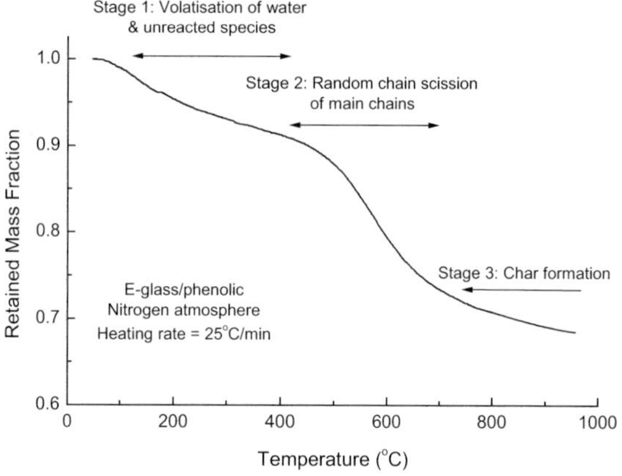


*Figure 2.25. TGA curve for a glass/phenolic composite, showing different regions of decomposition and condensation.*

The final stage involves the further fusion of aromatic rings into a carbonaceous char, with some further evolution of volatiles. The temperature at which char begins to form increases with the heating rate, as shown in Fig. 2.26. This suggests when a phenolic matrix composite is exposed to fire then the char formation temperature will decrease quickly with distance below the hot surface.  Due to the high aromatic content, 40-60% of the original mass of phenolic resins is transformed into char, resulting in a much lower yield of flammable volatiles compared to many other polymers: hence the superior fire performance and low flammability of phenolic composites.  The fire reaction properties and flammability of phenolic laminates are described in next chapter.

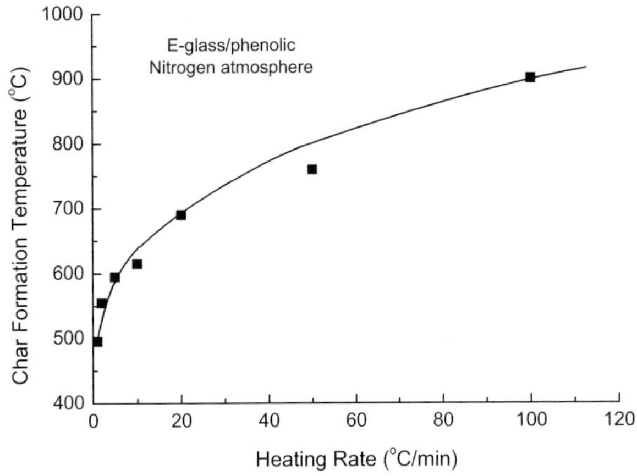


*Figure 2.26.  Effect of heating rate on the formation temperature of char in a glass/phenolic composite.  Data from Ninan [23].*

### 2.4.3    THERMOPLASTIC RESINS FOR COMPOSITES

Thermoplastic matrix composites presently account for a lower tonnage of composites than thermosets, but this proportion is increasing, driven by the need for recyclability, faster cycle times and cleaner processing technologies.  Many types of thermoplastic are used including polypropylene (PP), polyethylene terephthalate (PET), thermoplastic polyester, polyamide (PA), poly ether ether ketone (PEEK) and polyphenylene sulphide (PPS) (Fig. 2.27).  For brevity, however, the decomposition reactions of three of the commonly-used thermoplastics are discussed – PP, PEEK and PPS.

*Polypropylene (PP)*
Polypropylene (PP) is an olefin that is widely used for cost-effective composite parts that can be rapidly processed. It is presently the most widely used thermoplastic matrix material; being used in preference to polyethylene (PE) because low viscosity grades are readily available which aids impregnation.  Another advantage is its very low polarity, that is a characteristic of all olefins, which can be easily modified by the addition of coupling agents to promote strong bonding with inorganic fibres.  PP is

generally less thermally stable than PE because of the 'tertiary' carbon atom that constitutes a weak point in the backbone chain.

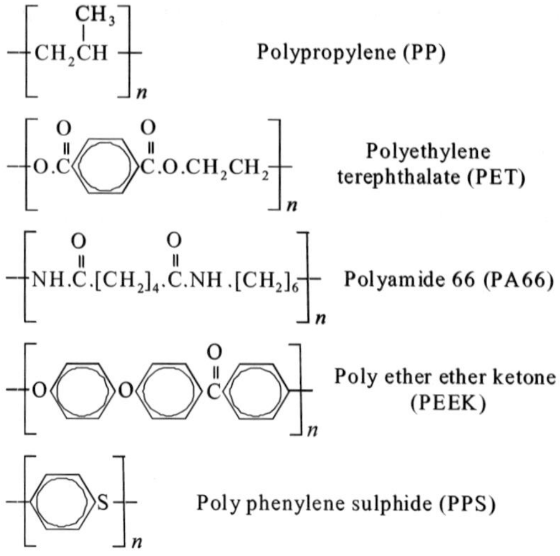

Polypropylene (PP)

Polyethylene terephthalate (PET)

Polyamide 66 (PA66)

Poly ether ether ketone (PEEK)

Poly phenylene sulphide (PPS)

*Figure 2.27.  Thermoplastic resins used in composite materials.*

The first step in the thermal decomposition of olefin polymers is generally 'homolytic dissociation', i.e. main chain scission at a tertiary carbon atom in the case of PP.  This reaction forms two free radicals which go through a chain reaction mechanism to attack the chain at further random points resulting in the formation of a range of olefinic fragments.  All olefin polymers tend to decompose completely into volatile products, leaving no char. In the presence of oxygen, free radical formation is accelerated which reduces the temperature at which the onset of decomposition takes place.  Oxygen also modifies the composition of the decomposition products to give aldehydes in addition to olefinic products.

The poor thermal stability of PP can be altered by the addition of free radical absorbing stabilisers. These also improve resistance to degradation during high temperature processing and to UV attack at ambient temperature, the latter being a problem with all olefin polymers.  The stabiliser compounds are rendered ineffective during the process of absorbing free radicals so, unfortunately, their effect on hindering decomposition is not permanent.

*Poly ether ether ketone (PEEK)*
PEEK has good thermal stability and a high melting point for a thermoplastic (380°C), and for these reasons is used in elevated temperature applications in aircraft and other

demanding performance applications.  Despite its high processing temperature, the melt viscosity of PEEK is relatively high which necessitates special processing methods for PEEK composites, the commonest example of which is known as aromatic polymer composite (APC) [25].  The monomer unit to PEEK is characterised by three aromatic units in the main chain, with both the ether and ketone linkages having a high level of thermally stably.

Day et al. [26] and Moulinie et al. [27] found that the decomposition of PEEK begins at about 500°C by a primary random chain scission reaction of the ether bonds and then, under more severe pyrolysis conditions, random scission of the ketone linkages.  While decomposition occurs by two scission reactions, PEEK appears to degrade in nitrogen in a single-stage process between 500 and 640°C.  The major volatile decomposition product is phenol, with smaller amounts of other organic gases including benezene, dibenzofuran, diphenoxybenzene, bis(phenoxyl)benzophenone and naphthalene.  Many of these gases decompose into lower molecular weight volatiles as they diffuse from the reaction zone through the hot char region towards the fire.  An important feature of the decomposition reaction is the high yield of char (~60% of the original mass) due to the high concentration of aromatic rings in the polymer chain.  Chain fragments containing aromatic rings fuse via a condensation reaction at high temperature to form a stable carbonaceous char.  The high char yield of PEEK, with the consequent reduction in combustible volatiles, results in flammability resistance superior to most other thermoplastics as well as the styrenic or solvent monomer-based thermosets.

*Polyphenylene Sulphide (PPS)*
PPS is also used in high performance thermoplastic matrix composites because, unlike PEEK, it forms a melt phase of relatively low viscosity which assists impregnation and processing.  PPS decomposes in a single-stage reaction over the temperature range 380 to 500°C [28].  Decomposition occurs by random scission of the C-S bonds along the PPS chain followed by depolymerisation and cyclisation reactions. The major volatile pyrolysis products are cyclic tetramer, linear trimers and dimers, benzene and benzenethiol. The char yield of PSS is about 60%, which is due to the high density of aromatic ring compounds in the molecular chain. A distinct feature of many thermoplastics used in composites is their high char yield.  PPS both yield a high fraction of char that is comparable with phenolic resin [29].  The fire reaction properties of thermoplastic composites are described in great detail in the next chapter.

2.4.4    THERMAL DECOMPOSITION OF FIBRE REINFORCEMENTS

Various types of fibres are used to reinforce polymer composites, most notably glass, carbon/graphite, aramid (eg. Kevlar$^{TM}$) and, less commonly, extended-chain polyethylene (eg. Spectra®, Dyeema®). The thermal stability and fire response of these fibres is described in this section.  Other types of reinforcement, including polyesters, rigid rod polymer fibres, basalt, silica, alumina, silicon carbide and natural fibres (eg. jute, sisal) are only used in relatively small amounts, and therefore their thermal stability is not considered here.

Glass is without doubt the most widely used reinforcement, accounting for more than 90% of all composite materials containing fibreglass. Glass fibres are chemically inert in fire and retain chemical and physical stability at high temperature and heat flux. E-glass fibre, which is the most used type of fibreglass, remains unaffected by fire until heated to ~830°C when softening and viscous flow starts and melting occurs at ~1070°C. However, the mechanical properties such as strength and creep resistance decrease over a range of temperatures well below the softening temperature. The other type of fibreglass often used in composites is S-glass, which has superior high temperature properties to E-glass. The softening and melting temperatures of S-glass fibres are ~1050 and 1500°C, respectively. The temperature of most fires is typically within the range of 500-1100°C, and therefore E-glass fibres have excellent fire resistance. It is only in extremely high temperature fires, such as gas-jet fires, when the fibres will soften and melt [30]. Upon cooling the molten glass can fuse which can slow the rate of heat conduction and act as a barrier against the release of flammable volatile gases. Under these conditions, fused glass fibres can reduce the flammability of composite materials.

While glass fibre possesses excellent fire resistance, the organic agents contained in the size coating to the fibres are not immune to fire. Glass fibres are usually covered with a thin layer of organic sizing agents, such as organosilanes, film formers, antistatic agents and lubricants, to provide chemical adhesion with the polymer matrix as well as binding, anti-static and abrasion resistant properties required during fibre processing and handling. These sizing agents chemically degrade in fire and release flammable volatiles [31]. However, since the film is very thin and the mass fraction is small (<2% of total fibre mass) the size has little influence on the fire properties of fibreglass composites.

Carbon fibres are available in various types, depending on their method of production. The two most common types are polyacrylonitrile-based (PAN) and pitch-derived (PDF) carbon fibres. Despite some morphological differences both types respond to fire in a similar way. When carbon fibres are exposed directly to fire their surface can be oxidised. The threshold temperature for oxidation of PAN- and pitch-based fibres is about 350 and 450°C, respectively, and the rate of oxidation increases rapidly above these temperatures. Sussholz [32] and McKee and Memeaulth [33] report that trace impurities within carbon fibre (such as sodium) act as a catalyst to the oxidation process, and this can cause thinning of the fibre. Sussholz [32] has also shown that metallic impurities together with flaws such as submicron-sized voids in the graphitic structure can cause fibrillation of carbon fibres in fire. The axial splitting of fibres into small fibrils can become a health hazard when released from a burning composite into the smoke plume where they can be inhaled, as described in chapter 12. It is important to note, however, that in most types of fire the extent of oxidation is small because most carbon fibres within a composite are protected by char. It is usually only fibres at the hot composite surface exposed directly to the fire and an oxygen-rich environment that experience significant oxidation, and only then when the heat flux is high.

The most common type of organic reinforcement is aramid fibre. 'Aramid' is short for 'aromatic polyamide' or more specifically poly($p$-phenylene terephthalamide), as shown in Fig. 2.28. Aramid fibres (Kevlar®, Twaron®) are based on rigid rod polymers possessing high tensile stiffness and strength in the chain direction. The thermal stability of aramid has been studied in a variety of atmospheres (eg. air, oxygen, nitrogen, vacuum) [34-39], and TGA curves for a fibre tested in air and nitrogen are shown in Fig. 2.29. Decomposition of aramid fibres is characterised by a two-stage process. A small loss in mass occurs in the first stage when the fibres reach about 100-120°C, and this is due to the loss of absorbed water. Aramid fibres have a glass transition temperature of about 300°C, when significant softening and loss in strength occurs. The TGA curve remains relatively flat until the main decomposition reaction occurs at about 450 and 500°C in air and nitrogen, respectively. In an inert atmosphere, the main pyrolytic transformation occurs over a narrow temperature range (500-575°C) involving a substantial break-down of the main polymer network structure by a random scission process. Due to the high aromatic content of the polymer chain, the fibres yield a high amount of char (~40% of the original mass) as a by-product of the decomposition reaction when heated in nitrogen. The char is composed of polyaromatic compounds that will eventually decompose in an oxidising atmosphere. In addition to char, aramid fibre yield various gases, with a large amount of carbon dioxide and smaller amounts of carbon monoxide, aromatic compounds, hydrogen cyanide, nitrogen oxides and other gases, depending on the combustion conditions. Despite the thermal instability, aramid fibres are inherently flame resistant with a limiting oxygen limit value of 29.

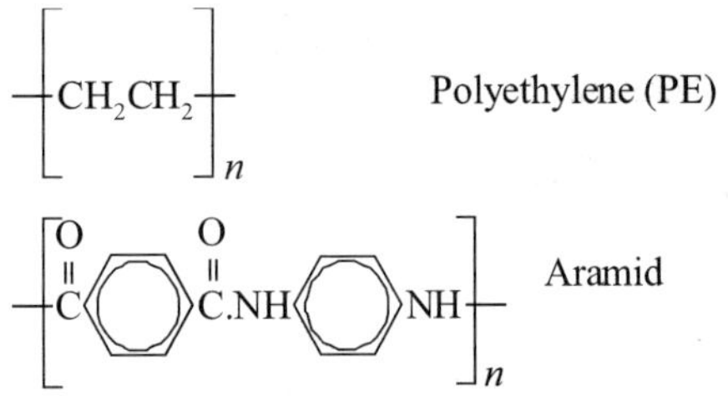

$$\left[ CH_2CH_2 \right]_n \qquad \text{Polyethylene (PE)}$$

$$\left[ \begin{array}{c} O \\ \| \\ C \end{array} \bigcirc \begin{array}{c} O \\ \| \\ C \end{array}.NH \bigcirc NH \right]_n \qquad \text{Aramid}$$

*Figure 2.28. Chemical structure of aramid and polyethylene fibres.*

Unlike aramid fibres, polyethylene fibres melt at a temperature well below the decomposition range. Ultra-high molecular weight polyethylene fibres are highly flammable and decompose more rapidly than aramid at elevated temperature [40,41]. The molecular structure of polyethylene is shown in Fig. 2.28, and consists of repeating -CH$_2$- units. Polyethylene fibres melt at about 145°C and therefore when polyethylene composites are exposed to fire the surface fibres usually melt before they and the surrounding polymer matrix have experienced significant decomposition. Polyethylene remains chemically stable until ~290°C when the molecular weight begins to decrease due to scission of the main chain.  More extensive degradation by random scission reactions occurs at temperatures above 370-390°C.  Polyethylene decomposes into paraffinic and olefinic compounds, particularly high amounts of linear *n*-alkanes and *n*-alkenes, which volatise leaving little or no residual char.

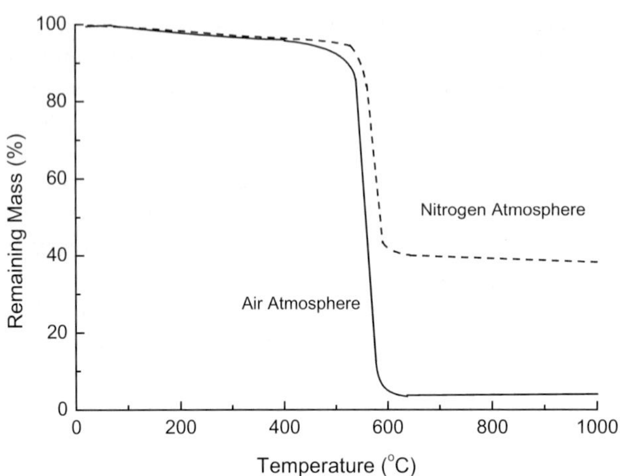


*Figure 2.29.  TGA curve for aramid fibre tested in oxygen and nitrogen.*

The decomposition of polyethylene into organic volatiles makes the fibres much more flammable than aramid fibres, which yield a lower amount of flammable gases.  Brown et al. [41] compared the fire reaction properties of extended-chain polyethylene and aramid fibres when exposed to an external heat flux, and found significant differences in their fire performance.  Table 2.1 compares the time-to-ignition, heat release rates, smoke density, yields of CO and $CO_2$ gases, and mass loss of polyethylene and aramid fibres when fire tested in air at an incident heat flux of 50 kW/m$^2$.  It is seen that the polyethylene fibres ignite more rapidly, have much higher peak and average heat release rates, and yield higher amounts of smoke and CO and $CO_2$ gases, which clearly demonstrates their inferior fire performance.

*Table 2.1. Fire reaction properties of polyethylene and aramid fibres.  (Incident heat flux of 50 kW/m$^2$).  Data from Brown et al. [41].*

| Fire Reaction Property | Polyethylene Fibres | Aramid Fibres |
|---|---|---|
| Time-to-ignition | 31 s | 185 s |
| Peak Heat Release Rate | 691 kW/m$^2$ | 63 kW/m$^2$ |
| Average Heat Release Rate | 275 kW/m$^2$ | 50 kW/m$^2$ |
| Average Smoke Extinction Area | 426 m$^2$/kg | 66 m$^2$/kg |
| Average Carbon Monoxide Yield | 0.0190 kg/kg | 0.0018 kg/kg |
| Average Carbon Dioxide Yield | 2.11 kg/kg | 1.09 kg/kg |

## 2.5        Fire Damage to Composites

### 2.5.1     INTRODUCTION

The damage caused to laminates and sandwich composites in fire has been a topic of intensive investigation in recent years [42-49].  Most attention has been given to determining the damage experienced by thermoset matrix laminates because of their usage in aerospace and ship structures, and much less given to thermoplastic composites. The major types of fire damage suffered by laminates are char formation, softening and degradation of the matrix and organic fibres, and delamination and matrix cracking. Figure 2.30 shows a through-thickness view of a composite exposed to one-sided heating in a fire, and the damage is seen to occur in different zones through the material. Also shown is the variation in resin content in the through-thickness direction.  The hot surface exposed directly to the fire is the first region to thermally decompose to char, which appears as a black layer in the figure.  The polymer content in this region is negligible because the matrix has completely degraded, and any residual organic material has condensed into char.  Below the char zone is a thin region called the decomposition zone where the polymer matrix has been heated to above the decomposition reaction temperature but below the char formation temperature.  In this region the matrix is partially degraded, usually by scission of the chains into high molecular weight fragments that are too heavy to vaporise.  However, the decomposition process is not complete and therefore the matrix has not been reduced to char and combustion gases.  Below the decomposition zone the composite contains delamination cracks between the plies and fine matrix cracks within the plies.  The region nearest the cold surface to the laminate has not been affected by the fire because the temperature is too low to cause any softening or decomposition of the matrix.  With increasing exposure time to a fire, the decomposition zone and char zone move progressively towards the unexposed surface, and eventually the polymer matrix is completely degraded to char.

The fire damage experienced by sandwich composite materials is somewhat different to that incurred by laminates due to the core material [48].  Figure 2.31 shows a cross-section view of a sandwich composite after being exposed to one-sided heating for various times. This material, which consists of glass/vinyl ester face skins and PVC foam core, was exposed to a heat flux of 50 kW/m$^2$ for various times up to 30 minutes. Other types of sandwich composites with combustible skins and core exhibit fire damage generally similar to that shown in Fig. 2.31. The face skin exposed to the heat flux experiences char formation, resin softening and degradation, delamination and matrix cracking. Degradation of the core occurs once the face skin has become severely degraded and is unable to provide significant thermal protection. Decomposition of the core causes it to detach from the charred face skin, and with increasing exposure time the decomposition and char zones move towards the unexposed face skin. The PVC core has experienced severe decomposition and volatilisation that leaves little residual char between the skins, which is a problem commonly experienced with low char yielding core materials, including Nomex paper honeycomb and polyurethane foam.

Certain types of core material, such as phenolic foam and balsa wood, yield a relatively large amount of char and therefore provide better structural and dimensional stability in fire.

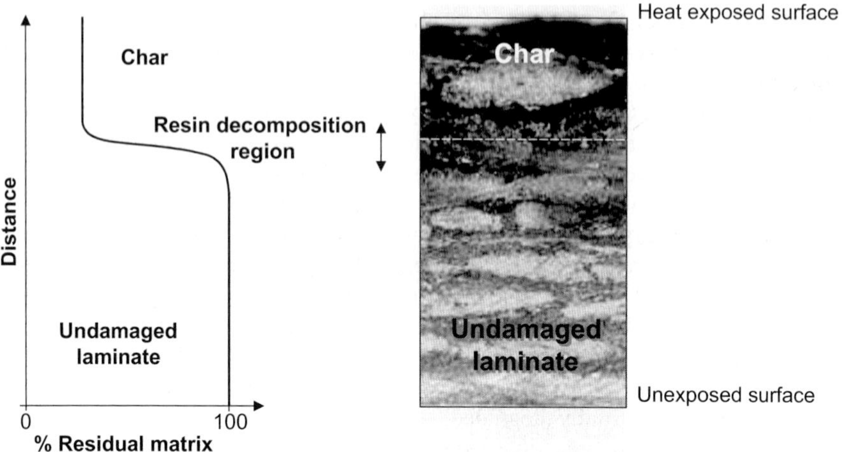


*Figure 2.30. Section of a fire-damaged laminate showing the different damage zones. The variation in resin content through the laminate is shown schematically.*

## 2.5.2    CHAR FORMATION

The formation of a char layer is an important process because it can promote significant flame retardation. Polymers with high char yield generally possess longer ignition times, lower heat release rates, slower flame spread rates, and generate less smoke and toxic gases than low char-forming polymers [50,52]. For example, Fig. 2.33 shows a plot of limiting oxygen index (LOI) against char yield for various polymers and polymer composites. The LOI is a measure of the minimum amount of oxygen in the atmosphere needed to sustain flaming combustion of a material, and this parameter is often used to rank the relative flammability of different materials. An increase in the LOI value is an indirect measure of the increased flammability resistance of a material. It is seen in the figure that the LOI values increase with the char yield, and this is because char reduces flammability in several ways. Firstly, in certain cases the char acts as a thermal insulation layer because the thermal conductivity of char can be lower than the conductivity of the virgin composite material. For example, Fanucci [53] reports that the thermal conductivity of solid char at room temperature is 0.17 $Wm^{-1}K^{-1}$, whereas the in-plane conductivity for a virgin carbon/epoxy laminate is many times higher at about 8-12 $Wm^{-1}K^{-1}$, depending on fibre content. Low density and highly porous chars tend to provide the best thermal insulation.   In cases of reduced conductivity, the char layer reduces the conduction of heat to the underlying virgin material and thereby slows the decomposition reaction rate of the polymer matrix and organic fibres.  As the char layer increases in thickness the decomposition reaction rate becomes progressively slower, and eventually the composite may self-extinguish when insufficient heat is conducted through the char to the underlying material.  However, chars produced from aromatic polymers generally have a dense, continuous network structure that can accelerate the rate of heat conduction and therefore the decomposition reaction rate.

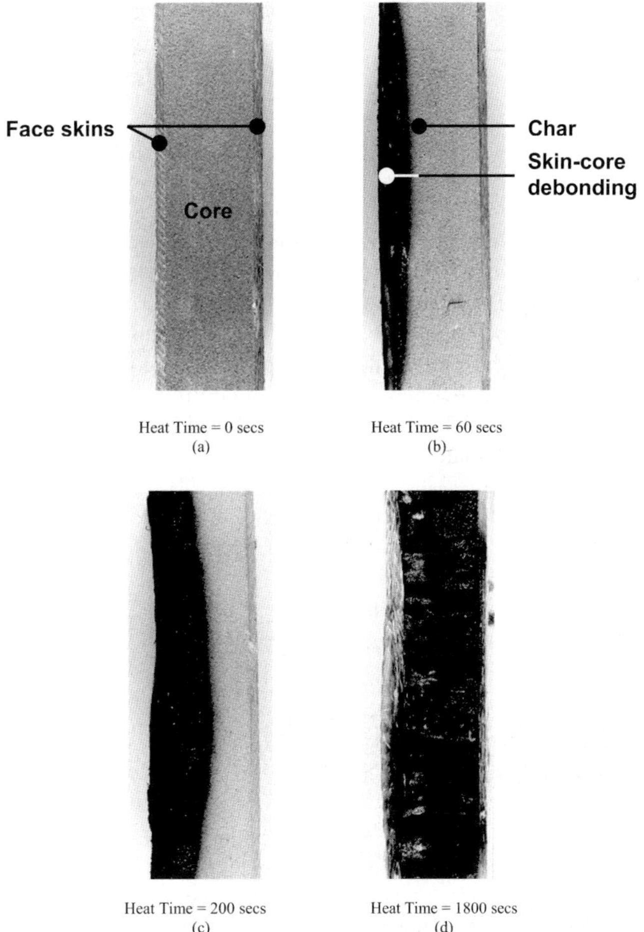


*Figure 2.31.  Section of a fire-damaged sandwich composite showing the different damage zones. The composite was exposed to a heat flux of 50 kW/m$^2$ for different times.*

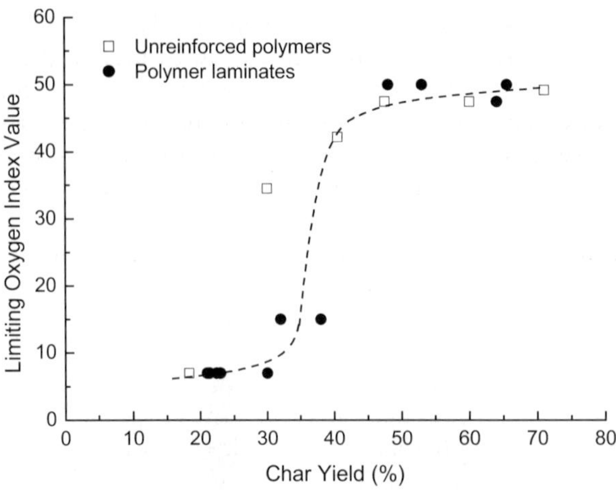



*Figure 2.32.  Effect of char yield on the limiting oxygen index (LOI) of polymers and their composites.*

Char can also improve fire resistance by limiting the access of oxygen from the atmosphere to the region of the composite undergoing decomposition, which again can slow the combustion rate.  Furthermore, char can act as a barrier against the flow of volatiles from the decomposition zone, thereby delaying ignition, slowing flame spread, and reducing the heat release rate.  In some types of composite materials, the volatiles may become trapped as gas bubbles, which upon cooling solidify into a highly porous char structure [52].  Figure 2.33 shows voids produced by volatile evolution within the char of a carbon/epoxy laminate.

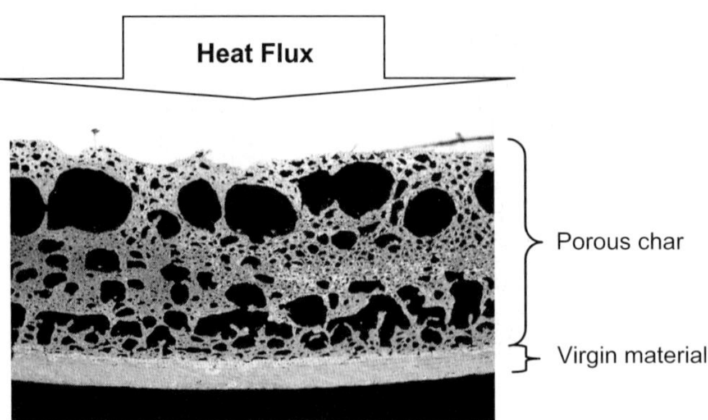



*Figure 2.33. Section of a fire-damaged carbon/epoxy laminate showing a porous char structure.*

Finally, char can help retain the structural integrity of a fire-damaged composite by holding the fibres in place after the polymer matrix has been degraded. However, for char to be effective in providing fire retardation it must form a continuous network structure that possesses low thermal conductivity and gas transport properties. Furthermore the char must adhere strongly to the underlying composite, otherwise it can flake off and expose virgin material directly to the fire. A discontinuous char structure containing cracks and fissures provides a pathway for the escape of flammable volatiles into the flame, and thereby reduces the effectiveness of the char layer to provide fire protection.

The formation and growth of char has been investigated for a variety of thermoset polymer composites under different fire conditions [15,42-48,54]. Figure 2.34 shows the typical increase in char layer thickness with time for polyester and phenolic matrix composites. In this figure the char layer thickness ($d_c$) is normalised to the total thickness of the composite ($d$). It is seen that the formation and growth of char in both materials occurs in three stages. An initial delay in the onset of char growth occurs because of the time required to heat the composite surface to the char formation reaction temperature. Once the char has formed, it grows in thickness at a fairly steady rate with increasing time until the front of the char layer approaches the rear face of the composite. At this point the remaining portion of virgin composite becomes 'thermally thin', which accelerates the growth rate of the char layer until the entire material is degraded. Various models are available to predict the formation and growth of char in composites exposed to fire, and these models are described in chapter 5.

The rate of char formation can be highly dependent on the orientation of the fibres, particularly in composites with a large difference in the thermal properties of the fibre reinforcement and polymer matrix. Table 2.2 shows approximate thermal conductivity values for common reinforcing fibres, along with resins for comparison.

The axial thermal conductivity values for common engineering fibres are higher than those for most polymers, which are typically in the range 0.2 - 0.4 $Wm^{-1}K^{-1}$. While the thermal properties of glass fibres are nearly isotropic, carbon fibre and all the organic fibres have anisotropic thermal conductivities with the higher value in the axial direction. As a result, faster heat conduction occurs along the fibre direction when composites are exposed to fire. Carbon fibres, in particular, have fairly high axial thermal conductivity, and it is with carbon reinforced composites where the effect of fibre orientation on char formation is the most apparent. Kucner and McManus [43] compared the char growth process in the through-thickness and in-plane directions of a carbon/epoxy laminate exposed to fire, and found a significant difference. Figure 2.35 shows plots of char growth with time in the two directions, and it is seen that the in-plane growth rate is much higher.

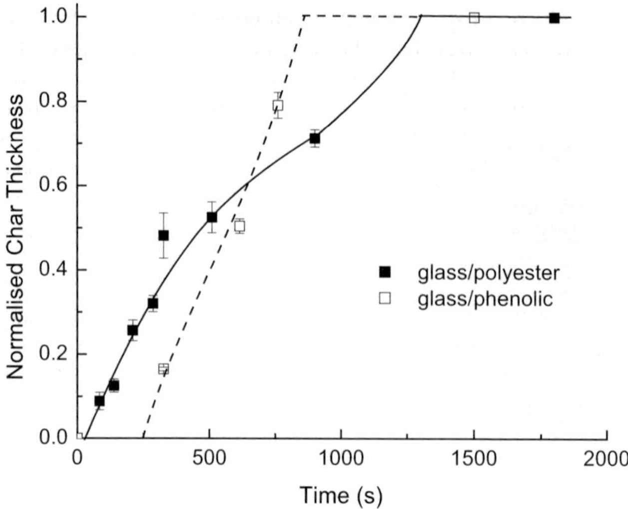

Figure 2.34. Effect of time  on the char layer thickness for glass/polyester and
glass/phenolic composites.

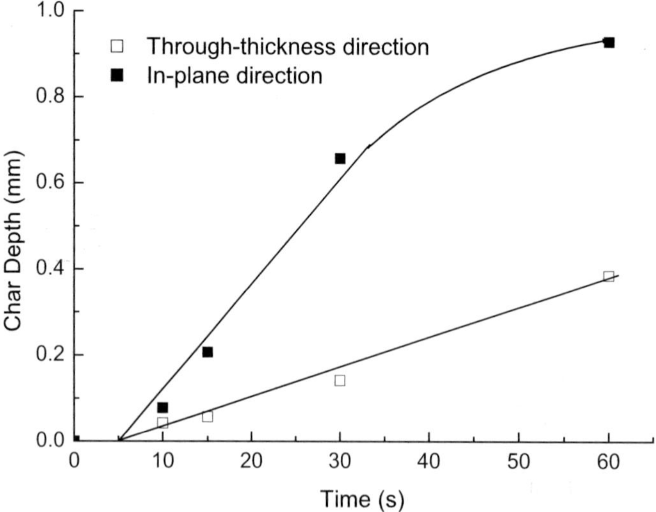

Figure 2.35.  Anisotropic char growth in the in-plane and through-thickness directions of a
carbon/epoxy laminate in fire.  Reproduced from Kucner and McManus [41].

*Table 2.2.   Thermal conductivity of reinforcing fibres and resins.*

|  | $k_{parallel}$<br>$Wm^{-1}K^{-1}$ | $k_{perpendicular}$<br>$Wm^{-1}K^{-1}$ |
|---|---|---|
| E-glass fibre | 1.13 | 1.13 |
| PAN-based carbon fibre | ~20 | 0.32 |
| Aramid (Kevlar 49) fibre | 0.52 | 0.16 |
| Polyethylene fibre | ~20 | 0.35-0.5 |
| Polyester resin | 0.19 | |
| Vinyl ester resin | 0.19 | |
| Phenolic resin | 0.25 – 0.38 | |
| Epoxy resin | 0.23 | |
| Polypropylene resin | 0.18 | |
| PEEK resin | 0.25 | |

## 2.5.3    DELAMINATION DAMAGE

Delamination cracking between the ply layers and matrix cracking within the plies often occurs ahead of the char zone when composites are exposed to fire [31,42,46,49,55-60]. This damage may be confined solely to the reaction zone, or may be spread through the reaction zone and underlying virgin material.  Figure 2.36 shows typical examples of delamination and matrix cracking caused by fire.  It is believed the cracking is due in part to the internal pressure build-up in the material due to volatile formation and, in some cases, vaporisation of trapped moisture [57-59,61,62].  The temperature in the delamination cracking region is well above the glass transition temperature to most polymer systems used in composites, so the phenomenon can be readily explained by the combination of pressure build-up and matrix softening.

The steep temperature gradient that can occur through composites may also exacerbate delamination because of the thermally-induced strain gradient between plies.  Cracking is usually more severe when there is a large difference in fibre orientation between neighbouring plies, which increases the thermal expansion mismatch. The depth of delamination cracking increases with time as the heat is conducted through the composite [42,49,52-60]. Figure 2.37, for example, shows how the penetration depth of delaminations in woven glass/polyester and chopped glass/polyester laminates increase with fire exposure time.

Delamination, and to a certain extent matrix cracking, can be expected to have a significant affect on fire behaviour due to the formation of unbonded interfaces between plies.  Although these interfaces may result in a significant increase in thermal resistance, which will slow the thermal decomposition rate [61], the magnitude of the effect has yet to be quantified. However, Bates and Solomon [54] also pointed out that matrix cracks provide a pathway for the rapid release of combustion gases. Furthermore, when laminates are located above a fire source or in a vertical orientation, delaminations can cause plies to fall off, thereby exposing fresh material to fire.

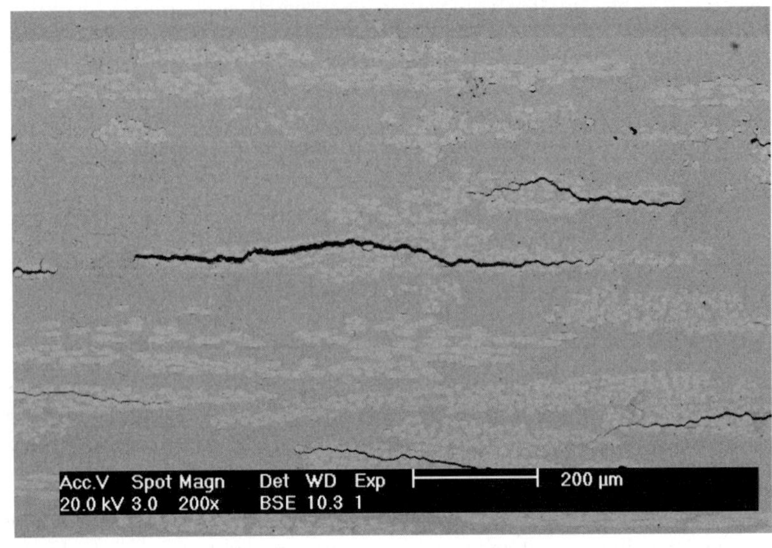

*(a)*

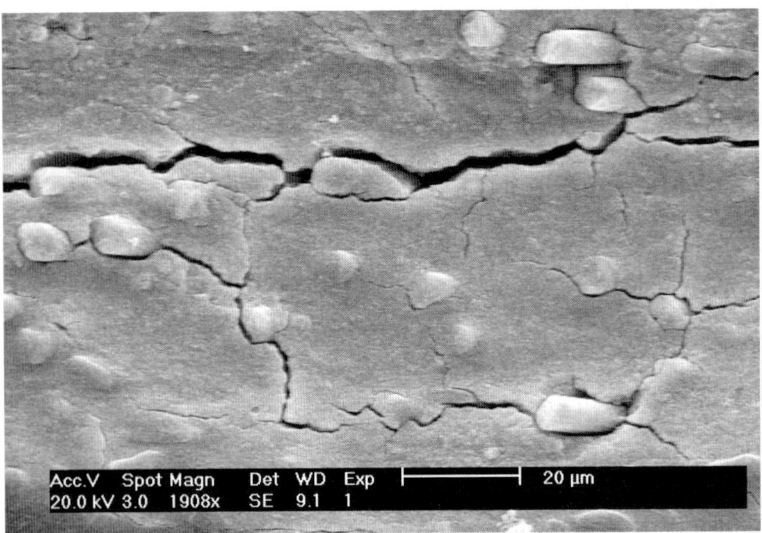

*(b)*

*Figure 2.36. (a) Delamination and (b) matrix cracking within composite laminates exposed to fire.*

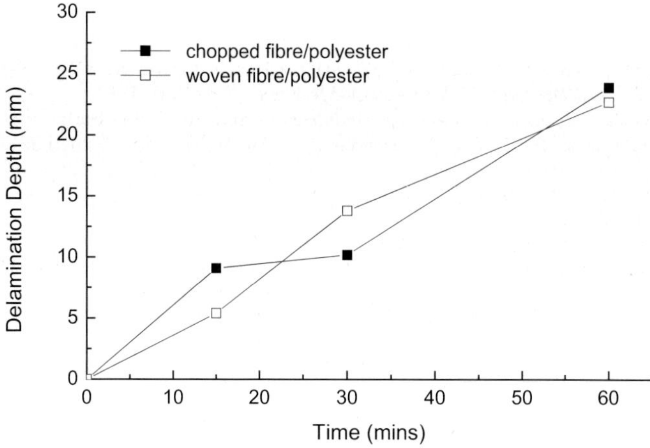


*Figure 2.37. Variation in penetration of delamination cracks in glass/polyester composites with fire exposure time. Adapted from Allison et al. [42].*

## 2.6     Concluding Remarks

A general description of the thermal decomposition and damage to polymer composite materials has been given in this chapter. The decomposition reactions, volatile gases and char yield resulting from the combustion of the polymer systems and organic fibres used in composites have been described. Despite the large amount of research and analysis into the decomposition reactions and the types and amounts of combustion gases produced by these reactions, there is not a complete understanding of the thermal decomposition behaviour of polymers and their composites for various fire scenarios. Furthermore, models for calculating the decomposition temperatures and composition of volatiles are under-developed. Despite this, the information provided in this chapter provides the basis for understanding the behaviour of composite materials in fire, particularly the fire reaction properties (eg. time-to-ignition, heat release rate, smoke density, flame spread) that are described in the next chapter and the potential health hazards of the combustion products (eg. toxic fumes, fibre fragments) that are outlined in Chapter 12. The information given in this chapter also provides the background to flame retardant polymers and nanopolymer composites for use in composites, which are discussed in Chapters 8 & 9. This chapter has also described the physical damage to composites in fire, such as charring, delamination and matrix cracking. The damage has a major influence on both the fire resistive properties (eg. burn-through resistance) and structural properties (eg. stiffness, strength). The fire resistive properties of composites are described in Chapter 3 whereas the structural properties of composites during and after fire are described in Chapters 6 & 7, respectively.

## References


1. S.L. Madorsky. *Thermal Degradation of Polymers*, Robert E. Kreiger, New York, 1985.
2. T. Kelen. *Polymer Degradation*, Van Nostrand Reinhold, New York, 1983.
3. W.L. Hawkins. *Polymer Degradation and Stabilization*, Springer Verlag, Berlin, 1984.
4. N. Grassie and G. Scott. *Polymer Degradation and Stabilisation*, Cambridge University Press, Cambridge, UK, 1985.
5. C.F. Cullis and M.M. Hirschler. *The Combustion of Organic Polymers*, Oxford University Press, Oxford, UK, 1981.
6. L. Pauling. *The Nature of the Chemical Bond*, Cornell University Press, New York, 1960.
7. S. Levchik and C.A. Wilkie. Char formation. In: *Fire Retardancy of Polymeric Materials*, ed. A.F. Grand and C.A. Wilkie, Marcel Dekker Inc, New York, 2000, pp. 171-215.
8. J.A. Parker and D.A. Kourtides. New fireworthy composites for use in transportation vehicles. *Journal of Fire Science*, 1983; 1:432-458.
9. I. Pektas. High-temperature degradation of reinforced phenolic insulator. *Journal of Applied Polymer Science*, 1998; 68:1337-1342.
10. G. Tipping. Composite solutions to the fire performance needs of mass transit. In: *Proceedings of the 3rd Conference on Composites in Fire*, 9-10 Sept 2003, Newcastle, UK.
11. M. Woodward. Modar composites: Achieving very low smoke and toxicity without compromising structural performance. In: *Proceedings of the 3rd Conference on Composites in Fire*, 9-10 Sept 2003, Newcastle, UK.
12. R.J. Bansal, J. Mittal and P. Singh. Thermal stability and degradation studies of polyester resins. *Journal of Applied Polymer Science*, 1989; 37:1901-1908.
13. A.G. Gibson and J. Hume. Fire performance of composite panels for large marine structures. *Plastics, Rubber and Composites, Processing and Applications*, 1995; 23:175-183.
14. T. Arii, S. Ichihara, H. Nakagawa and N. Fujii. A kinetic study of the thermal decomposition of polyesters by controlled-rate thermogravimetry. *Thermochimica Acta*, 1998; 319:139-149.
15. A.G. Gibson, P.N.H. Wright, Y.-S. Wu, A.P. Mouritz, Z. Mathys and C.P. Gardiner. Modelling residual mechanical properties of polymer composites after fire. *Plastics, Rubbers & Composites*, 2003; 32:81-90.
16. S. Knop and W. Krieger. Flame retardant gelcoats on composite laminates. In: *Proceedings of the 3rd Conference on Composites in Fire*, 9-10 Sept 2003, Newcastle, UK.
17. N. Regnier and B. Mortaigne. Analysis by pyrolysis/gas chromatograhy/mass spectrometry of glass fibre/vinylester thermal degradation products. *Polymer Degradation & Stability*, 1995; 49:419-428.
18. J. Vogt. Thermal analysis of epoxy-resins: identification of decomposition products. *Thermochemica Acta*, 1985; 85:407-410.
19. H.W. Lochte, E.L. Strauss and R.T. Conley. The thermo-oxidative degradation of phenol-formaldehyde polycondensates: thermogravimetric and elemental composition studies of char formation. *Journal of Applied Polymer Science*, 1965; 9:2799-2810.
20. G.P. Sullivan and H.W. Lochte. Thermal degradation of polymers. II. Mass spectrometric thermal analysis of phenol-formaldehyde polycondensates. *Journal of Applied Polymer Science*, 1996; 10: 619-635.
21. A. Knop and L.A. Pilato. *Phenolic Resins*. Springer-Verlag, Berlin, 1985.
22. A. Knop and W. Scheib. *Chemistry and Application of Phenolic Resins*. Springer-Verlag, Berlin, 1979.
23. K.N. Ninan. Effect of heating rate on thermal decomposition kinetics of fibreglass phenolic. *Engineering Notes*, 1983; 23:347-348.
24. C.P. Reghunadhan Nair, R.L. Bindu and K.N. Ninan. Thermal characteristics of addition-cure phenolic resins. *Polymer Degradation & Stability*, 2001; 73:251-257.
25. F.N. Cogswell. Thermoplastic aromatic polymer composites: a study of structure, processing and properties of carbon fibre reinforced PEEK and related materials, Butterworth-Heineman, Oxford, 1992.
26. M. Day, J.D. Cooney and D.M. Wiles. The thermal degradation of poly(aryl-ether-ether-ketone) (PEEK) as monitored by pyrolysis-GC/MS and TG/MS. *Journal of Analytical & Applied Pyrolysis*, 1990; 18:163-173.

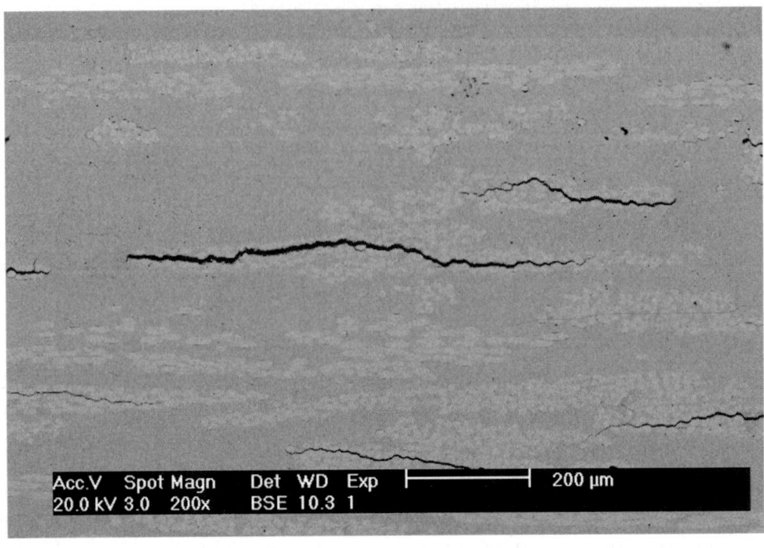

27. P. Moulinie, R.M. Paroli, Z.Y. Wang, A.H. Delgado, A.L. Guen, Y. Qi and J.-P. Gao. Investigating the degradation of thermoplastics by thermogravimetry-fourier transform infrared spectroscopy (TG-FTIR). *Polymer Testing*, 1996; 15:75-89.

28. M. Day and D.R. Budgell. Kinetics of the thermal degradation of poly(phenlyene sulfide). *Thermochemica Acta*, 1992; 203:465-474.

29. D.A. Kourtides, W.J. Gilwee and J.A. Parker. Thermochemical characterization of some thermally stable thermoplastic and thermoset polymers. *Polymer Engineering & Science*, 1979; 19:24-29.

30. P.S. Hill and G.C. White. Jet fire testing and performance of composite materials. In: *Proceedings of the 1st Conference on Composites in Fire*, Sept 1999, Newcastle, UK.

31. J.R. Brown and Z. Mathys. Reinforcement and matrix effects on the combustion properties of glass reinforced polymer composites. *Composites*, 1997; 28A:675-681.

32. B. Sussholz. Evaluation of Micron Sized Carbon Fibers Released From Burning Graphite Composites, NASA CR-159217.

33. D.W. McKee and V.J. Memeault. Surface properties of carbon fibers. *Chemistry & Physics of Carbon*, 1981; 17:1.

34. J.R. Brown and D.K.C. Hodgeman. An e.s.r. study of the thermal degradation of Kevlar 49 aramid. *Polymer*, 1982; 23:365-368.

35. L. Penn and F. Larsen. Physicochemical properties of Kevlar 49 fibre. *Journal of Applied Polymer Science*, 1979; 23:59-73.

36. J.R. Brown and B.C. Ennis. Thermal analysis of Nomex® and Kevlar® fibres. *Textile Research Journal*, 1977; 47:62-66.

37. S. Bourbigot, X. Flambard and F. Poutch. Study of the thermal degradation of high performance fibres – application to polybenzazole and p-aramid fibres. *Polymer Degradation & Stability*, 2001; 74:283-290.

38. S. Bourbigot, X. Flambard, F. Poutch and S. Duquesne. Cone calorimeter study of high performance fibres – application to polybenzazole and p-aramid fibres. *Polymer Degradation & Stability*, 2001; 74:481-486.

39. S. Bourbigot and X. Flambard. Heat resistance and flammability of high performance fibres: a review. *Fire and Materials*, 2002; 26:155-168.

40. J.A. Conesa, A. Marcilla, R. Font and J.A. Caballero. Thermogravimetric studies on the thermal decompostion of polyethylene. *Journal of Analytical & Applied Pyrolysis*, 1996; 36:1-15.

41. J.R. Brown, P.D. Fawell and Z. Mathys. Fire-hazard assessment of extended-chain polyethylene and aramid composites by cone calorimetry. *Fire & Materials*, 1994;18:167-172.

42. D.M. Allison, A.J. Marchand and R.M. Morchat. Fire performance of composite materials in ships and offshore structures. *Marine Structures*, 1991; 4:129-140.

43. L.K. Kucner and H.L. McManus. Experimental studies of composite laminates damaged by fire. In: *Proceedings of the 26th International SAMPE Technical Conference*, 17-20 October 1994, pp. 341-353.

44. A.P. Mouritz and Z. Mathys. Post-fire mechanical properties of marine polymer composites. *Composite Structures*, 1999; 47:643-653.

45. A.P. Mouritz and Z. Mathys. Mechanical properties of fire-damaged glass-reinforced phenolic composites. *Fire & Materials*, 2000; 24:67-75.

46. A.P. Mouritz & Z. Mathys. Post-fire mechanical properties of glass-reinforced polyester composites. *Composite Science & Technology*, 2001; 61:475-490.

47. A.P. Mouritz. Post-fire flexural properties of fibre-reinforced polyester, epoxy and phenolic composites. *Journal of Materials Science Letters*, 2002; 37:1377-1386.

48. A.P. Mouritz and C.P. Gardiner. Compression properties of fire-damaged polymer sandwich composites. *Composites*, 2002; 33A:609-620.

49. P.G.B. Seggewiss. Properties of fire-damaged polymer matrix composites. In: *Proceedings of the 3rd Conference on Composites in Fire*, 9-10 Sept 2003, Newcastle, UK.

50. D. Price, G. Anthony and P. Carty. Introduction: polymer combustion, condensed phase pyrolysis and smoke formation. In: *Fire Retardant Materials*, ed. A.R. Horrocks and D. Price, Woodhead Publishing Ltd, Cambridge, 2002, pp. 1-30.

51. J.P. Fanucci. Thermal response of radiantly heated Kevlar and graphite/epoxy composites. *Journal of Composite Materials*, 1987; 21:129-139.

52. G.A. Pering, P.V. Farrell and G.S. Springer. Degradation of tensile and shear properties of composites exposed to fire or high temperature. *Journal of Composite Materials*, 1989; 14:54-66.

53.   J. Hume. Assessing the fire performance characteristics of GRP composites. In: *International Conference on Materials and Design Against Fire*, London, 1992, pp. 11-15.

54.   S.C. Bates and P.R. Solomon. Elevated temperature and oxygen index apparatus and measurements. *Journal of Fire Sciences*, 1993; 11:271-284.

55.   H.L.N. McManus and G.S. Springer. High temperature thermomechanical behaviour of carbon-phenolic and carbon-carbon composites, I. Analysis. *Journal of Composite Materials*, 1992; 26: 206-229.

56.   H.L.N. McManus and G.S. Springer. High temperature thermomechanical behaviour of carbon-phenolic and carbon-carbon composites, II. Results. *Journal of Composite Materials*, 1992; 26: 230-255.

57.   H.L. McManus. Prediction of fire damage to composite aircraft structures. In: *Proceedings of the 9th International Conference on Composite Materials*, Madrid, Spain, July 1993, pp. 929-936.

58.   J.A. Milke and A.J. Vizzini. The effects of simulated fire exposure on glass-reinforced thermoplastic materials. *Journal of Fire Protection Engineering*, 1993; 5:113-124.

59.   H. Ramamurthy, F.L. Test, J. Florio and J.B. Henderson. Internal pressure and temperature distribution in decomposing polymer composites. In: *Proceedings of the 9th Heat Transfer Conference*, August 1990, Jerusalem, Israel.

60.   Y.I. Dimitrienko. Thermomechanical behaviour of composite materials and structures under high temperatures: 1. Materials. *Composites*, 1997; 28A:453-461.

61.   T. Ohlemiller and J. Shields. One and two-sided burning of thermally thin materials. *Fire & Materials*, 1993; 17:103-110.

# Chapter 3

# Fire Reaction Properties of Composites

## 3.1    Introduction

The fire reaction properties of fibre reinforced polymer composites are described in this chapter.  The properties that are described are time-to-ignition, heat release rate, mass loss, extinction flammability index, thermal stability index, limiting oxygen index, smoke density, smoke toxicity and surface spread of flame.  The fire resistance properties of burn-through resistance and resistance to jet-fire attack are also described, although other resistance properties such as structural capacity in fire and post-fire mechanical properties are discussed in separate chapters later in this book.

The fire reaction properties of a diverse variety of polymer laminates and sandwich composites are discussed in this chapter.  Special attention is given to polyester-, vinyl ester-, epoxy- and phenolic-matrix laminates because of their widespread use in aircraft, ships, offshore platforms, civil structures and rail carriages, although the fire reaction of advanced thermoset and thermoplastic matrix composite materials are also described. This chapter is confined to describing the fire response of composites containing unmodified resins. The fire reaction properties of materials containing flame retardant resins (eg. ATH additives, bromated resins, nanoclay polymers) are outlined in Chapter 8.

## 3.2    Time-to-Ignition

Ignition is an important fire reaction property of flammable materials because it defines the start of flaming combustion.  The organic resins commonly used in composites (eg. polyesters, vinyl esters and epoxies) can ignite within a very short time of being exposed to a hot fire.  Following ignition, composites often burn with large, high-temperature flames that contribute to the rapid spread of fire.  For this reason, ignition is an important property in describing the fire hazard of composite materials.

Ignition usually occurs when the surface of a composite exposed to fire is heated to about the endothermic decomposition temperature of the polymer matrix.  The thermal decomposition reaction of the matrix produces flammable volatile gases that flow from the composite into the fire.  When the amount of volatiles at the composite/fire interface reaches a critical concentration then ignition and combustion flaming will occur.  Most of the volatiles are generated by the endothermic decomposition of the polymer matrix, and depending on the type of resin may include a mixture of flammable components such as carbon monoxide, styrene vapour, aromatic compounds and other low molecular weight hydrocarbons.  Smaller quantities of volatiles can be produced by the decomposition of organic sizing and binding agents that coat the fibre reinforcement.  Volatiles can also be released by organic (combustible) fibres such as aramid and UHMW polyethylene.

The ease of ignition is generally characterised by the time-to-ignition, which is the minimum time required to promote ignition and continuous flaming of a combustible material when exposed to an external heat flux.  The ignition time depends on a variety of factors such as oxygen availability, temperature and the chemical and thermo-physical properties of the polymer matrix and fibre reinforcement.  In this section the effects of resin composition, resin content, fibre composition, fibre sizing agents, specimen thickness, heat flux and the fire atmosphere on the ignition response of composite laminates are described.  The ignition properties of sandwich composite materials are also discussed.

The time-to-ignition of composites is determined by experimental fire testing using techniques such as cone calorimetry (that is described in chapter 11) and the ISO ignitibility test.  Ignition times can be measured in two conditions called spontaneous and piloted ignition. In spontaneous ignition, flaming occurs spontaneously within the flammable vapour/air mixture at the hot composite surface.  Piloted ignition is initiated in a flammable vapour/air mixture by an external pilot or ignition source, such as an electrical spark or flame.  In this section we will principally discuss piloted ignition of composite materials.

Most interest in ignition has centred on composite laminates used in aerospace, marine and infrastructure applications.  For this reason, the ignition times of laminates with polyester, vinyl ester, epoxy and phenolic resin matrices have been determined for a wide variety of fire conditions [1-25].  Figure 3.1 shows the typical effect of external heat flux on the ignition times for fibreglass laminates with different resin matrices. The composites do not ignite below a threshold heat flux, even after exposure to fire for a very long time.  The threshold heat flux below which piloted ignition will not occur for polyester, vinyl ester and epoxy composites is ~13 kW/m$^2$ and for phenolic laminate is ~25 kW/m$^2$.  Ignition cannot occur because the heat flux is too low to heat the composite to the decomposition temperature of the resin matrix.  Above the threshold value, the ignition times decrease rapidly with increasing heat flux. A log-log relationship between decreasing ignition time and increasing heat flux is generally observed for most types of polymer composites. The rapid reduction in the ignition times with increasing heat flux is due to the large increase in the pyrolysis rate. An increase in heat flux (or temperature) increases rapidly the production and flow rates of volatiles to the composite/fire interface, thereby lowering the time-to-ignition [3].

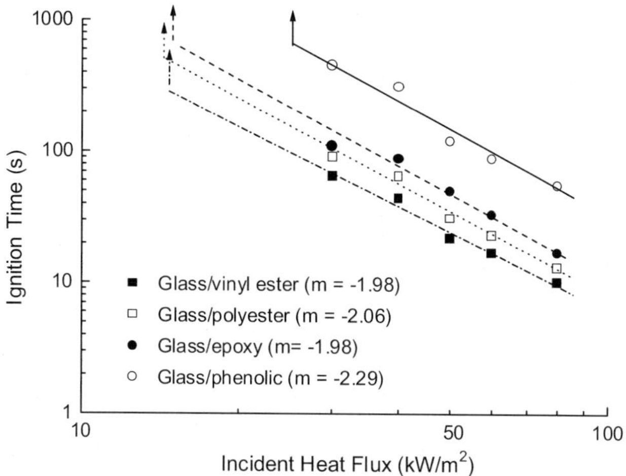


*Figure 3.1. Log-log plot of ignition time against incident heat flux for fibreglass laminates with a polyester, vinyl ester, epoxy and phenolic matrix. The m values define the slope of the curves. (Gibson A.G.; Hume, J. Plast. Rubb. Comp., 23, 1995, 175-183, reference 12, Reproduced with permission of Institute of Metals, Minerals and Mining.)*

Figure 3.1 also shows the ignition times for the phenolic laminate is considerably longer than for the other materials. Delayed ignition is one of the outstanding fire reaction properties of phenolic composites, and is due to the high decomposition temperature, high charring tendency on decomposition, and low yield of combustible volatiles from the phenolic resin matrix [1,6,7,8,10,11,13,15,18,21-23].

One of the main reasons for the common usage of phenolic composites in high fire-risk applications is their excellent ignition resistance. However, there are other high-temperature thermoset resins that provide composites with long ignition times, most notably bismaleimides, polyimides, cyanate esters and phthalonitriles [4,6,14,17,21,26,27]. In addition, Sorathia and colleagues have shown that many thermoplastic composites have excellent ignition times, even at high fluxes, including polyphenylene sulfide (PPS), polyether ether ketone (PEEK) and polyether ketone ketone (PEKK) laminates [4,6,21]. For example, Fig. 3.2 compares the ignition time of a standard glass/vinyl ester laminate against various high-temperature thermoset and thermoplastic composites when fire tested at a heat flux of 75 kW/m$^2$. Compared to the glass/vinyl ester the other composites have much longer ignition times, particularly the phthalonitrile- and cyanate ester-based materials. The excellent ignition resistance of these materials is due to several factors, most importantly high thermal stability and low release of flammable volatiles because the resins yield a high amount of char during decomposition.

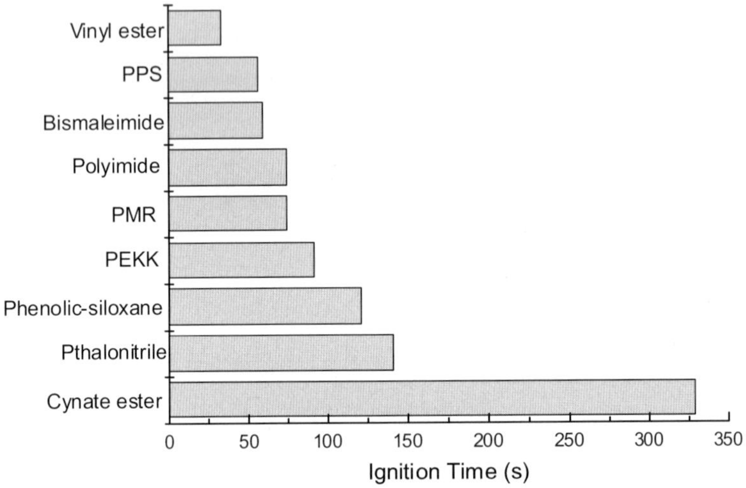


*Figure 3.2. Ignition times for various fibreglass composites with advanced thermoset and thermoplastic matrices at the heat flux of 75 kW/m$^2$. Data from Sorathia [6].*

While the ignition properties of many types of polymer laminates have been determined, there is little published information on the ignition times of sandwich composite materials [23,28,29]. Most interest has focussed on the ignition behaviour of sandwich composites used in boat and ship construction, and relatively little information is available for non-marine sandwich materials. Ignition of sandwich composites can be a complex process because of the influence of the core material. Figure 3.3 is a log-log plot showing the effect of incident heat flux on the ignition times of two types of marine sandwich composites. It is seen that the GRP skin-balsa core composite shows a linear relationship between ignition time and heat flux when plotted on a log-log scale, and this behaviour is identical to single-skin laminate (eg. Fig. 3.1). The GRP skin-PVC foam core composite, on the other hand, does not show a linear relationship at high heat fluxes. Furthermore, the ignition times for this material are considerably shorter at heat flux values under ~50 kW/m$^2$, and this is due to the influence of the PVC core on the ignition process.  When exposed to fire the PVC core melts and decomposes, and this causes an air-gap to form between the skin and core.  This air-gap reduces the rate of heat conduction through the skin, causing it to heat-up and ignite more rapidly.  In the case of the balsa sandwich, on the other hand, heat transfer is not interrupted because the core remains in contact with the skin, resulting in longer ignition times.  Core materials with a low thermal conductivity, such as polymer foams, honeycombs and syntactic polymers, can also lower the ignition time.  These types of core materials are relatively inefficient in conducting heat away from the face skin that is exposed to fire.  This leads to the rapid heating of the face skin that causes early ignition.  This is illustrated in Fig. 3.3 where the thermal conductivity of the PVC core is substantially lower than the balsa core, and this may be partly responsible for the shorter ignition times of the GRP skin-PVC foam core sandwich composite.

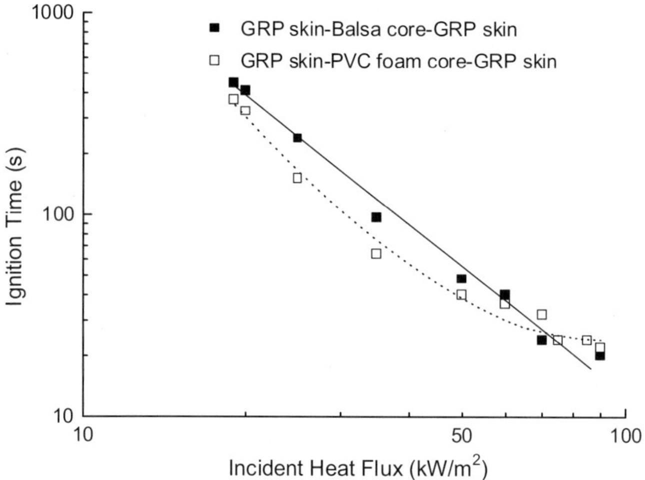


*Figure 3.3.  Effect of heat flux on the ignition times of two sandwich composites.  The difference in the two curves is due to the influence of the core material on the ignition process.  Data from Grenier et al. [28].*

The fibre reinforcement can also influence the ignition time.  Glass and carbon fibres are inert to fire when the heat flux is below 100-125 kW/m$^2$, although the reinforcement can still affect the ignition process.  Le Bras et al. [30] and Kootsookos and Mouritz [24] found that the amount of reinforcement changes the ignition resistance of fibreglass composites.  Figure 3.4 shows the effect of fibre content on the ignition time of a glass/polyester composite.  It is seen that raising the fibre volume content from 0% (ie. neat resin) to nearly 70% causes a substantial increase in the ignition time.  Increasing the fibre content obviously reduces the amount of resin in a composite, and therefore less combustible material is available to produce flammable volatiles needed for ignition.

The fibres can contribute in various other ways to the ignition process. While glass and carbon fibres remain inert in fires with temperatures below ~1000-1200°C, the sizes, emulsion binders and other organic agents applied to fibres during manufacture to promote adhesion, binding, anti-static properties and abrasion resistance will thermally decompose. The fibres are coated with a thin film of organic agents such as film formers, antistatic agents and lubricants to a weight content of ~1-2%. In a fire this coating decomposes and releases flammable volatiles that can reduce the ignition time. The film of organic agents coating the fibres is usually very thin (~0.1 μm), and therefore produces a much smaller amount of volatiles than the resin matrix. However, certain types of reinforcement – most notably chopped strand mats – contain a significant amount of organic emulsion binding agent to bind the fibre strands, and in a fire this can produce volatiles that lower the ignition time. For example, Fig. 3.5 compares the ignition times of woven glass/polyester and chopped glass/polyester composites at different heat fluxes [15]. The glass fibres in both composites are coated with a similar sizing agent, although the chopped glass fibres are also coated with an organic emulsion binder to hold the short fibre strands together as a mat. It can be seen that the chopped glass composite has substantially lower ignition times, particularly at low to intermediate heat fluxes (<50 kW/m$^2$). The generation of flammable volatiles in the decomposition of the emulsion binder is believed responsible for the inferior ignition resistance of the chopped fibre composite [7,15].

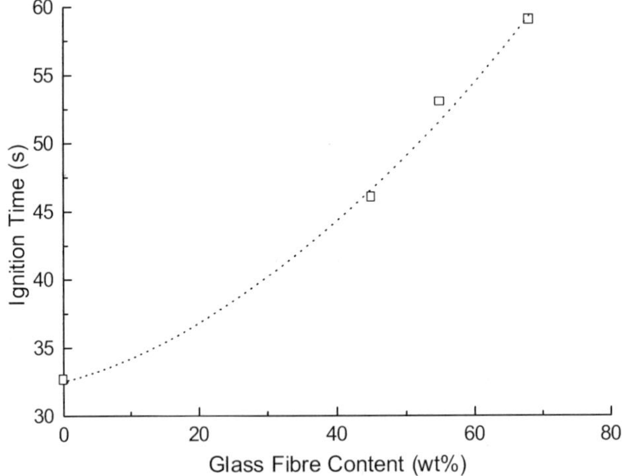

*Figure 3.4. Effect of glass fibre content on the ignition time of a glass/polyester composite. (Le Bras, M.; Bourbigot, S.; Mortaigne, B; Cordellier, G., Poly. & Poly. Comp., 6, 1998, 535-539, reference 30, Reproduced with permission Rapra Technology Limited).*

The ignition resistance can also be reduced by combustion of the fibre reinforcement itself. Combustible organic fibres such as aramid and Spectra (UHMW polyethylene) can thermally decompose along with the organic matrix in a fire, and thereby reduce the ignition time. The ignition behaviour of composite materials containing combustible reinforcing fibres in addition to a combustible organic matrix has not been thoroughly investigated, and is a topic that requires further investigation.

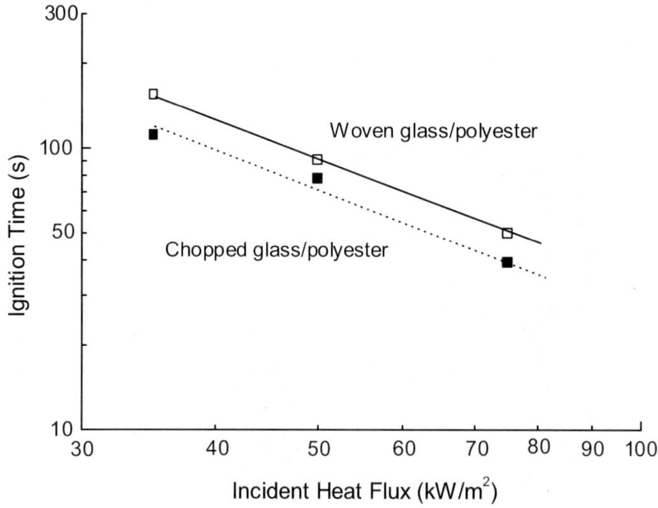


*Figure 3.5. Effect of heat flux on the ignition times of polyester laminates reinforced with woven glass or chopped glass fibres. Data from Brown and Mathys [15].*

In one of few reported studies, Brown et al. [9] examined the ignition properties of composites reinforced with aramid or extended-chain polyethylene fibres. It was discovered that the ignition times of these composites are shorter than composites reinforced with non-combustible fibres such as glass. Figure 3.6 compares the ignitions times for vinyl ester-based composites containing woven glass, aramid or polyethylene fibres at two heat flux levels. It is seen that ignition occurs more readily when organic fibres are present, particularly when polyethylene reinforcement is present. The polyethylene and, to a lesser extent, aramid fibres thermally degrade with the polymer matrix in a fire, resulting in increased volatile release that causes the ignition time to be reduced.

The time-to-ignition is also dependent on thickness [5,7,25]. Figure 3.7 shows the effect of specimen thickness on the ignition time of a glass/phenolic laminate tested at different heat fluxes. It is seen that the thickness-sensitivity is most marked at low heat fluxes. At 35 kW/m² the ignition time increases rapidly and continuously with thickness. At higher heat fluxes, however, the ignition time increases gradually with thickness up to ~3 mm, and above this value the time is almost independent of thickness. Hume [5] suggests that the increase in ignition time with thickness is due to an increase in the thermal capacity of the material, prolonging the time taken to heat the laminate to the resin decomposition temperature.

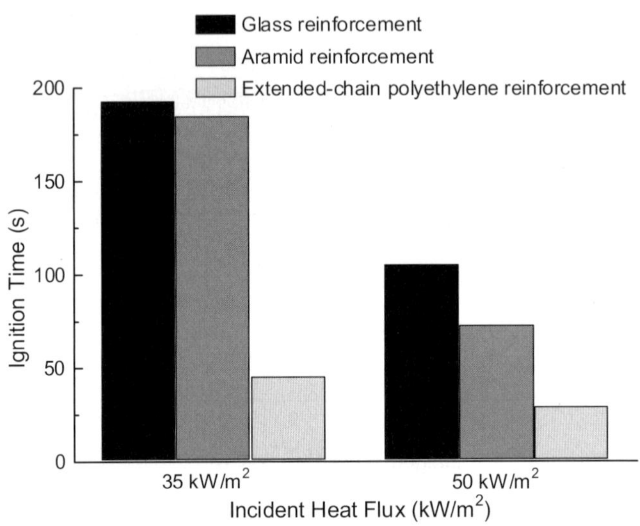


*Figure 3.6. Effect of heat flux on the ignition times of vinyl ester-based laminates containing glass, aramid or extended chain polyethylene fibres. Data for the glass laminate from Brown and Mathys [15] and for the aramid and polyethylene reinforced composites from Brown et al. [9].*

The fire environment also has a large influence on the ignition properties of polymer composites. Most studies into the ignition behaviour are performed under standard atmospheric conditions (eg. 21%O₂/78%N₂). However the atmosphere to some fires can be substantially different. For example, when a fire occurs within an enclosed space without ventilation, the oxygen content drops with time and can reach as low as ~10-12% before the flames are extinguished from the lack of oxygen. Space agencies such as NASA, on the other hand, are concerned about ignition of combustible materials within the oxygen-rich atmosphere of manned spacecraft. Studies into the ignition properties under non-standard atmospheric conditions are limited, although the few studies that have been performed show that the oxygen content of the atmosphere can affect the ignition time.

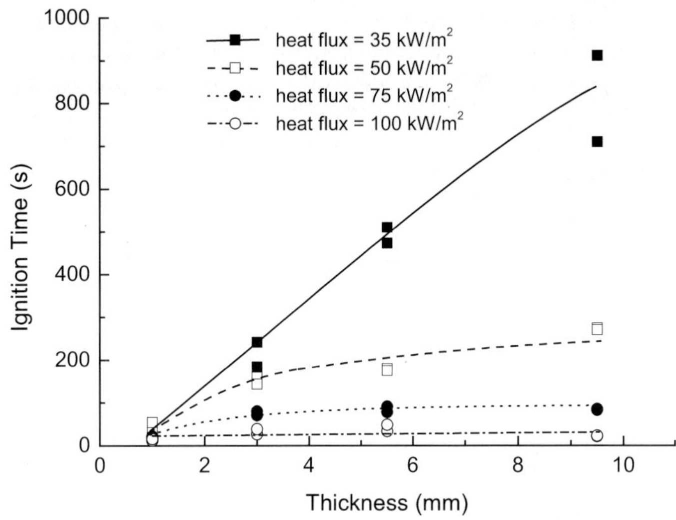


*Figure 3.7. Effect of thickness on the ignition time of a glass/phenolic laminate. (Scudamore,M.J., Fire & Materials, 18, **1994**, 313-325,reference 7. Copyright John Wiley & Sons Limited, Reproduced with permission).*

Hshieh and Beeson [16] investigated the effect of oxygen concentration between 18% and 30% in the fire atmosphere on the ignition time of epoxy and phenolic composites. This investigation was performed using an atmosphere-controlled cone calorimeter that allows the fire properties to be determined under different atmospheric conditions, and this technique is described in chapter 11. Figure 3.8 shows the effect of oxygen content on the time-to-ignition values for carbon, glass and aramid reinforced epoxy laminates. It is seen the ignition times decrease slowly with increasing oxygen content, and this is because more oxygen is available to react in the flame with the flammable volatiles released in the decomposition of the epoxy matrix and aramid fibres. Figure 3.8b shows the influence of oxygen content on the ignition times for several phenolic composites, and in this case different relationships are observed. The carbon/phenolic shows a rapid reduction in the time-to-ignition with increasing oxygen content, the glass/phenolic shows a small increase in ignition time with oxygen concentration, while the aramid/phenolic shows no clear correlation between the oxygen level and ignition time. The reason for the different behaviour is unclear, and further research into the effect of atmospheric conditions on the ignition properties of composites is needed.

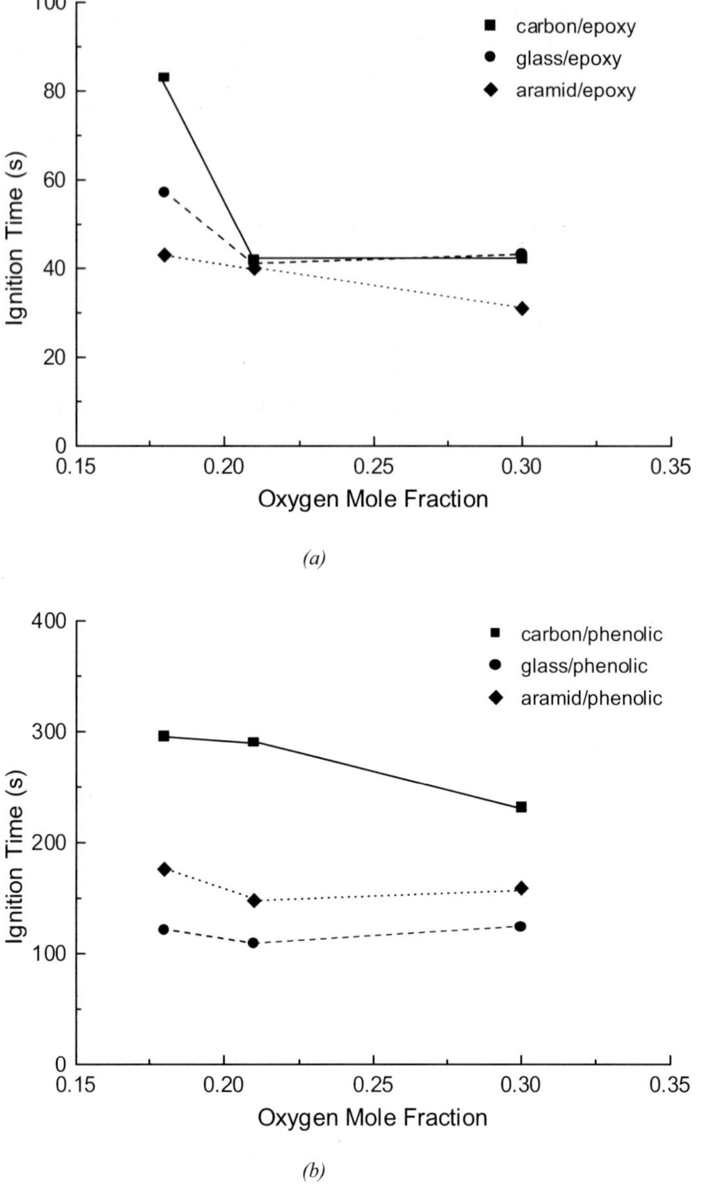



*Figure 3.8. Effect of oxygen content of the ignition times for (a) epoxy-based and (b) phenolic-based composites. (Hshieh, F.-Y.; Beeson, H.D., Fire & Materials, 21, **1997**, 41-49, reference 16. Copyright John Wiley & Sons Limitied, Reproduced with permission).*

Ignition times are generally determined experimentally, because of the difficulty in modelling the ignition mechanisms of composites.  Many theories have been proposed for calculating the ignition times of combustible materials, although in almost all cases the models have only been proven accurate for materials such as woods and plastics and have not been validated for polymer composites.  It is difficult to theoretically model the ignition (and other fire reaction properties) of composites due to the complex decomposition reactions of the resin matrix and combustible fibres, the anisotropic heat flow behaviour, the mass flow of volatiles, and other events that occur in a fire. When composites are exposed to fire they can selectively burn, chemically degrade, char, and release volatile gases, smoke, fumes and heat as well as crack and delaminate.  The complexity of these processes does not permit easy analytical prediction. Nevertheless, several ignition models have been found to give good estimates of the time-to-ignition in certain cases.

Lyon et al. [31] report that ignition will occur when there is sufficient thermal energy at a polymer surface to convert it from a solid to gaseous fuel.  The heat of gasification per unit mass of solid polymer (hg) can be calculated using:

$$h_g = c(T_s - T_o) + (1 - \mu)h_v \tag{3.1}$$

where $T_i$ is the ignition temperature, $T_o$ is the ambient temperature, $h_v$ is the heat of vapourisation of the decomposition products, and $\mu$ is the mass fraction of non-combustible material in the polymer, which can include char, filler particles and/or fibre reinforcement.  The temperature dependence of the heat capacity is $c = c_o T/T_o$, where $c_o$ is the heat capacity at ambient temperature.  Based on this analysis, the heat of gasification is related to the ignition temperature of a polymer via the expression:

$$h_g = \int_{T_o}^{T_{ign}} c(T)dT + (1-\mu)h_v = \frac{c_o T_{ign}^2}{2T_o} + \left[(1-\mu)h_v - \frac{c_o T_o}{2}\right] \approx \frac{c_o T_{ign}^2}{T_o} \tag{3.2}$$

or

$$T_{ign} \approx \left[\frac{T_o h_g}{c_o}\right]^{1/2} \tag{3.3}$$

Knowing the values of $h_g$ and $c_o$ for a polymer, it is possible to estimate the ignition temperature.  Figure 3.9 compares the calculated and measured ignition temperatures for several polymer systems, and good agreement is observed.

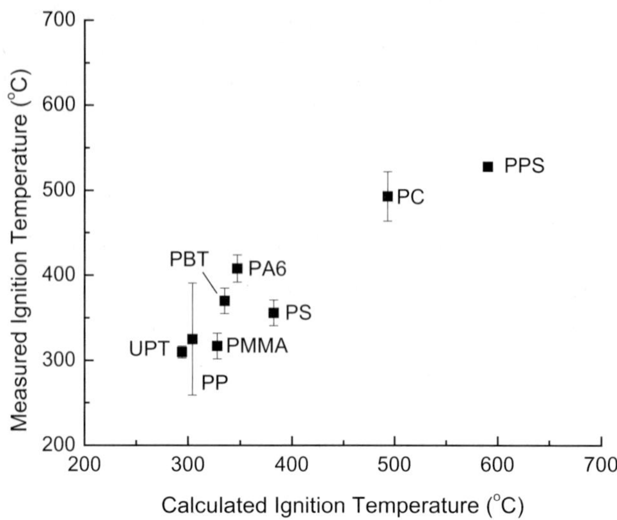



*Figure 3.9. Comparison of calculated and measured ignition temperatures for several polymers. Data from Lyon et al. [31].*

As discussed, one of the critical factors influencing the ignition of combustible materials, including polymer composites, is their thermal thickness (as shown in Fig. 3.7). A material is classified as 'thermally thin' when the heat from the fire (otherwise known as the thermal wave) is absorbed so rapidly that there is no significant temperature gradient through it.

In a 'thermally thick' material, by contrast, a significant temperature gradient exists through the material. Mikkola and Wichman [32] propose that the thermal thickness of a material can be defined by its characteristic thermal conduction length ($\Delta$), which is calculated using:

$$\Delta = \sqrt{\alpha t_i} = \sqrt{\frac{k t_i}{\rho c}} \tag{3.4}$$

where $\alpha$ is thermal diffusivity, $t_i$ is ignition time, $k$ is thermal conductivity, $\rho$ is material density and $c$ is specific heat. This equation is only valid for 'simple' radiative thermal ignition, and does not consider heat losses such as the emissivity of the material. When the characteristic thermal conduction length is greater than the sample thickness, $L_o$, then it is considered to be thermally thin. In practice, this means that most composite materials are thermally thin when less than ~1-2 mm thick. This being the case, then time-to-ignition of a homogenous material can be determined by [32,33]:

$$t_i = \rho c L_o \frac{(T_i - T_o)}{q_{net}}$$

(3.5)

where $T_i$ is the surface temperature at ignition, $T_o$ is ambient temperature, and $q_{net}$ is net heat flux to the surface (including heat losses). This equation shows the ignition time of a thermally thin material decreases linearly with increasing net heat flux.

Composites used in most engineering structures are much thicker than the characteristic conduction length, and therefore when exposed to fire behave as a thermally thick material. Carslaw and Jaeger [34] report that the ignition time of a thermally thick material is proportional to the inverse square of net heat flux to the surface according to the model:

$$t_i = \rho c k \frac{\pi}{4} \left( \frac{T_i - T_o}{q_{net}} \right)^2$$

(3.6)

In this relationship it is assumed that heat losses are negligible and the substrate is an inert, thermally thick and opaque solid. This equation has proven accurate in the theoretical determination of the ignition times for thermally thick specimens of wood (eg. birch, oak) and plastics (eg. polyurethane, polyvinyl chloride, polypropylene), but not polymer composites. Equation 3.6 suggests that a log-log plot of ignition time against heat flux should give a straight line with a slope of -2 for an inert, thermally thick material. The plots in Fig. 3.3 for the fibreglass composites shows values for the slope ($m$) of about -2, confirming this relationship. However, calculating the ignition time using Eqn. 3.6 is difficult because the values of $\rho$, $c$ and $k$ change with increasing temperature, and these need to be empirically determined at the ignition temperature.

There is also the intermediate case between thermally thin and thermally thick. Here, the ignition time can be approximated by:

$$t_i \sim \rho c \sqrt{kL_o} \left( \frac{T_i - T_o}{q_{in} - q_{out}} \right)^{3/2}$$

(3.7)

Further information on models for calculating the time-to-ignition of combustible materials is given in Chapter 4.

Another approach to modelling the ignition times of certain types of composite laminates was developed recently by Gibson et al. [35]. Recent research has shown that the mass flux at ignition of fibreglass composites is about 0.005 kg/m$^2$s [31,35]. Using the thermal model [36] described in Chapter 5, it is possible to calculate the mass loss rate of a composite when exposed to fire, and then relate this to the ignition time.  The mass loss rate controls the concentration of flammable volatiles produced and released by the composite that in turn determines the ignition time.  Gibson et al. [35] observed that when polyester, vinyl ester and epoxy-based laminates are exposed to medium to high heat fluxes (>25 kW/m$^2$) then ignition occurred at the time when the theoretical mass loss rate calculated using the thermal model reached ~0.0075 kg/m$^2$s.  It appears that this mass loss rate is required to produce sufficient flammable volatiles to cause ignition.  Table 3.1 compares the calculated and measured ignition times for several composites, and it is seen the thermal model can accurately determine the ignition times.

*Table 3.1.  Comparison of measured and calculated ignition times for several composite materials.  Data from Gibson et al. [35].*

| COMPOSITE | HEAT FLUX (kW/m$^2$) | THEORETICAL IGNITION TIME (s) | MEASURED IGNITION TIME (s) |
|---|---|---|---|
| Glass/polyester | 50 | 39 | 34 |
|  | 75 | 11 | 12 |
|  | 100 | 6 | 7 |
| Glass/vinyl ester | 50 | 56 | 53 |
|  | 75 | 14 | 18 |
|  | 100 | 6 | 5 |
| Carbon/epoxy | 50 | 79 | 83 |
|  | 75 | 26 | 37 |
|  | 100 | 13 | 18 |

## 3.3    Heat Release Rate

Heat release rate (HRR) is the single most important fire reaction property because the heat released by a burning material can provide the additional thermal energy required for the growth and spread of fire [37,38].  No other reaction property has such a dominant influence on the fire behaviour of composites.  Furthermore, several other reaction properties, such as surface spread of flame, smoke generation and CO emission, are dependant on or related to the HRR [39,40].

Heat release is the thermal energy produced, per unit area of surface, when flammable decomposition products ignite and burn in the vicinity of a material in fire, or subjected to heat flux.  The peak HRR occurs over a very short period of time and often shortly after ignition, and is usually a good indication of the maximum flammability of a material.  The average HRR is the total heat released averaged over the combustion period, and is considered the most reliable measure of the heat contribution to a sustained fire.

A model for estimating the HRR of composite materials is described in Chapter 4, although it is usually determined by experiment because of the difficulty in accurately calculating the value. Various instruments have been developed to measure the HRR properties of composite materials, with the most popular techniques being the cone calorimeter and Ohio State University heat release calorimeter. These instruments are limited to testing small specimens of a composite material. The HRR properties of components, assemblies and structures made of composite material can be measured using large-scale fire tests such as the single-burning item test, intermediate-scale cone calorimeter and room fire calorimeter test (see chapter 11).

The HRR properties have been determined for a diverse variety of composites, including polyester, vinyl ester and epoxy matrix laminates that are used in aerospace, marine and construction applications [3,5,6,10,12,14-16,21,30,39,41]. An example of the HRR response over time for a typical flammable composite material (glass/vinyl ester) that is exposed to a constant heat flux is shown in Fig. 3.10. The HRR was measured using a cone calorimeter and is defined as the heat released per unit time by the composite sample divided by the exposed surface area of the sample. The HRR profile fluctuates considerably over time due to various chemical and thermal events that occur to the composite when exposed to fire, and these events are designated as *A* to *D* in Fig. 3.10. The figure shows an initial induction period (designated *event A*) during which the composite does not release any heat. No heat is released during this period because the exposure time to the external heat flux is insufficient to heat the composite to the decomposition reaction temperature. This delay is followed by a rapid rise in the HRR spectrum (*event B*) due to the sudden, short-term release of heat from the ignition of flammable volatiles released from the resin-rich surface film on the composite. The curve continues to rise to a peak HRR, after which the heat release rate decreases progressively with time due to the formation and growth of char at the hot surface (*event C*). The char reduces the HRR in two ways: (i) it acts as a thermal insulator which retards heat transfer to the underlying virgin material and thereby slows the decomposition reaction rate, and (ii) it limits the supply of combustible gases to the flame front. In some materials, the HRR curve may rise again, and at this stage the specimen is nearly completely degraded by the fire. Only a small portion of virgin composite remains near the unexposed surface of the sample, and behaves in the manner of a thermally thin material. This causes the rapid decomposition of the last remaining amount of resin matrix, resulting in an increase to the HRR. Eventually the HRR becomes negligible when the polymer matrix has completely degraded (*event D*). The profile shown in Fig. 3.10 for the vinyl ester matrix composite is typical of the HRR response of composites that yield low amounts of char, and similar behaviour is found for polyester and epoxy laminates.

A somewhat different HRR response is observed for more fire resistant composites, such as phenolic laminates [10,12,15,41]. The HRR profile for a glass/phenolic composite is shown in Fig. 3.11, and the rate that heat is released by the phenolic composite is substantially lower because of the higher thermal stability and lower release rate of flammable volatiles from the polymer matrix. The start of the release of heat is delayed for a longer time with the phenolic laminate and the amount of heat released is much lower than for the vinyl ester laminate shown in Fig. 3.11. This behaviour is often observed for advanced thermoset and high temperature thermoplastic composites that contain thermally stable resins that yield a high amount of char and generate low levels of combustible volatiles.

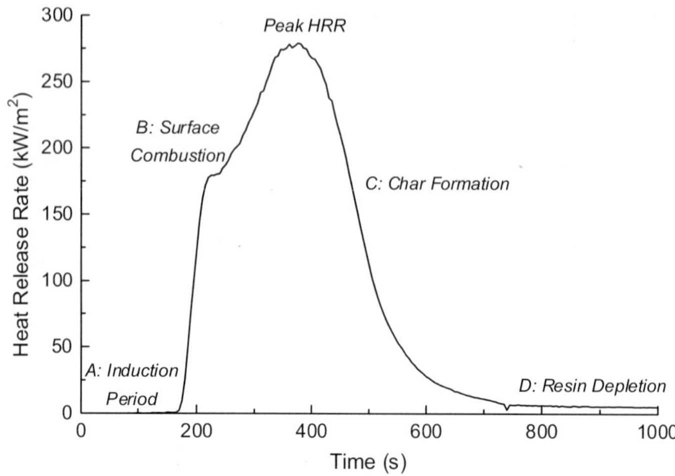


Figure 3.10. Heat release rate profile for a glass/vinyl ester composite exposed to a heat flux of 50 kW/m².

The heat flux of a fire has a significant influence on the rate of heat release from a composite material [10,12,15,17,27,42,43]. Figure 3.12 shows the effect of increasing heat flux on the peak HRR values for a variety of glass/thermoset laminates. It is seen that the heat released by a highly flammable composite material (ie. glass/epoxy) rises rapidly with heat flux due to increases in the decomposition reaction rate and yield of combustibles. In comparison, the amount of heat released by high-performance thermoset composites (eg. phenolic, cyanate ester, phthalonitrile) is much lower at all heat flux levels. In many cases the heat released by these composites is just a small fraction of that released by the epoxy-based laminate. Furthermore, the peak and average HRRs of the high-performance thermoset laminates are less sensitive to increases in the external heat flux.

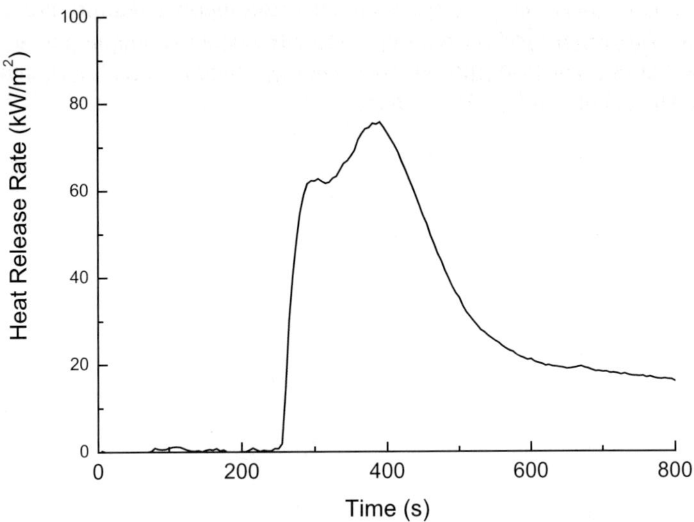

*Figure 3.11. Heat release rate profile for a glass/phenolic composite exposed to a heat flux of 50 kW/m².*

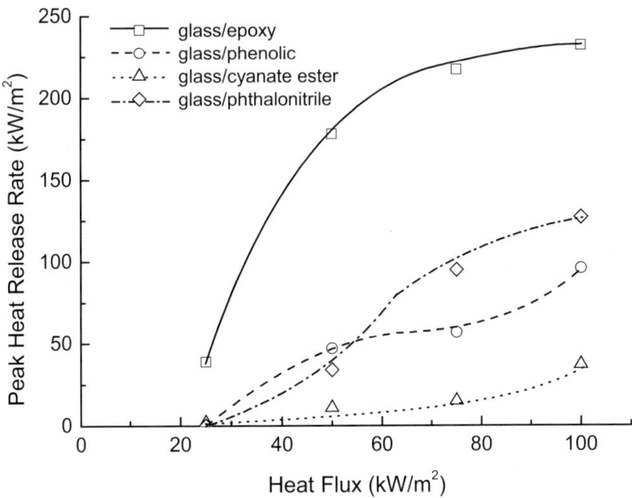


*Figure 3.12. Effect of heat flux on the peak HRR for various glass/thermoset polymer laminates.*

The superior fire performance of these composites is attributed to the highly aromatic composition of the polymer matrix that confers high thermal stability and an ability to yield a high amount of char on burning. High-temperature thermoplastic composites also possess much better fire performance than styrene-based polymer materials [21]. This is shown in Fig. 3.13, which compares the peak and average HRR values for a standard carbon/epoxy laminate against values for a variety of carbon fibre reinforced composites with high-temperature thermoplastic matrices. As with high-performance thermoset laminates, the excellent fire performance of these thermoplastic composites is due to the thermal stability, low release rate of flammable volatiles and high char yield of the polymer matrix.

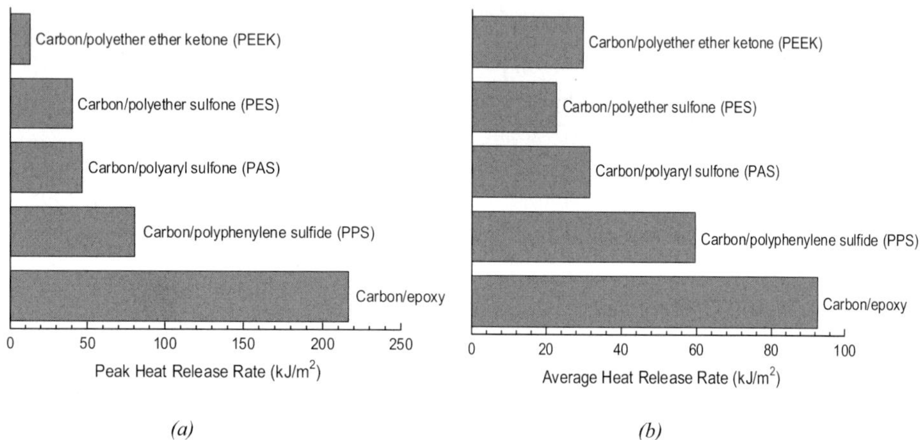


*Figure 3.13. Comparison of the (a) peak HRR and (b) average HRR for a carbon/epoxy laminate against various carbon/high-temperature thermoplastic polymer composites. Data from Sorathia [6] and Sorathia [21].*

The heat released by composite materials can be affected by the fibre reinforcement in several ways. Le Bras et al. [30] and Mouritz and Kootsookos [24] have shown that increasing the fibre content causes a reduction in the amount of heat released by a burning composite. For example, Fig. 3.14 shows a rapid decline in the peak HRR of a glass/polyester composite with increasing fibre content. This behaviour occurs because less polymer matrix material is available to generate heat during thermal decomposition. Numerous studies have found that the type of fabric used to reinforce a composite can also affect the heat release properties [5,7,12,13,15]. In particular, significant differences can occur in the heat release response of composites reinforced with woven roving or chopped mat fabrics. The HRR profile for the chopped fibre composite is characterised by a broad peak, and this is indicative of relatively steady-state decomposition throughout most of the material. The profile for a woven composite, on the other hand, can fluctuate with a series of peaks and troughs over time that is indicative of erratic burning, as shown in Fig. 3.15. These differences in the HRR responses are attributed to the different distribution of glass fibre and resin in the composites. Chopped fibre laminates have a reasonably even distribution of resin between the glass layers, and this facilitates the stable formation and combustion of degradation products when exposed to fire. In contrast, woven fabric composites have a distinct layered construction, with resin-rich regions between the ply layers. When exposed to fire the burning rate through the laminate varies considerably, with higher amounts of volatiles released during decomposition of the resin-rich layers that results in the peaks to the HRR spectrum. As combustion progresses through each glass layer, where less resin is present, the heat release rate is reduced that causes the troughs in the HRR curve.

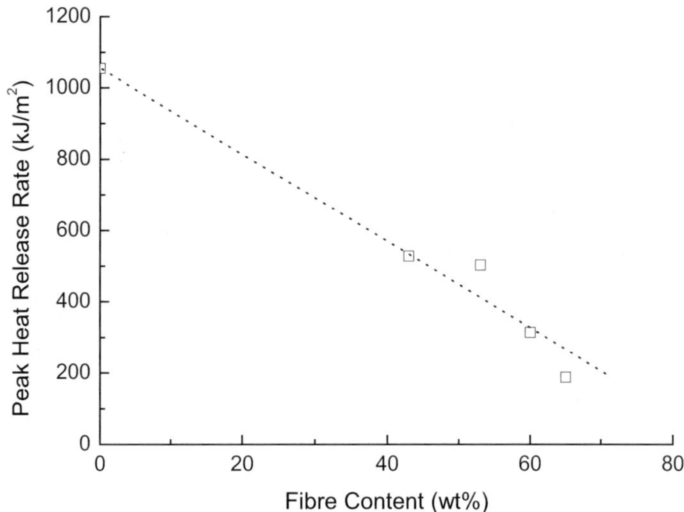

*Figure 3.14. Effect of increasing glass fibre content on the peak HHR of a polyester composite exposed to an incident heat flux of 50 kW/m². (Le Bras, M.; Bourbigot, S.; Mortaigne, B; Cordellier, G., Poly. & Poly. Comp., 6, 1998, 535-539, reference 30, Reproduced with permission Rapra Technology Limited).*

It was described earlier that the thickness of a composite material affects the time-to-ignition. The ignition time generally increases with thickness as the thermal response transforms from thermally thin to thermally thick behaviour. Likewise, the HRR properties are dependent on thickness with numerous fire studies reporting large reductions to the peak and average HRRs with increasing thickness [7,12,25,41]. Figure 3.16 shows the effect of increasing thickness on the heat released by four types of fiberglass laminates.  In this figure the HRR values are expressed as the heat release *per unit volume of composite*, and all the materials have been fire tested at a constant incident heat flux of 60 kW/m$^2$.  It can be seen that the HRR decreases rapidly with increasing thickness to about 8 mm, but above this value the HRR is insensitive to further increases in thickness.  A similar relationship between peak HRR and thickness was measured by Scudamore [7] for glass/phenolic laminates when fire tested at different heat fluxes, and again the HRR was only dependent on thickness at values under 8 mm.  When composites are very thin the heat penetrates rapidly through the material, causing the complete decomposition of the resin matrix within a short period that accelerates the heat release rate.  Composites behave increasingly as a thermally thin material as their thickness is reduced below 8 mm, and this is the cause of the progressive increase in HRR.

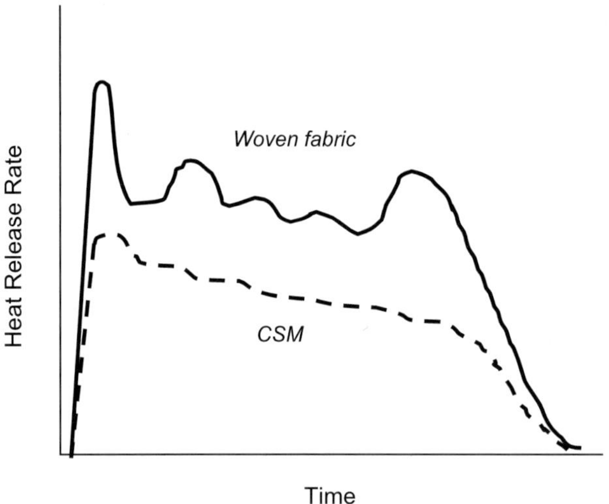


*Figure 3.15.  Comparison of the HHR profiles for a composite reinforced with chopped glass mat and woven glass fabric.*

Fire tests conducted by Ohlemiller and Shields [44] reveal that the dependence of HRR on thickness can be quite complex, particularly when the composite is thermally thin.  Most fire reaction studies on composite materials are conducted as single-sided fire tests, and only rarely are double-sided tests performed.   In the only reported study into the double-sided fire properties, Ohlemiller and Shields [44] compared that heat release behaviour of polymer composites when subjected to one-sided or two-sided burning.  It was found that when a thermally thick composite is exposed to fire attack of equal intensity (ie. identical heat flux) from both sides then the HRR simply doubles because twice as much surface area is involved.    The response of a thermally thin laminate to two-sided burning is more complex than this, and usually the HRR is more than doubled. For example, Fig. 3.17 compares the peak HRR of a thermally thin glass/epoxy laminate when subjected to one-sided and two-sided burning.  In this case the HRR is over two times higher for two-sided burning.  Ohlemiller and Shields [44] report that the thermal process responsible for this effect is not well understood, although they suggest that when a thermally thin composite experiences two-sided burning the thermal waves from the two surfaces merge at a point, resulting in an acceleration in the decomposition reaction rate that is much greater than two.

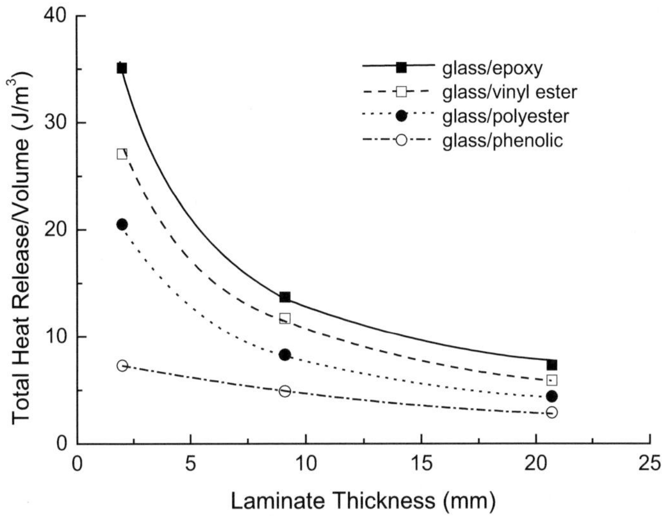


*Figure 3.16.  Effect of thickness on the HRR of various fiberglass composites. The HRR is expressed as the total heat released per unit volume of composite material. (Gibson A.G.; Hume, J. Plast. Rubb. Comp., 23, **1995**, 175-183, reference 12, Reproduced with permission of Institute of Metals, Minerals and Mining.)*

## 3.4      Extinction Flammability Index & Thermal Stability Index

The extinction flammability index (ESI) is a useful quantitative measure of the flammability of a composite material, although it is a property that is rarely determined because of the greater attention given to heat release rate [3,5,15].   The ESI is determined by plotting the average HRR against the incident heat flux, as shown in Fig. 3.18.  The HRR value at which the line intercepts the y-axis (ie. heat flux = 0) is taken to be the ESI.  A negative ESI implies that a material will self-extinguish soon after the heat flux is removed, and an example of this behaviour is shown in Fig. 3.18 for the phenolic laminate.  Other examples of composites with negative ESI values include advanced thermoset resins (eg. polyimide, phthalonitrile) and high temperature thermoplastics (eg. PEEK, PPS, PES).  When a composite has a positive ESI value this indicates that combustion will continue after the external fire has been extinguished, and the glass/polyester laminate shown in Fig. 3.18 is such as material.  In addition to polyester laminates, other composites with positive ESI values are epoxy- and vinyl ester-based materials.  Generally, composite materials containing a resin matrix that has low thermal stability and a high yield of flammable volatiles will have a positive ESI value.

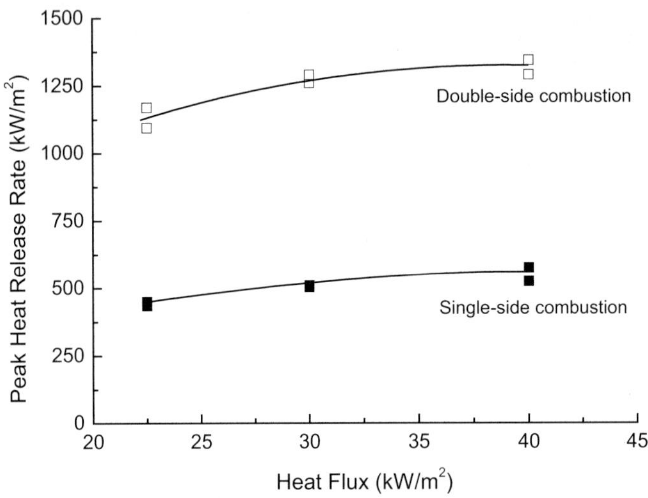


*Figure 3.17. Effect of one- and two-sided burning on a thermally thin glass/epoxy laminate (1.6 mm thick) on the peak HHR. (Ohlemiller, T.;Shields, J.; Fire & Materials, 17, **1993**, 103-110, reference 44. Copyright John Wiley & Sons Limited, Reproduced with permission).*

Another fire reaction property that is used to describe flammability is the thermal stability (or sensitivity) index (TSI).  The TSI provides a value by which the fire performance of materials may be ranked and compared over a range of external heat fluxes, simulating different fire environments [3].  The intercept of such a plot in principle indicates whether flaming combustion of a material is self-sustaining in the absence of an external heat flux.  This index is determined from the same plot as the ESI, but it is the slope of the line (see Fig. 3.18).  A low TSI value indicates that a material will burn near its maximum rate even at low heat fluxes.  The higher the TSI the greater must be the external heat flux for the material to burn at its maximum rate.

Again, high temperature thermoset and thermoplastic composites have relatively high TSI values compared against flammable materials.

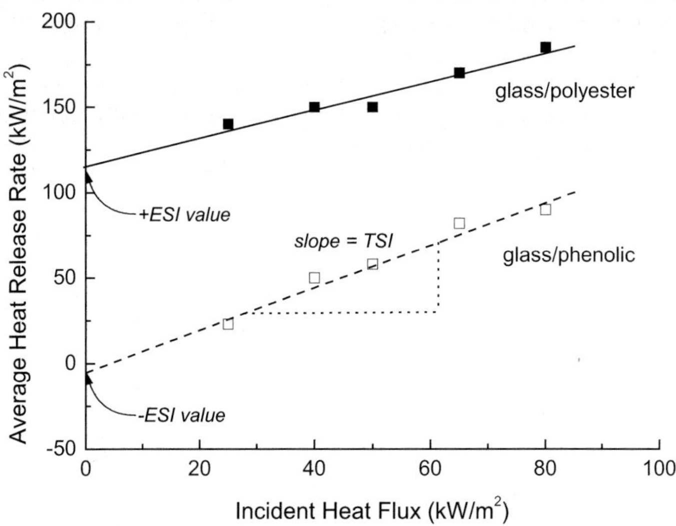


*Figure 3.18. Plots of incident heat flux against heat release rate showing the determination of ESI and TSI for glass/polyester and glass/phenolic laminates. Data from Egglestone and Turley [10].*

## 3.5    Mass Loss

Mass loss is another important fire property because it gives a quantitative measure of the amount of materials that will decompose in fire. The amount and rate of decomposition of the organic constituents of a composite material can be determined over the course of a fire by measuring the weight change of a composite sample using test instruments such as the cone calorimeter. The mass loss of a wide variety of composite materials has been determined under different fire conditions [3,15,45], and two examples of the mass loss behaviour of composites are given in Fig. 3.19. Shown is the weight change over time for glass/vinyl ester and glass/phenolic laminates exposed to a constant heat flux of 50 kW/m². The mass loss curves show four distinct regions that are identified as stages I to IV, and each stage represents a different event in the fire response of the material. Stage I represents the short delay period when the composite shows no change in weight when first exposed to fire. The delay occurs because the composite has not reached the decomposition reaction temperature of the polymer matrix. Stage II is characterised by a rapid loss in mass with increasing time due to endothermic decomposition of the matrix. During stage III a change in the mass loss rate of the composite occurs because most of the polymer matrix has degraded and only a small region of virgin material near the back-face remains unaffected by the fire. In this stage the composite behaves as a thermally thin material, and this usually accelerates the mass loss rate. In stage IV the mass loss curve reaches a constant minimum value because the polymer matrix has been consumed, and this represents the final mass of the degraded laminate (ie. mass of the fibre reinforcement and residual char).

The total mass loss of a composite is determined largely by the char yield of the organic constituents.  Composites that yield only a small amount of char (eg. polyesters, epoxies) experience a greater mass loss than materials with a high char yield  (eg. phenolics, phthalonitriles, thermoplastics).  The mass loss is also higher when combustible fibres (eg. aramid) are used in the composite because the reinforcement decomposes along with the resin matrix.  It has been found that even the organic sizing and binding agents on the fibres can affect the mass loss, with chopped fibre composites that contain a high amount of fibre binding agent experiencing a greater mass loss than composites that do not contain a binding agent [15].

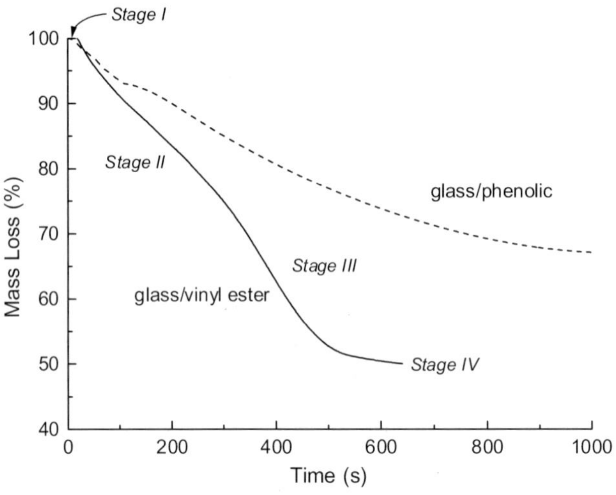


*Figure 3.19.  Mass loss curves for glass/vinyl ester and glass/phenolic laminates.*

Mass loss is one of the few fire reaction properties that can be calculated by thermo-chemical modelling.  Gibson et al. [35] recently showed that the mass loss of laminates with a polyester, vinyl ester or epoxy matrix can be modelled.  For example, Fig. 3.20 compares the measured mass loss curve for a fibreglass/polyester composite against the theoretical curve determined using a thermal model [36]. It is seen the model can predict the change in weight over time with good accuracy. As yet, the mass loss behaviour of composites containing combustible fibres or a highly aromatic polymer matrix cannot be accurately predicted, although research into this problem is in progress.

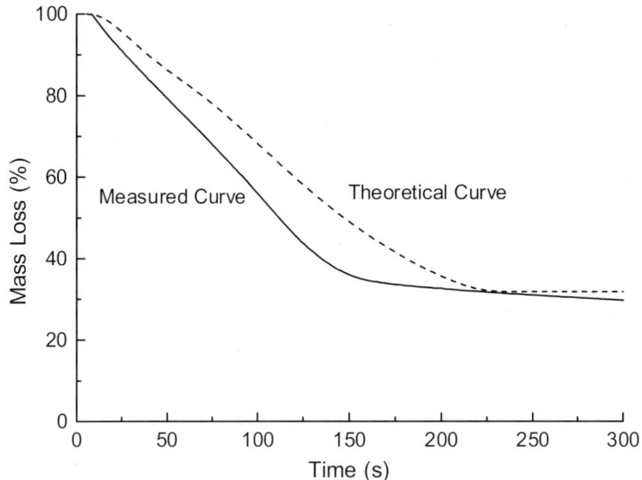


*Figure 3.20. Measured and calculated mass loss curves for a glass/polyester laminate.*

Figure 3.19 shows that the mass loss rate of a burning composite is not constant. This is because the mass loss is controlled by several thermal processes that change with time, including ignition, combustion, char formation, and the transition from thermally thick to thermally thin behaviour as burn-through occurs. Therefore, it is important to characterise the change in mass loss rate of composites with increasing exposure time to fire. The change in mass loss rate of glass/vinyl ester and glass/phenolic composites with time is shown in Fig. 3.21. It is seen the mass loss rate varies considerably as the thermal response of these materials change with time, particularly the more volatile vinyl ester-based material. The curve for the glass/vinyl ester laminate shows an initial spike in mass loss rate due to the rapid decomposition of the resin-rich surface layer. Following this initial peak the mass loss rate decreases steadily with increasing time (as the decomposition reaction rate slows due to the increasing thermal insulation provided by the growth of a char surface layer. After this, the mass loss rate increases again with time, and this rise is due to an increase in the decomposition reaction rate as the composite becomes thermally thin. Finally, the mass loss rate decreases with time as the last of the polymer matrix is degraded. These events also occur with the glass/phenolic shown in Fig. 3.21, although they are not as obvious due to the slower mass loss rate of this material.

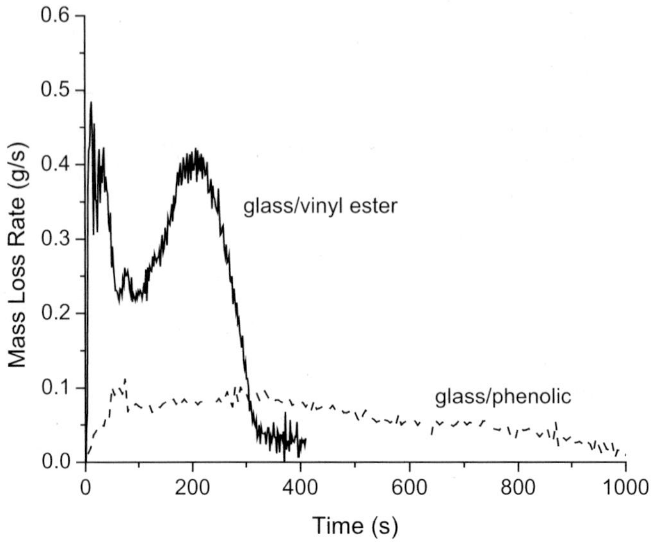


*Figure 3.21.  Mass loss rate curves for glass/vinyl ester and glass/phenolic laminates.*

The thermal model used by Gibson [35] is capable of describing the mass loss rate behaviour of various types of composite material.  As an example, Fig. 3.22 compares the measured and theoretical mass loss rate curves for a glass/polyester composite.  The model is able to predict with good accuracy the complex series of oscillates in the mass loss rate curve as the thermal response of the composite changes with time.

### 3.6      Smoke

One of the main safety concerns with polymer composites is the generation of dense smoke in a fire.  The smoke produced by a burning composite is a mix of small fragments of fibre and ultra-fine carbon (soot) particles.  The short-term exposure of people to smoke released from a burning composite is usually not considered a serious health hazard, as described in Chapter 12.  However, the smoke can be extremely dense and thereby reduce visibility, cause disorientation and make it difficult to fight the fire. For these safety reasons, the smoke properties of many composite materials have been characterised.

Numerous fire studies report that the smoke produced by highly flammable polymer composites (eg. polyesters, vinyl esters, epoxies) is much more dense than smoke from phenolic laminates [4-6,10,12,15,16,46-48]. Figure 3.23 compares the specific extinction area (SEA) for glass/vinyl ester and glass/phenolic composites when exposed to an identical fire. The SEA is a measure of how effectively a given mass of flammable volatiles released by a combustible material is converted into smoke, and is often used to quantitatively define the smoke density. The SEA of the vinyl ester laminate increases rapidly at the onset of combustion, and then remains high until the polymer matrix is completely consumed after about 500 seconds. The phenolic composite produces much less smoke, which is one of the main reasons for their common use in high fire-risk applications.

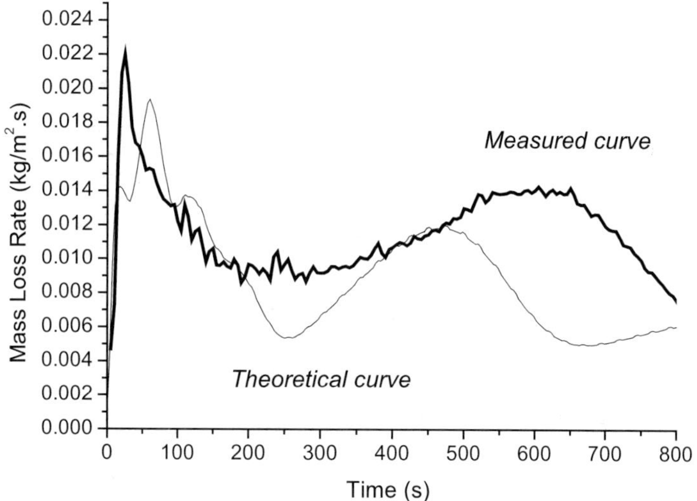


*Figure 3.22. Measured and calculated mass loss rate curves for a glass/polyester laminate.*

In addition to phenolic resins, other highly aromatic thermoset polymers and many types of high temperature thermoplastics have low smoke emission. Examples include aromatic thermosets such as polyimides, cyanate esters and phthalonitriles as well as thermoplastics such as PPS, PES and PEEK [4,6,17,43,46,48,49]. Figure 3.24 shows the maximum smoke density values for various carbon fibre composites, and the smoke released by the thermoplastic materials is generally very low. Price et al. [50] report that polymers with aliphatic backbones, or those that are largely aliphatic and oxygenated, have a tendency to yield low levels of smoke, while polyenic polymers and those with pendant aromatic groups generally produce more smoke. Polymers with high thermal stability or which form small amounts of flammable pyrolyzates generally produce little visible smoke. Increasing char formation is one way of minimising the yield of pyrolyzates and hence smoke production. For this reason, composites containing resins that yield a high amount of char, such as phenolics, PEEK and PES, generate less smoke.

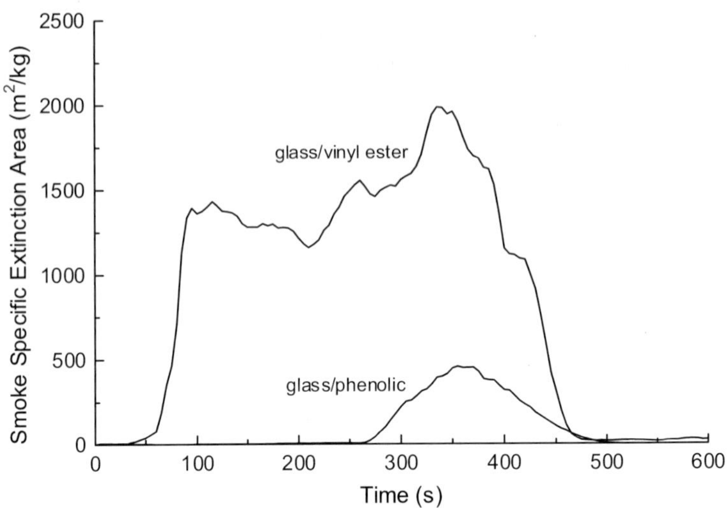


*Figure 3.23. Smoke generation (specific extinction area) verses time for glass/vinyl ester and glass/phenolic composites tested at the heat flux of 50 kW/m².*

The production of char can also reduce the smoke density by impeding the release of ultra-small fragments of fibre into the smoke. The continuous char structure formed in high char yield composites is effective in eliminating the environmentally hazardous release of fibres into a smoke plume [51,52]. For example, Gilwee [51] compared the amount of fibres released from a carbon/polystyrylpyridine composite - that yields a high amount of char (~68%) – against a conventional carbon/epoxy laminate – that has a low char yield (~15%). It was found that under impact loading, the charred carbon/polystyrylpyridine composite released less than 0.2% of its fibres whereas the charred carbon/epoxy lost between 1.2-1.4%. The strong, continuous char in the polystyrylpyridine composite restrained any loose fragments of carbon fibres, whereas the open, discontinuous char of the epoxy laminate allowed fibre fragments to escape more easily.

The fibre reinforcement can also influence the amount of smoke released from a burning composite. Increasing the fibre content lowers the maximum and total amounts of smoke because less organic material is available to produce smoke [19,24,30]. Combustible fibres, such as aramid and polyethylene, and the organic sizing and binder agents used on glass and carbon fibres will however increase the smoke density [7,11,15].

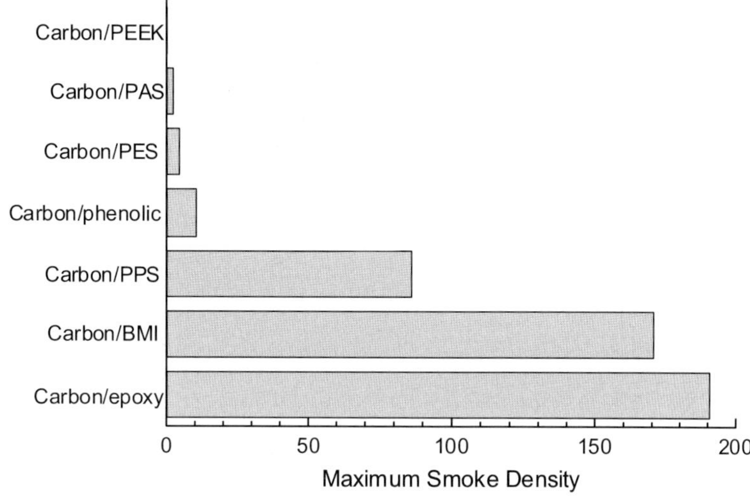


*Figure 3.24. Comparison of the maximum smoke density produced by various thermoset and thermoplastic carbon fibre composites. Data from Sorathia [6].*

The density of smoke released from a burning composite is also controlled by the intensity of the fire. The smoke density usually increases slightly with the external heat flux due to an increase in the release rate of soot particles [7,12,15,46]. For example, Figure 3.25 shows that the smoke density of fibreglass laminates increase with the incident heat flux. However, in some highly aromatic polymer composites the smoke density can decrease with increasing heat flux in most cases. This behaviour is attributed to the high char yield of these composites, which inhibits the transport of volatiles and soot particles to the surface and thereby slows the smoke production rate [10].

It is evident from the information presented above that the smoke released from a burning composite is dependent on a variety of factors, including the amount and type of resin and fibre reinforcement together with the heat flux of the fire. It also appears that smoke production is related to the heat release rate properties of the composite material [40].  Figure 3.26 shows the relationship between average heat release rate and smoke extinction area (SEA) for a variety of thermoset and thermoplastic composites. The smoke data was determined for composites tested at various heat flux levels between 25 and 100 kW/m$^2$.  It can be seen that a strong correlation exists between the SEA and average heat release rate.  It is believed that this relationship exists because the endothermic decomposition reaction rate of the organic matrix determines both the heat release rate and smoke density.  An increase in the reaction rate results in an increase to both the heat release rate and smoke density, and for this reason a strong correlation exists between these two fire reaction properties.

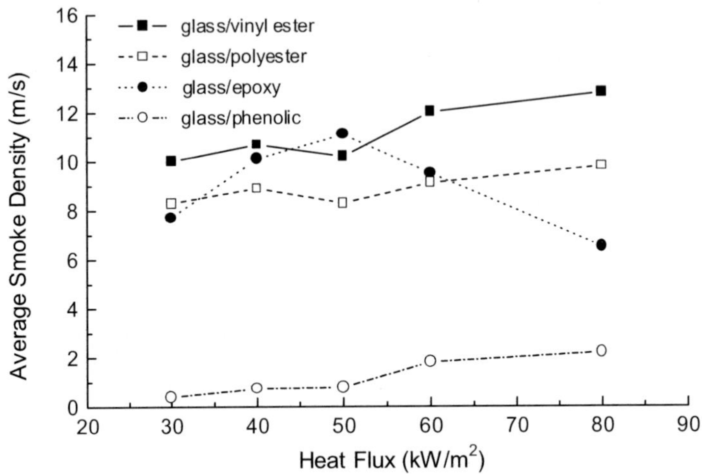


*Figure 3.25.  Effect of incident heat flux on the smoke yield of various fibreglass laminates. (Gibson A.G.; Hume, J. Plast. Rubb. Comp., 23, **1995**, 175-183, reference 12, Reproduced with permission of Institute of Metals, Minerals and Mining.)*

## 3.7      Smoke Toxicity

While the most important fire reaction property is heat release rate, it is often the toxic gases released during combustion that pose the greater health hazard.  It is well recognised that the main cause of death in fires is the toxicity of  combustion products, and the gas that generally has the greatest individual hazard is carbon monoxide.  The amount of CO produced by a burning composite depends on the composition of the organic constituents, the temperature of the fire, and oxygen availability, but even very low levels of CO can cause incapacitation or death [53].  Death in humans will occur within one hour when the CO concentration in air reaches about 1500 ppm.  In comparison, the $CO_2$ content must exceed 100,000 ppm for death to occur within the same time.  In combination with CO, a variety of other gases can be produced during combustion of composite materials [5,6,8,17,27].  The type of gases is determined by the composition of the organic constituents.  For example, polyester laminates release CO, $CO_2$, low molecular weight organic volatiles such as propylene, benzene, toluene and styrene, and higher molecular weight ring compounds including aromatic C-H and aromatic C-H-O [54].  As another example, phenolic laminates produce CO, $CO_2$, toluene, methane, acetone, propanol, propane, benzene, benzaldehyde and volatile aromatic compounds [8,55].  Corrosive and toxic gases can also be released, including HCl, HCN and aromatic halogenated species [6,8].

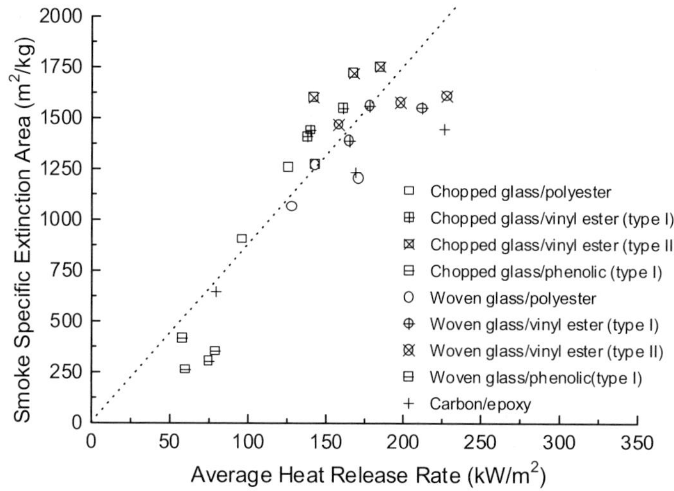


*Figure 3.26.  Effect of average heat release rate (after 300 seconds) on the smoke extinction of composite materials.  (Mouritz, A.P.; Mathys, Z., Proc. SAMPE Symp. & Exhib., reference 40. With permission of SAMPE).*

Table 3.2 lists the concentration of combustion gases measured by Sorathia et al. [4] and Sastri et al. [17] for various thermoset and thermoplastic laminates.  The gases were measured by analysing a sample of combustion products in a Drager calorimeter tube.  It is seen the CO level is substantially higher for the thermoset laminates, and even aromatic resins such as phenolic and polyimide that have excellent flammability resistance still produce a high amount of CO.  Certain types of advanced thermoset resins, such as phthalonitrile, cyanate ester and phenolic/siloxane, and thermoplastics yield small amounts of CO and $CO_2$ in a fire due to their exceptionally high char yield that retains most of the carbon within the composite [17,27].  The concentration of gases is also dependent on the mode of combustion.  Hunter and Forsdyke [47] found that a phenolic composite that was smouldering when exposed to fire released about 50 ppm

CO and 300 ppm $CO_2$. However, when the same material burnt in a flaming mode the gas concentrations increased to 100 ppm CO and 5000 ppm $CO_2$. The quantity of gas also tends to vary over the course of a fire, with the CO yield usually increasing in the later stages of the combustion process when the polymer matrix is extensively carbonised.

Organic fibres also generate toxic gases in a fire [9,11]. For example, Sorathia et al. [11] measured the gases released by various types of phenolic composites reinforced with organic or non-combustible fibres. The composites reinforced with aramid or Spectra fibres released a higher concentration of CO (700 ppm) compared to materials containing glass (190-330 ppm) or carbon (500 ppm) fibres. A similar amount of $CO_2$ was released by the different composites, and therefore it appears that combustible fibres only have a significant influence on the production of CO. It is worth noting that fragments of damaged fibres can also be released by burning composites, and while these are not toxic at low concentrations they can cause irritation to the respiratory tract and affect breathing [56,57].

*Table 3.2. Combustion gases released by burning composite materials. Data from Sorathia et al. [4] and Sastri et al. [17].*

| COMPOSITE | CO (ppm) | $CO_2$ (vol%) | HCN (ppm) | HCl (ppm) |
|---|---|---|---|---|
| Glass/vinyl ester | 230 | 0.3 | not detected | not detected |
| Glass/epoxy | 283 | 1.5 | 5 | not detected |
| Glass/BMI | 300 | 0.1 | 7 | trace |
| Glass/phenolic | 300 | 1.0 | 1 | 1 |
| Glass/polyimide | 200 | 1.0 | trace | 2 |
| Glass/PPS | 70 | 0.5 | 2 | 0.5 |
| Glass/phthalonitrile | 40 | | | |
| Carbon/PEEK | trace | trace | not detected | not detected |

The amount of CO produced by a composite is influenced by the heat release rate properties [40]. Figure 3.27 shows a plot of average yield of CO against average heat release rate for various types of glass and carbon fibre laminates. A linear correlation exists between the CO level and heat release rate, which suggests that the toxic hazard caused by the release of CO can be minimised by using composite materials that have low heat release rate properties. Figure 3.27 also shows the effect of average heat release rate on $CO_2$ yield, although in this case there is no clear correlation.

## 3.8      Limiting Oxygen Index

The limiting oxygen index (LOI) is often used to quantify the flammability of organic polymers and composite materials. The LOI is defined as the minimum percentage of oxygen needed to sustain flaming combustion, and thus may be considered as a measure of the ease of self-extinguishment of a burning material [58]. The LOI is experimentally determined using the oxygen index test that is described in chapter 11.

In brief, the test involves subjecting a sample material to an ignition flame in atmospheres having different oxygen levels, and from this determining the minimum oxygen content that allows the sample to burn with a candle-like flame. Unfortunately the method does not test the sample in a realistic fire environment, and therefore the LOI index cannot be used to accurately quantify the fire reaction behaviour of a material. However, the oxygen index test is often used to rank the relative flammability of polymer composite materials [8,30,42,46,48,59,60,61].

The LOI values for a range of thermoset and thermoplastic composites are presented in Fig. 3.28. It is seen the LOI values for highly flammable composites, such as polyester-, vinyl ester- and epoxy-based materials, are below about 30. Composites with highly stable or aromatic polymers have much higher index values. It is generally recognised that the LOI values for polymers and polymer composites increase with their ability to yield char in a fire [55,60,62,63]. This is because the formation of char occurs at the expense of combustible volatiles, which in turn increases the oxygen level required to sustain flaming combustion. In addition to the type of polymer matrix, the LOI value can be affected by other factors, most notably the degree of resin cure, fibre content, and the flammability of the fibre reinforcement [8,30,48,64,65].

The LOI index values shown in Fig. 3.28 were determined at room temperature. However, a composite material will reach a much higher temperatures in a fire. LOI studies have shown the index values of composites are dependent on the test temperature [8,59,61]. The values can change dramatically with temperature, usually decreasing with increasing temperature, and often changing the relative ranking of some materials. It is therefore questionable to use LOI values measured at room temperature to assess the flammability of composite materials. Figure 3.29 shows the effect of test temperature from $25°C$ to $300°C$ on the LOI of two fibreglass composites. The index values increase with temperature up to $100°C$, but at higher temperatures there is a steady reduction in the values because less heat is needed to sustain decomposition and burning.

While the LOI is often used to characterise the fire performance of composites, there is no clear correlation between the index value and other fire reaction properties (eg. heat release rate) [58]. Therefore, it is not valid to use the index value as a quantitative measure of fire resistance, although it can be used (with caution) to rank the relative flammability of different composite materials.

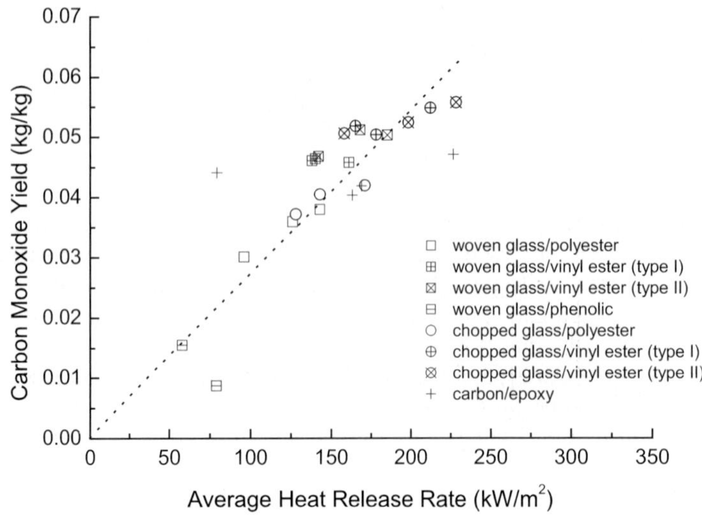



*(a)*

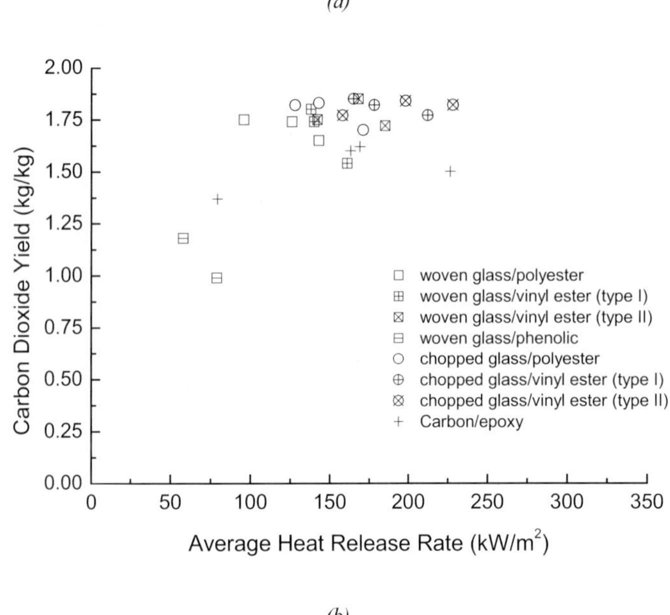



*(b)*

*Figure 3.27. Effect of average heat release rate on the yields of (a) carbon monoxide and (b) carbon dioxide from burning composite materials. (Mouritz, A.P.; Mathys, Z., Proc. SAMPE Symp. & Exhib., reference 40. With permission of SAMPE).*

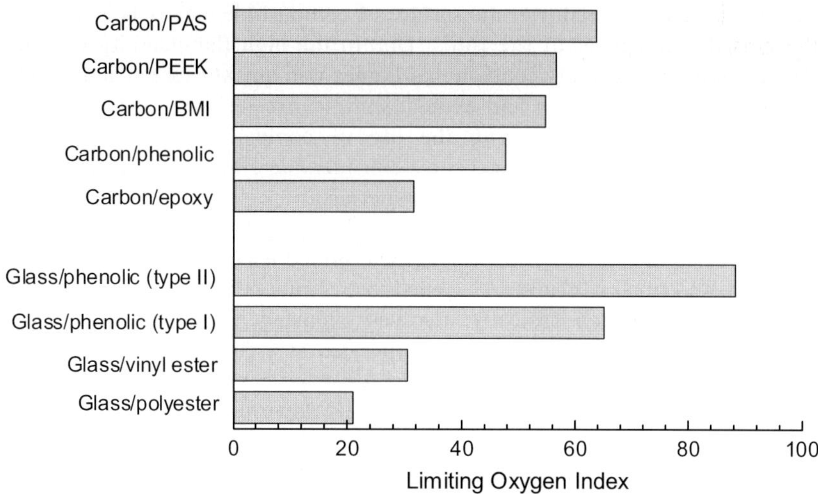


*Figure 3.28. LOI values for various thermoset and thermoplastic composite materials at room temperature. Data for the fibreglass and carbon fibre laminates are from Allison et al. [48] and Sorathia et al. [6], respectively.*

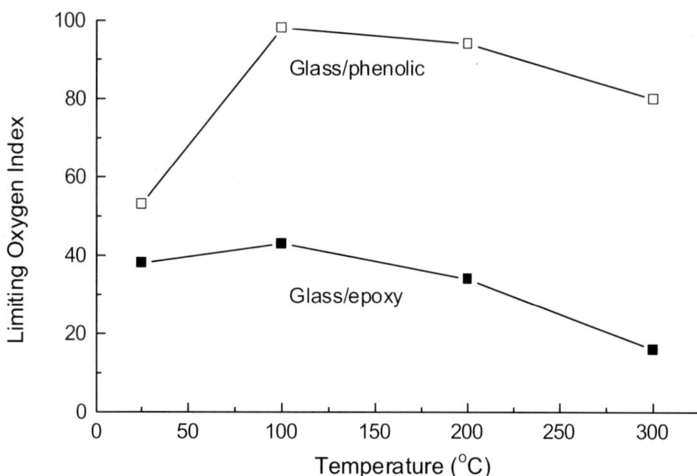


*Figure 3.29. Effect of increasing temperature on the LOI values for two fibreglass composites. Data from Tewarson and Macaione [8].*

## 3.9      Surface Spread of Flame

The speed at which flames spread over the surface of combustible materials is a critical factor in the growth and spread of fire [66]. Due to the high flammability of many composites, there is a serious safety concern that flames will quickly spread and thereby increase the difficulty in containing and extinguishing a fire. This is one of the key concerns of fire safety authorities with the use of composites in high fire-risk applications, and for this reason there has been a considerable effort over many years to characterise the flame spread properties of FRP materials.

The flame spread of composite materials can be determined using several experimental techniques that are described in chapter 11. The most common technique is the radiant panel flame spread test, which basically involves exposing a flat composite panel inclined at an angle of $45^{\circ}$ to a radiant heater operated at a constant heat flux. The heater is placed at the higher end of the panel to force the composite to ignite at the upper edge. The speed at which the flame front travels down the inclined specimen is measured during a test. In this respect, the radiant panel flame spread test is unrealistic because the flame front is required to travel downwards, whereas in actual fires it is the more rapid upward movement of flames that is responsible for the spread of fire. Despite this, it is a standard test for determining the flame spread properties of composites and other combustible materials.

Most investigations into the flame spread rates of composites have been performed on polyester-, epoxy- and phenolic-based laminates due to their use in aircraft, ships, buildings, oil rigs and rail carriages [4,5,7,11,12,16,41,53]. Typical downward flame spread speeds of these composites are shown in Fig. 3.30. This figure shows the time taken for the flame front to travel down the composite from the ignition point on the specimen panel in the flame spread test. It is seen that the flame propagates readily down the surface of glass/polyester and glass/epoxy laminates, and this is due to the high flammability of these materials. However, the flame is unable to spread down the glass/phenolic laminate, and this material can be regarded as self-extinguishing. Numerous studies have reported that phenolic laminates have excellent resistance of flame spread, and this is another outstanding fire reaction property of these materials that makes them suited for many high fire risk applications [4,5,7,11,12,16,53]. In addition to phenolic resins, Sorathia et al. [4] have shown that composites with a bismaleimide, polyimide or high-temperature thermoplastic (eg. PEEK, PPS) matrix display excellent flame spread resistance. The combustion behaviour of the fibre reinforcement can also have a major impact on flame spread. It is generally found that composites have much higher flame spread speeds when reinforced with combustible fibres (eg. aramid, polyethylene) rather than non-combustible fibres (eg. glass, carbon) [11].

The greatest influence on the downward flame spread rate of composite materials is their heat release rate. Sorathia et al. [39] have shown the flame spread rate of both thermoset and thermoplastic matrix composites is highly dependent on their peak heat release rate, as shown in Fig. 3.31.  In this figure the peak heat release rates of the composites have been normalised to their ignition times.  The flame spread index is a parameter often used to quantify the downward flame spread rate, and the higher the index value the faster is the average flame spread speed.  The figure shows a clear correlation between the flame spread index and the peak heat release rate parameter.

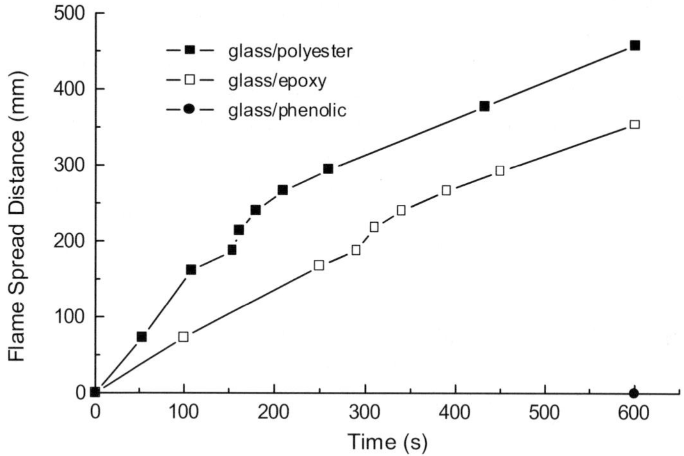


*Figure 3.30.  Surface spread of flame verses time.  (Gibson A.G.; Hume, J. Plast. Rubb. Comp., 23, **1995**, 175-183, reference 12, Reproduced with permission of Institute of Metals, Minerals and Mining.)*

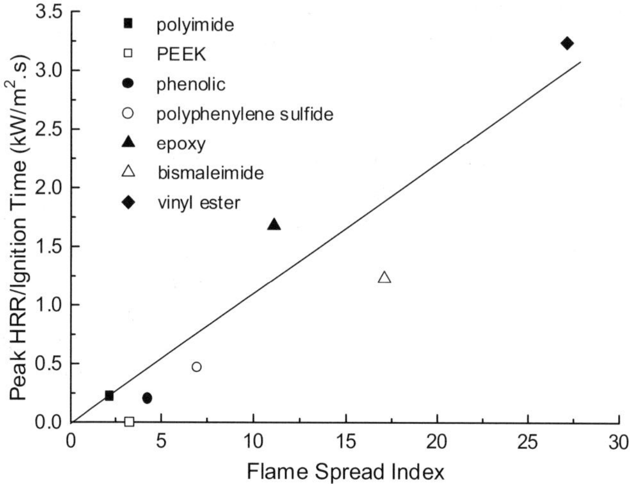


*Figure 3.31.  Relationship between normalised peak heat release rate and flame spread index for various thermoset and thermoplastic composites.  Data from Sorathia et al. [11] and  Sorathia et al. [39].*

Several models have been proposed for calculating the upward flame spread speed in composite laminates, including models by Cleary and Quintiere [67], and Brehob and Kulkarni [68,69]. Unfortunately, the accuracy of these models in the prediction of the upward flame spread rate for the many types of thermoset and thermoplastic composites has not been rigorously assessed [41].

## 3.10     Fire Resistance

This chapter is principally focussed on the fire reaction of polymer composites, although some mention will be made here about their fire resistance. Fire resistance describes the ability of a material or structure to restrict the spread of fire and to retain mechanical and physical integrity. In this section the fire resistive properties of heat transmission and burn-through resistance of laminates and sandwich composites are briefly described. The effect of fire on the mechanical integrity of composites is described in chapters 6 & 7. These chapters describe the degradation to the load-bearing properties of composite materials during and after a fire.

Various test methods are used to characterise the fire resistance of composite materials, and these are outlined in chapter 11. One of the most common methods for assessing the resistance to heat transmission is the furnace test. This test involves exposing a composite panel to a furnace fire for an extended period of time, during which the transmission of heat is measured by thermocouples located through the material. Composite materials that heat-up slowly are less likely to spread a fire via heat conduction to neighbouring rooms, and are therefore considered to have good fire resistance. The furnace test is usually performed using the cellulosic fire condition, which has a maximum temperature of about $850°C$ and is representative of a room fire, or a hydrocarbon fire, with a maximum temperature of $\sim 1100°C$ that is typical of a fuel fire (see Fig. 11.10).

The fire resistance of many types of laminates and sandwich composites exposed to cellulosic and hydrocarbon fires have been determined [5,12,42,48,70,71]. Examples of the fire resistance of glass/polyester and glass/phenolic laminates are presented in Fig. 3.32. The curves show the temperature increase to the back (unexposed) face of the laminate specimens with increased exposure time to a cellulosic fire. It is seen that the temperature rise to the back-face of the phenolic laminate over the initial 30-40 minutes is lower than the polyester composite, and this is due to the superior flammability resistance of the phenolic matrix. However, at longer times the temperature of the phenolic laminate increases much more rapidly, and this is due mainly to explosive delamination damage caused by the internal pressure build-up from the vapourisation of entrapped water in the phenolic matrix.

The fire resistance of combustible materials is often defined by the time taken for the back-face temperature to reach $160°C$, at which point the fire is likely to spread to neighbouring rooms. A comparison of the fire resistance of different laminates when exposed to a hydrocarbon fire is shown in Fig. 3.33 [71].  As expected, the time to reach 160°C increases rapidly with the panel thickness, and when used in thick sections most types of composites are capable of withstanding severe temperatures for a considerable time.  This is achieved by virtue of their low thermal conductivity and the endothermic nature of the resin decomposition reaction that slows heat transmission through the laminate.  Figure 3.33 shows that the phenolic laminates have superior fire resistance, and this is due to their high char yield that acts as an insulator against the rapid transmission of heat.  Other studies performed on sandwich composites have shown that the fire resistance can be greatly improved by the use of a low density fire-resistant core material, such as phenolic foam or balsa, which can be highly effective in reducing heat transmission due to the low thermal conductivity of the core [12,21,70].  For this reason, sandwich composites are commonly preferred over single-skin laminates in applications requiring high fire resistance.

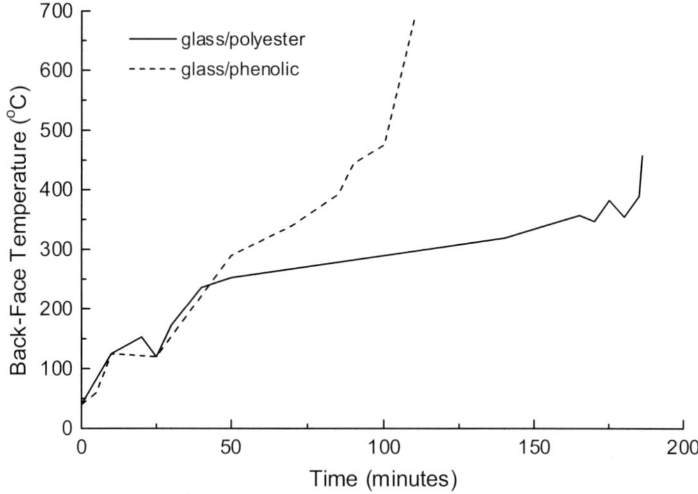


*Figure 3.32.  Average temperatures at the unexposed surface to glass/polyester and glass/phenolic laminates exposed to a cellulosic fire.  Reproduced with permission from Hume [5].*

Another important fire resistive property is burn-through resistance, which is the time taken for a fire or directed flame to penetrate a composite material.  The burn-through resistance is experimentally determined using the DTRC test (see chapter 11) that involves directing a jet flame onto a composite panel, although the furnace test has also been used to measure the burn-through resistance of composites [21].  The burn-through of laminates occurs in a step-wise process in which resin is depleted from the surface layers, leaving plies of reinforcement that eventually fall away, thereby exposing underlying virgin material to the flame. Under extremely high temperature jet fire attack conditions, however, the fibres in glass reinforced laminates can fuse together at the surface that reduces ablation and provides protection to the underlying material [72].

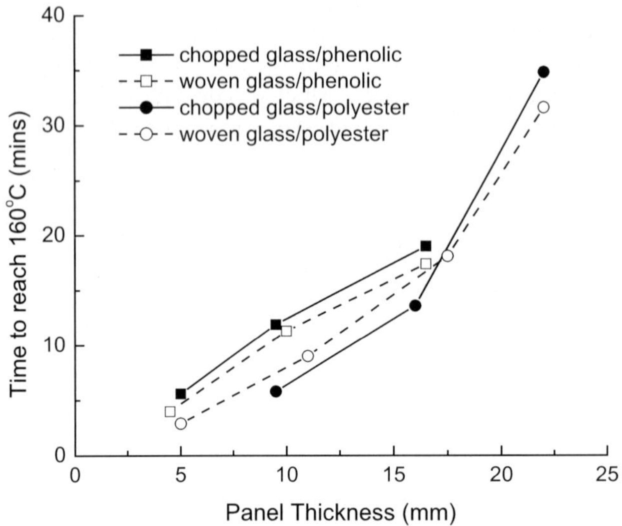


*Figure 3.33. A comparison of fire resistance (defined as the time for the back-face to reach 160°C) vs. panel thickness for various fibreglass/thermoset laminates. The laminates were exposed to the hydrocarbon fire curve. (Dodds, N; Gibson, A.G.; Dewhurst, D.; Davies, J.M., Comp. 31A, 2000, 689-702, reference 71, with permission from Elsevier).*

## References


1.   J.G. Quintiere, V. Babrauskas, L. Cooper, M. Harkleroad, K. Steckler and A. Tewarson. The role of aircraft panel materials in cabin fires and their properties. DOT/FAA/CT-84/30, Federal Aviation Administration, 1985.

2.   V. Babrauskas and W.J. Parker. Ignitability measurements with the cone calorimeter. *Fire & Materials*, 1987; 11:31-43.

3.   J.E. Brown, E. Braun and W.H. Twilley. Cone calorimetry evaluation of the flammability of composite materials. NBS Report NBSIR 88-3733, 1988.

4.   U. Sorathia, T. Dapp and J. Kerr. Flammability characteristics of composites for shipboard and submarine internal applications. In: *Proceedings of the 36th International SAMPE Symposium*. San Diego, CA, 15-18 April 1991, p. 1868-1878.

5.   J. Hume. Assessing the fire performance characteristics of GRP composites. In: *International Conference on Materials and Design Against Fire*, London, 1992, p. 11-15.

6.   U. Sorathia. Flammability and fire safety of composite materials. In: *Proceedings of the 1st International Workshop on Composite Materials for Offshore Operations*. Houston, Texas, 26-28 Oct 1993, p. 309-317.

7.    M.J. Scudamore. Fire performance studies on glass-reinforced plastic laminates. *Fire and Materials*, 1994; 18:313-325.

8.    A. Tewarson and D.P. Macaione. Polymers and composites – an examination of fire spread and generation of heat and fire products. *Journal of Fire Sciences*, 1993; 11:421-441.

9.    J.R. Brown, P.D. Fawell and Z. Mathys. Fire-hazard assessment of extended-chain polyethylene and aramid composites by cone calorimetry. *Fire & Materials*, 1994; 18:167-172.

10.   G.T. Egglestone and D.M. Turley. Flammability of GRP for use in ship superstructures. *Fire and Materials*, 1994; 18:255-260.

11.   U. Sorathia, H. Telegadas and M. Bergen. Mechanical and flammability characteristics of phenolic composites for naval applications. In: *Proceedings of the 39th International SAMPE Symposium*. 11-14 April 1994, p. 2991-3002.

12.   A.G. Gibson and J. Hume. Fire performance of composite panels for large marine structures. *Plastics, Rubbers & Composites Processing and Applications*, 1995; 23:175-183.

13.   J.R. Brown and N.A. St John. Fire-retardant low-temperature-cured phenolic resins and composites. *TRIP*, 1996; 4:416-420.

14.   S.B. Sastri, J.P. Armistead and T.M. Keller. Phthalonitrile-carbon fiber composites. *Polymer Composites*, 1996; 17:816-822.

15.   J.R. Brown and Z. Mathys. Reinforcement and matrix effects on the combustion properties of glass reinforced polymer composites. *Composites*, 1997; 28A:675-681.

16.   F.-Y. Hshieh and H.D. Beeson. Flammability testing of flame-retarded epoxy composites and phenolic composites. *Fire and Materials*, 1997; 21:41-49.

17.   S.B. Sastri, J.P. Armistead, T.M. Keller and U. Sorathia. Flammability characteristics of phthalonitrile composites. In: *Proceedings of the 42th International SAMPE Symposium*. 4-8 May 1997, p.1032-1038.

18.   A.P. Mouritz and Z. Mathys. Post-fire mechanical properties of marine polymer composites. *Composite Structures*, 1999; 47:643-653.

19.   P. Van Dine and M.D. Smith. The effect of manufacturing method on the fire performance of composites. In: *Proceedings of the 31st International SAMPE Technical Conference*. 26-30 October, 1999, p. 177-187.

20.   A.P. Mouritz and Z. Mathys. Mechanical properties of fire-damaged glass-reinforced phenolic composites. *Fire and Materials*, 2000; 24:67-75.

21.   U. Sorathia, T. Gracik, J. Ness, M. Blum, A. Le, B. Scholl and G. Long. Fire safety of marine composites. In: *Proceedings of the 8th International Conference on Marine Applications of Composite Materials*. Florida, 14-16 March, 2000.

22.   A.P. Mouritz. Post-fire flexural properties of fibre-reinforced polyester, epoxy and phenolic composites. *Journal of Materials Science*, 2002; 37:1377-1386.

23.   A.P. Mouritz and C.P. Gardiner. Compression properties of fire-damaged polymer sandwich composites. *Composites*, 2002; 33A:609-620.

24.   A. Kootsookos and A.P. Mouritz. Fire reaction of polymer composites: effect of resin matrix content. *Composites A* (to be submitted).

25.   A. Kootsookos and A.P. Mouritz. Fire reaction of polymer composites: effect of specimen thickness. *Composites A* (to be submitted).

26.   J.E. Brown. Combustion characteristics of fiber reinforced resin panels. NBS Report No. FR3970, 1987.

27.   J.H. Koo, B. Muskopf, S. Venumbaka, R. Van Dine, B. Spencer and U. Sorathia. Flammability properties of polymer composites for marine applications. In: *Proceedings of the 32nd International SAMPE Technical Conference*, 5-9 November 2000, Paper No. 136.

28.   A.T. Grenier, N.A. Dembsey and J.R. Barnett. Fire characteristics of cored composite materials for marine use. *Journal of Fire Safety*, 1998; 30:137-159.

29.   N.A. Dembsey and D.J. Jacoby. Evaluation of common ignition models for use with marine cored composites. *Fire and Materials*, 2000; 24:91-100.

30.   M. Le Bras, S. Bourbigot, B. Mortaigne and G. Cordellier. Comparative study of the fire behaviour of glass-fibre reinforced unsaturated polyesters using a cone calorimeter. *Polymers and Polymer Composites*, 1998; 6:535-539.

31.   R.E. Lyon, J. Demario, R.N. Walters and S. Crowley, 'Flammability of glass fiber-reinforced polymer composites', *Proceedings of the Fourth Conference on Composites in Fire*, 15-16 September 2005, Newcastle-upon-Tyne, UK.

32. E. Mikkola and I.S. Wichmann. On the thermal ignition of combustible material. *Fire & Materials*, 1989; 14:87-96.

33. E. Mikkola. Ignitability of solid materials. In : *Heat Release in Fires*. Ed. V. Babrauskas and S.J. Grayson. London: Elsevier Applied Science, 1992, pp. 225-232.

34. H. Carslaw and J. Jaeger. *Conduction of Heat in Solids*. Oxford: Clarendon Press, 1959.

35. A.G. Gibson, A.P. Mouritz, Y. Wu, C.P. Gardiner and Z. Mathys. Validation of the Gibson model for the fire reaction properties of fibre-polymer composites. In: *Proceedings of Composites in Fire 3*, Newcastle, 2003.

36. A.G. Gibson, Y.S. Wu, H.W. Chandler, J.A.D. Wilcox and P. Bettess. A model for the thermal performance of thick composite laminates in hydrocarbon fires composite materials in the petroleum industry. *Revue de I'Institute Francais du Petrole*, 1995; 50:69-74.

37. V. Babrauskas and R.D. Peacock. Heat release rate: the single most important variable in fire hazard. *Fire Safety Journal*, 1992; 18:255-272.

38. V. Babrauskas. Why was the fire so big? HHR: The role of heat release rate in described fires. *Fire & Arson Investigator*, 1997; 47:54-57.

39. U. Sorathia, R. Lyon, T. Ohlemiller and A. Grenier. A review of fire test methods and criteria for composites. *SAMPE Journal*, 1997; 33(July/August):23-31.

40. A.P. Mouritz and Z. Mathys. Heat release of polymer composites in fire. In: *Proceedings of the SAMPE Symposium and Exhibition*, 16-20 May 2004.

41. T.J. Ohlemiller and T.G. Cleary. Upward flame spread on composite materials. *Fire & Polymers II*, American Chemical Society, 1995, p. 422-434.

42. D.A. Kourtides, W.J. Gilwee and J.A. Parker. Thermal response of composite panels. *Poly. Eng. & Sci.*, 1979; 19:226-231.

43. A. Lin. Cyanate esters with improved fire resistance. In: *Proceedings of the 44th International SAMPE Symposium*. 23-27 May 1999, p. 1424-1430.

44. T. Ohlemiller and J. Shields. One and two-sided burning of thermally thin materials. *Fire & Materials*, 1993; 17:103-110.

45. L.K. Kucner and H.L. McManus. Experimental studies of composite laminates damaged by fire. In: *Proceedings of the 24th International SAMPE Technical Conference*, 17-20 October 1994, p. 341-353.

46. D.A. Kourtides, W.J. Gilwee and J.A. Parker. Thermochemical characterisation of some thermally stable thermoplastic and thermoset polymers. *Polymer Engineering & Science*, 1979; 19:24-29.

47. J. Hunter and K.L. Forsdyke. Phenolic glass fiber-reinforced plastic and its recent applications. *Polymer Composites*, 1989; 2:169-185.

48. D.M. Allison, A.J. Marchand and R.M. Morchat. Fire performance of composite materials in ships and offshore structures. *Marine Structures*, 1991; 4:129-140.

49. J.L. Sevart, O.H. Griffin, Z. Gürdal and G., A. Warner. Flammability and toxicity of materials for marine vehicles. *Naval Engineers Journal*, Sept 1990, p. 45-54.

50. D. Price, G. Anthony and P. Carty. Introduction: polymer combustion, condensed phase pyrolysis and smoke formation. In: *Fire Retardant Materials*, ed. A.R. Horrocks and D. Price, Cambridge: Woodhead Publishing Ltd, 2002, p. 1-30.

51. W.J. Gilwee. Fire retardant materials in aviation. In: *Proceedings of the 4th Annual Plastics Symposium*, 29 May 1975.

52. J.A. Parker and D.A. Kourtides. New fireworthy composites for use in transportation vehicles. *Journal of Fire Sciences*, 1983; 1:432-458.

53. M.M. Hirschler. Fire hazard and toxic potency of the smoke from burning materials. *Journal of Fire Sciences*, 1987; 5:289-307.

54. E. Braun and B.C. Levin. Polyesters: a review of the literature on products of combustion and toxicity. *Fire & Materials*, 1986; 10:107-123.

55. B.K. Kandola and A.R. Horrocks. Composites. In: *Fire Retardant Materials*, ed. A.R. Horrocks and D. Price, Cambridge: Woodhead Publishing Ltd, 2002, p. 182-203.

56. F. Barthorpe. Damage: fibres on fire. *Professional Engineering*, June 1995, p. 10-11.

57. S. Gandhi, R. Lyon and L. Speitel. Potential health hazards from burning aircraft composites. *Journal of Fire Sciences*, 1999; 17:20-41.

58. E.D. Weil, M.M. Hirschler, N.G. Patel, M.M. Said and S. Shakir. Oxygen index: correlations to other fire tests. *Fire & Materials*, 1992; 16:159-167.

59. D.P. Macaione, R.P. Downing and P.R. Berguist. Summary Report AMMRC TR 83-53. Army Materials and Mechanics Research Laboratory, Watertown, MA, 1983.

60. D.A. Kourtides. Processing and flammability parameters of bismaleimide and some other thermally stable resin matrices for composites. *Polymer Composites*, 1984; 5:143-150.

61. S.C. Bates and P.R. Soloman. Elevated temperature oxygen index apparatus and measurements. *Journal of Fire Sciences*, 1993; 11:271-284.

62. D.W. Van Krevelan. Some basic aspects of flame resistance of polymeric materials. *Polymer*, 1975; 16:615-620.

63. W.J. Gilwee, J.A. Parker and D.A. Kourtides. Oxygen index tests of thermosetting resins. *Journal of Fire & Flammability*, 1980; 11:22-31.

64. A. Casu, G. Camino, M. De Giorgi, D. Flath, A. Laudi and V. Morone. Effect of glass fibres and fire retardant on the combustion behaviour of composites, glass fibres-poly(butylene terephthalate). *Fire & Materials*, 1998; 22:7-14.

65. A. Casu, G. Camino, M.P. Luda and M. De Giorgi. Mechanisms of fire retardance in glass fibre polymer composites. *Makromol. Chem. Macromol. Symp.*, 1993; 74:307-310.

66. J.G. Quintiere. Surface flame spread. In: *SFPE Handbook of Fire Protection Engineering*, ed. P.J. DiNenno, D. Drysdale, C.L. Beyler, W.D. Walton, R.L.P. Custer, J.R. Hall and J.M. Watts, Society of Fire Protection Engineers, 2000, p. 2-246 to 2-257.

67. T. Cleary and J. Quintiere. A framework for utilizing fire property tests. In: *Proceedings of the 3rd International Symposium of Fire Safety Science*. NY, 1991, p. 647.

68. E. Brehob. PhD Thesis. Penn State University, 1994.

69. A. Kulkarni, C. Kim and C. Kuo. NIST-GCR-91-597. National Institute of Standards and Testing, May 1991.

70. E. Greene. *Marine Composites*. www.marinecomposites.com.

71. N. Dodds, A.G. Gibson, D. Dewhurst and J.M. Davies. Fire behaviour of composite laminates. *Composites*, 2000; 31A:689-702.

72. P.S. Hill and G.C. White. Jet fire testing and performance of composite materials. In: *Proceedings of Composites in Fire*. Newcastle, 1999.

# Chapter 4

# Fire Modelling of Composites

B. Lattimer
*Hughes Associates*

T. Campbell
*Materials Sciences Corporation*

## 4.1    Introduction

Composite materials include an organic resin that will lose strength and thermally decompose when exposed to elevated temperatures. As with many other types of construction materials, the fire performance of composites is assessed through the material's propensity to spread flame and its fire resistance capabilities. Flame spread is the ignition of the composite material by an initiating fire and propagation of flame along the surface. Fire resistance is a measure of the ability of a wall or ceiling to prevent heat transmission through the assembly and structural integrity. Chapter 5 will address how to predict the heat transmission through the thickness of the composite when exposed to fire. This chapter will focus on modelling the flame spread over composite surfaces and the structural response of composites during fires.

One of the most important considerations in modelling the behaviour of composites during fires is the thermal exposure. The chapter will begin with a discussion of the thermal exposures that should be expected from a range of scenarios including local fires, local fires with hot gas layer heating, fully-developed compartment fires, and some standard tests. Following this section, an overview will be provided on modelling flame spread over composites with a focus on the contribution to the growth of fire inside a compartment. The last section of the chapter will focus on modelling the structural response of composites exposed to fires, which is a topic expanded upon in Chapter 6.

## 4.2      Thermal Exposure

### 4.2.1     BACKGROUND

The thermal exposure from a fire is typically modelled as a heat flux boundary condition. The heat flux from a fire is composed of both convection and radiation as shown in Fig. 4.1. Of interest in modelling is the net heat flux into the composite material, which is described by[1]:

$$-k\frac{dT}{dx} = q''_{net,f} = \varepsilon_f \sigma T_f^4 - \varepsilon_s \sigma T_s^4 + h\left(T_f - T_s\right)$$
(4.1)

where the first term on the right hand side is the radiation from the fire, the second term is the radiative loss from the composite surface, and third term is the convective heat transfer between the fire and composite surface. The heat flux from a fire can be calculated by knowing the local gas temperature, $T_f$, the emissivity of the gases, $\varepsilon_f$, and the local heat transfer coefficient. In many applications, the total heat flux at a boundary has been measured using a water cooled heat flux gauge typically close to ambient temperature. By setting the surface temperature to the ambient in Eqn. 4.1, the boundary condition at the total heat flux gauge is represented by the following equation:

$$q''_{hfg,f} = \varepsilon_f \sigma T_f^4 + h\left(T_f - T_\infty\right) - \varepsilon_s \sigma T_\infty^4$$
(4.2)

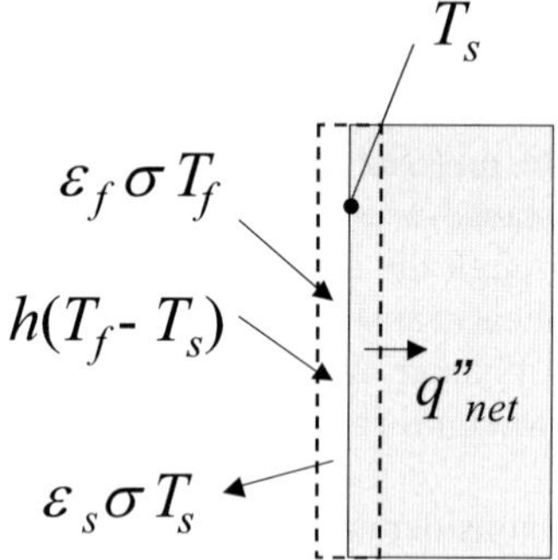


*Figure 4.1. Heat fluxes from a fire to an adjacent surface.*

---

[1] The nomenclature for equations are given at the end of the chapter.

Cooling the gauge surface maximizes the convective heat transfer and minimizes the radiative losses; thus, the cooled heat flux gauges measure the maximum total incident heat flux. The total incident heat flux measured using a heat flux gauge, Eqn. (4.2), is related to the actual heat flux through the following relation:

$$q''_{net,f} = \left[\varepsilon_f \sigma T_f^4 + h(T_f - T_\infty) - \varepsilon_s \sigma T_\infty^4\right] - h(T_s - T_\infty) - \varepsilon_s \sigma\left(T_s^4 - T_\infty^4\right) \quad (4.3)$$

or

$$q''_{net,f} = q''_{hfg,f} - h(T_s - T_\infty) - \varepsilon_s \sigma\left(T_s^4 - T_\infty^4\right) \quad\quad\quad (4.4)$$

Therefore, measuring the heat flux has removed the need to predict both the gas temperature and the emissivity of the gases, both of which are difficult to calculate. To determine the actual net heat flux into a surface from the measured heat flux, a surface temperature correction needs to be applied as done in Eqn. (4.4) which requires the local heat transfer coefficient and surface emissivity. A reasonable estimate of the composite surface emissivity is usually possible. The local heat transfer coefficient may be configuration dependent, and can vary from 0.010 kW/m.K for flat walls to as high as 0.050 kW/m.K where a fire is impinging on a ceiling. Since overestimating the convective heat transfer coefficient will make the heat flux boundary non-conservative, it is usually ignored except for simple configurations where the heat transfer coefficient can be readily calculated (e.g. flat walls).

### 4.2.2    LOCAL FIRE PLUMES

A fire plume consists of the fire and hot gases rising above a burning item. Fire plumes may be a burning object next to the composite, the burning composite itself, or both as depicted in Fig. 4.2. This section will discuss the heat fluxes transmitted by fire plumes back to a solid boundary such as a composite surface. The heat fluxes from these types of fires will be spatially and possibly time dependent. Heat fluxes from local fires have been experimentally measured using water cooled heat flux gauges for a variety of configurations and fuel types. All of the discussion and correlations in this section pertain to fires that are in direct contact with the surface. If fires are moved away from the surface so that the flame is not "attached" to the surface, the heat fluxes will be lower than predicted using correlations in this section.

Heat fluxes from local fires are typically related to the flame length of the fire. The flame length is related to the heat release rate of the fire and a characteristic length, usually the dimensional size of the burning area. The heat release rate is the amount of energy created by the burning material and can be calculated by:

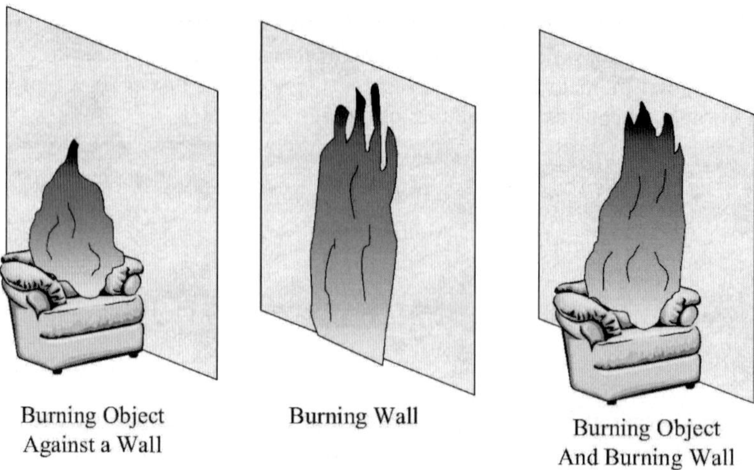



*Figure 4.2. Different types of local fires plumes.*

$$Q = \chi \dot{m}_f \Delta H_c \qquad (4.5)$$

The heat release rate of a material is usually taken from large-scale test data or determined using small-scale test data. See the section on fire growth modelling for details on predicting the heat release rate of burning composites. Correlations for predicting flame lengths in different configurations are provided in Table 4.1.

Tables 4.2 and 4.3 provide a list of empirical correlations for predicting heat fluxes from a burning boundary back onto itself and from burning items to a boundary, respectively, for a number of different configurations. Note that the appropriate flame length from Table 4.1 must be used to calculate the heat flux in each particular configuration.

Most of these correlations in Tables 4.2 and 4.3 have been developed using data for fires up to a particular physical size and heat release rate. However, the size of the fire can affect the peak heat fluxes produced by the fire. In Tables 4.2 and 4.3, maximum heat fluxes in many configurations were approximately 100 kW/m$^2$ with fires less than 1.0 MW and less than 1.0 m in diameter. With larger fires impinging on a ceiling and diameters up to 1.6 m, heat fluxes as high as 130 kW/m$^2$ were measured [8]. As the heat release rate and physical size of a fire increases, the radiative path length will increase causing the gas emissivity to go to 1.0. In addition, very large pool fires will have higher gas temperatures. The higher gas temperature is due to large fires generating more smoke, which blocks radiation from escaping the fire plume. McCaffrey [9] measured the gas temperature to be as low as 820°C, while Baum and McCaffrey [10] clearly showed the dependence of gas temperature on fire diameter, with measured gas temperatures as high as 1000°C for 6 m diameter fires and 1250°C for 30 m diameter fires.

*Table 4.1. Flame length correlations for different configurations.*

| Configuration | Correlation |
|---|---|
| Vertical Wall [1] | $L_f = 0.052\left(\dfrac{Q}{d}\right)^{2/3}$ |
| 90° Corner Walls With or Without Ceiling [2] | $L_f = 5.9\left(\dfrac{Q}{1100d^{5/2}}\right)^{1/2} d$ |
| In Open [3] | $L_f = 0.23Q^{2/5} - 1.02D$ |
| Vertical Wall Without Ceiling [4] | $L_f = 0.23Q^{2/5} - 1.02D$ |
| 90° Corner Walls With or Without Ceiling [2] | $L_f = 5.9\left(\dfrac{Q}{1100D^{5/2}}\right)^{1/2} D$ |
| Impinging on Unbounded Ceiling [5] | $L_f = 2.89\left(\dfrac{Q}{1100H^{5/2}}\right)^{1/3} H$ |

*Table 4.2. Correlations for the heat flux from a burning surface back onto itself.*

| Configuration | Correlation | |
|---|---|---|
| Vertical Wall | $q'' = 60$ | $\left(z/L_f\right) \le 0.53$ |
| | $q'' = 12.3\left(z/L_f\right)^{-2.5}$ | $\left(z/L_f\right) > 0.53$ |
| 90° Corner Walls [6] | $q''_{max} = q''_{peak}$ | $\left(z/L_f\right) \le 0.5$ |
| | $q''_{max} = q''_{peak} - 5\left(z/L_f - 0.5\right)\left(q''_{peak} - 27\right)$ | $0.5 < \left(z/L_f\right) \le 0.7$ |
| | $q''_{max} = 10.0\left(z/L_f\right)^{-2.8}$ | $\left(z/L_f\right) > 0.7$ |
| | where, | |
| | $q''_{peak} = 120\left[1 - exp\left(-0.1Q^{1/2}\right)\right]$ | |

*Table 4.3. Correlations for the heat flux from a burning object to an adjacent surface.*

| Configuration | Correlation | |
| --- | --- | --- |
| Vertical Wall [4] | $q''_{max} = q''_{peak}$ | $(z/L_f) \leq 0.4$ |
| | $q''_{max} = q''_{peak} - 5/3(z/L_f - 2/5)(q''_{peak} - 20)$ | $0.4 < (z/L_f) \leq 1.0$ |
| | $q''_{max} = 20(z/L_f)^{-5/3}$ | $(z/L_f) > 1.0$ |
| | where, | |
| | $q''_{peak} = 200\left[1 - exp\left(-0.09Q^{1/3}\right)\right]$ | |
| Corner Walls [7] | $q''_{max} = q''_{peak}$ | $(z/L_f) \leq 0.4$ |
| | $q''_{max} = q''_{peak} - 4(z/L_f - 2/5)(q''_{peak} - 30)$ | $0.4 < (z/L_f) < 0.65$ |
| | $q''_{max} = 7.2(z/L_f)^{-10/3}$ | $(z/L_f) \geq 0.65$ |
| | where, | |
| | $q''_{peak} = 120\left[1 - exp(-4.0D)\right]$ | |
| Ceiling and Top of Walls at Ceiling [7] | $q''_{max} = 120$ | $\left[(x+H)/L_f\right] \leq 0.58$ |
| | $q''_{max} = 18\left[(x+H)/L_f\right]^{-3.5}$ | $\left[(x+H)/L_f\right] > 0.58$ |
| Impinging on Unbounded Ceiling [5] | $q'' = 120$ | $\left[(r+H+z')/(L_f+z')\right] \leq 0.50$ |
| | $q'' = 682 exp\left\{-3.4\left[(r+H+z')/(L_f+z')\right]\right\}$ | $\left[(r+H+z')/(L_f+z')\right] > 0.50$ |
| | where, | |
| | $z' = 2.4D\left[Q_D^{*2/5} - Q_D^{*2/3}\right]$ | $Q_D^* < 1.0$ |
| | $z' = 2.4D\left[1 - Q_D^{*2/5}\right]$ | $Q_D^* \geq 1.0$ |
| | $Q_D^* = Q/\left(1100D^{5/2}\right)$ | |

The maximum heat fluxes that should be expected due to a local fire can be estimated from measurements conducted on items immersed in large pool fires, see Table 4.4. In general, heat fluxes were measured to be as high as 170 kW/m$^2$, with one exceptionally high data point at 220 kW/m$^2$. Based on these data, a bounding heat flux for local fires is expected be 170 kW/m$^2$.

*Table 4.4. Heat Fluxes to Objects Immersed in Large Pool Fires [11].*

| Test | Pool Size | Fuel | Peak Heat Flux (kW/m$^2$) |
|---|---|---|---|
| AEA Winfrith [11] | 1.6 ft x 31 ft | Kerosene | 150 |
| US DOT [11] | Not listed. | Kerosene | 138 |
| USCG [11] | Not listed. | Kerosene | 110-142 |
| US DOT [11] | Not listed. | Kerosene | 136-159 |
| Sandia [11] | Not listed. | Kerosene | 113-150 |
| HSE Buxton [11] | Not listed. | Kerosene | 130 |
| Shell Research [11] | 13 ft x 23 ft | Kerosene | 94-112 |
| Ref. [12] large cylinder | 30 ft x 60 ft | JP-4 | 100-150 |
| Ref. [12] small cylinder | 30 ft x 60 ft | JP-4 | 150-220 |
| Ref. [13] | 8 ft x 16 ft | JP-5 | 144 |

## 4.2.3    EFFECTS OF HOT GAS LAYERS

Local fire plumes that are located inside of an enclosure will cause a hot gas layer to develop in the upper part of the enclosure (upper-layer). This hot gas layer will preheat the boundaries of the enclosure. If the flames of the local fire plume are not optically thick, then a portion of the heat flux due to the hot gas layer will also contribute to the heat flux on the boundary.

There are a variety of fire models that can be used to predict the upper-layer temperature inside an enclosure containing a fire. Equally accurate is to use empirical correlations to predict the upper-layer temperatures for specific applications. Correlations exist for enclosures that have a door opening where air is naturally drawn into the enclosure, forced ventilation scenarios, and for completely closed compartments. The correlation for natural ventilation was developed by McCaffrey, Quintiere and Harkelroad [14]:

$$\Delta T = C \left( \frac{Q^2}{h_k A_T A_o \sqrt{H_o}} \right)^{1/3} \tag{4.7}$$

where,

$$h_k = \sqrt{\frac{k\rho C_p}{t}} \quad \text{for} \quad t < \frac{\alpha \delta^2}{4} \tag{4.8}$$

$$h_k = \frac{k}{\delta} \quad \text{for} \quad t > \frac{\alpha \delta^2}{4} \tag{4.9}$$

Karlsson and Magnusson [15] determined that the constant $C$ in Eqn. (4.7) was a function of the fire location inside the compartment.

$$C = \begin{cases} 6.83 \text{ for room-centered fires} \\ 9.22 \text{ for fires in a corner} \end{cases} \tag{4.10}$$

Fires in the corner of a room induced less air entrainment into the fire plume resulting in higher gas temperatures.

For forced ventilation, the correlation developed by Deal and Beyler [16] is most appropriate:

$$\Delta T = \frac{\dot{Q}}{\dot{m}_{ex} C_{p,air} + h_k A_T} \tag{4.11}$$

where the heat transfer coefficient is calculated using:

$$h_k = 0.4 \, max\left( \sqrt{\frac{k\rho C_p}{t}}, \frac{k}{\delta} \right) \tag{4.12}$$

This correlation can also be used to estimate the gas temperature in naturally ventilated compartments if the ventilation rate through the door is known. For enclosures with thermally thin boundaries (eg. steel), Peatross and Beyler [17] determined that the correlation developed Deal and Beyler [16] can be used with a modified heat transfer coefficient to quantify the heat losses to the wall. Therefore, for thermally thin boundaries Eqn. (4.12) is replaced with the following expression to determine the heat transfer coefficient:

$$h_k = 30 - 18\left[ 1 - exp\left( -\frac{50}{\rho \delta C_p} t \right) \right] \tag{4.13}$$

Tanaka et al. [18] measured the total heat flux to the boundaries of a 3.3 m wide, 3.3 m deep, 2.35 m high enclosure containing a fire with a door supplying natural ventilation. Figure 4.3 shows a plot of the water-cooled heat flux gauge measurements made on the upper-part of the inside walls to the enclosure with respect to the upper-layer gas temperatures inside the enclosure. The line in this plot corresponds to the black-body heat flux calculated using the upper-layer gas temperature, $\sigma T^4_{layer}$. In reality, there is both convective and radiative heat transfer occurring between the gas layer and boundary. However, assuming the gas layer behaves like a black-body it provides a good estimate of the total heat flux to the enclosure boundaries. Therefore, the heat flux contribution from hot gas layers that develop inside enclosures can be calculated using the following relation:

$$\dot{q}''_{hfg,layer} = \sigma T^4_{layer} - \varepsilon_s \sigma T^4_\infty \tag{4.14}$$

or

$$q''_{net,layer} = q''_{hfg,layer} - \varepsilon_s \sigma \left( T_s^4 - T_\infty^4 \right) \qquad (4.15)$$

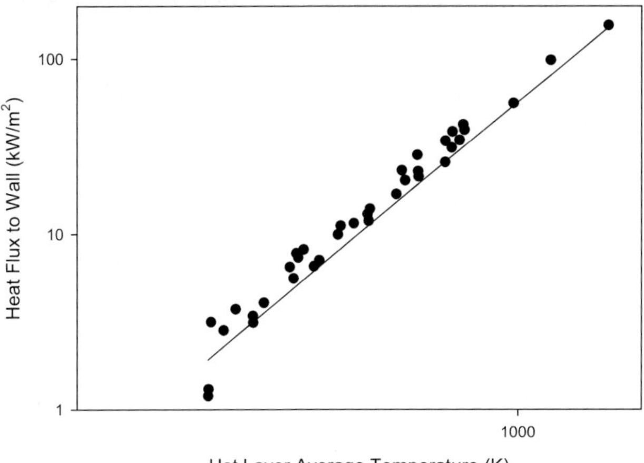



*Figure 4.3. Heat fluxes to the upper part of a wall inside an enclosure measured using a water-cooled heat flux gauge. The line is the black body heat flux. Reproduced from Tanaka et al. [18].*

In the area of the fire plume, the net heat flux to the surface is the sum of the heat flux from the fire (Eqn. (4.5)) and the heat flux from the layer that is transmitted through the fire plume:

$$q''_{net} = q''_{net,f} + \tau_f q''_{net,layer} \qquad (4.16)$$

Assuming combustion product attenuation is secondary to attenuation due to smoke and the fire plume behaves as a gray gas, then the gas transmissivity of heat through the flame to the surface will decrease with an increase in smoke levels and path-length (ie. flame thickness). Assuming the fire plume is optically thin ($\delta \sim 0$), none of the radiation from the layer is attenuated by the fire. Optically thick fires ($\delta \sim 1.0$) will not allow the radiation from the layer to penetrate through the gases, and therefore the heat flux to the boundary is only due to the fire itself. The heat fluxes for these three conditions are determined using:

$$q''_{net} = q''_{net,f} + \left[ exp(-\kappa d) \right] q''_{net,layer} \qquad \text{(gray gas fire plume)} \quad (4.17)$$

$$q''_{net} = q''_{net,f} + q''_{net,layer} \quad \text{(optically thin fire plume)} \quad (4.18)$$

$$q''_{net} = q''_{net,f} \quad \text{(optically thick fire plume)} \quad (4.19)$$

## 4.2.4    FULLY-DEVELOPED ENCLOSURE FIRES

The fire plume has been the dominant thermal exposure source up to this point in the chapter. However, a fire plume that is inside of an enclosure may grow large enough so that the gas layer conditions begin to dominate the exposure. Figure 4.4 presents a graphic depicting the different stages of an enclosure fire. During the pre-flashover stage, the fire plume, including layer heating effects, will dominate the exposure. However, if conditions are sufficient the fire can continue to grow and the compartment may reach flashover. Flashover for an enclosure fire occurs when the thermal conditions inside of the enclosure are sufficient to ignite all combustible items within the enclosure. Of more interest, from the thermal exposure point of view, are the conditions that exist during post-flashover or the fully-developed stage of a fire.

The thermal exposure from fully-developed enclosure fires has been primarily defined by the gas temperature inside of enclosure. The classical approach to analysing a fully-developed enclosure fire was developed by Babrauskas and Williamson [19]. In this analysis, the enclosure is considered as a well-stirred reactor with uniform temperatures. The first law of thermodynamics is used to balance the energy in and out of the enclosure to determine the gas temperature inside the enclosure, see Fig. 4.5. The governing equation for the enclosure control volume is:

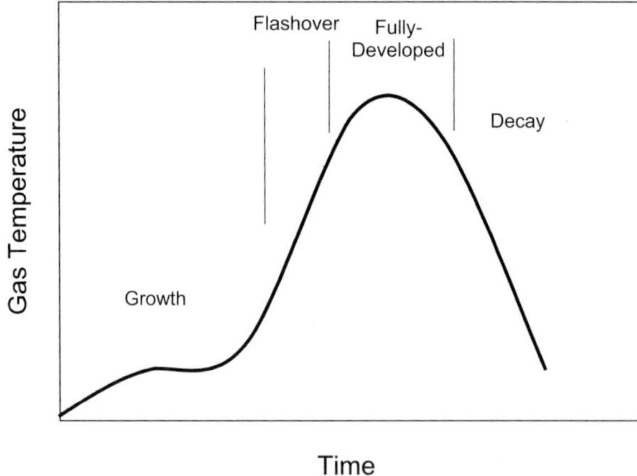


*Figure 4.4. The different stages of an enclosure fire.*

$$Q_{fire} - Q_{opening,flow} - Q_{opening,rad} - Q_{walls} \qquad (4.20)$$

where $Q_{fire}$ is the heat release rate of the fire, $Q_{opening}$ is the heat convected out of openings in the enclosure, $Q_{opening,rad}$ is the heat radiated out of the openings in the enclosure, and $Q_{walls}$ is the heat loss to the walls of the enclosure.

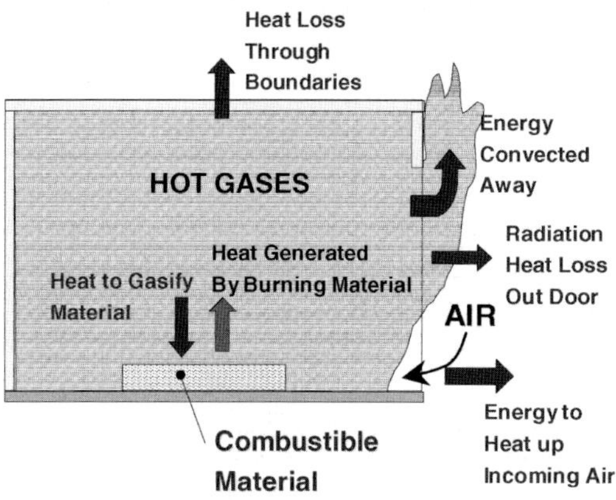


*Figure 4.5. Analysis of a fully-developed enclosure fire.*

The computer code COMPF2 contains this theory and is publicly available through the National Institute of Standards and Technology.

More commonly, empirical correlations are used to estimate the gas temperature inside an enclosure fire. Many of the correlations are based on the classic set of data developed for the Conseil International du Batiment (CIB) [20], shown in Fig. 4.6. Tests were conducted in 1.0 m high enclosures with the width and depth ranging from 1.0 to 4.0 m. In all test, the fuel was wood cribs with a fuel loading of 10-40 kg/(m$^2$ of floor area). From this data, the gas temperature inside the enclosure was determined to be a function of the surface area of the compartment excluding the floor and door opening, $A$, divided by the ventilation parameter, $A_o\sqrt{H_o}$. As shown in Fig. 4.7, this same factor was also found to be related to the wood mass burning rate measured in the tests when normalized relative to the ventilation parameter and the square root of the compartment aspect ratio to the one-half power.

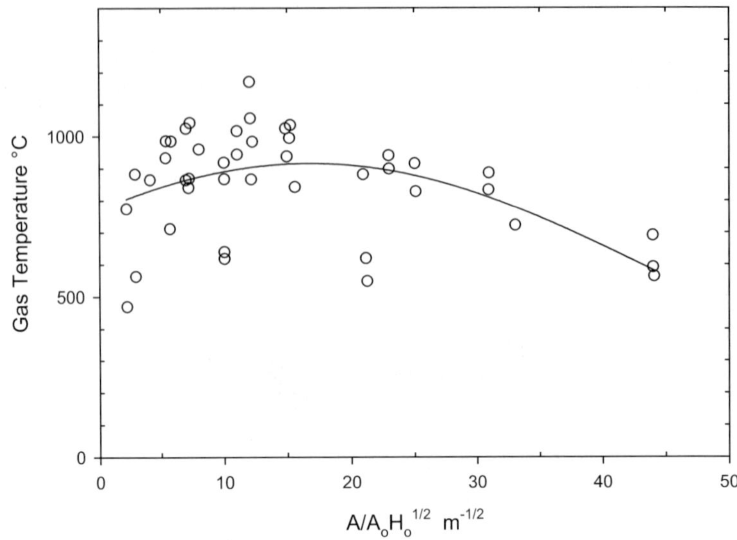


*Figure 4.6.  Average gas temperature during the fully-developed stage in the CIB enclosure fire tests with wood cribs as fuel.  Reproduced from Thomas and Heselden [20].*

Compartment fire tests have also been conducted with other fuels including thermoplastics. The gas temperatures and burning rate data for wood cribs as the fuel are also provided in Figs. 4.6 and 4.7. Comparison of the wood crib data with data from other fuels demonstrates that changing fuels has an impact on both the gas temperature and burning rate of the fuel. Changing the fuel type may change the energy required to gasify the fuel, the heat generated by the fuel, and the overall stoichiometry of the enclosure fire. This will ultimately affect the energy balance on the enclosure fire, which will in turn affect the gas temperature reached inside the enclosure during the fully-developed stage.

An estimate of the enclosure fire duration can be determined knowing the combustible fuel mass and fuel mass loss rate:

$$\tau = \frac{m_{comb}}{\dot{m}_f} = \frac{\eta m_t}{\dot{m}_f} \tag{4.21}$$

The combustible mass of fuel can be determined by multiplying the total mass of combustible fuel by the fraction of fuel that is expected to be remaining after the fire, which could be char, glass reinforcement, or other filler material in the case of a polymer matrix composite. As previously discussed, the mass loss rate of fuel is dependent on the thermal feedback from the enclosure fire and the stoichiometry of the compartment (ie. oxygen concentration).

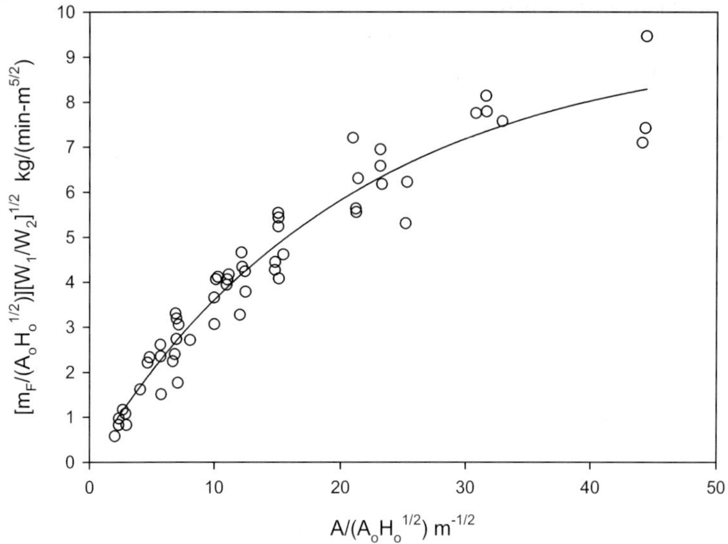

Figure 4.7.  Normalized burning rate of wood cribs during fully-developed stage in the CIB enclosure fire tests.  Reproduced from Thomas and Heselden [20].


## 4.3      Modelling Material Fire Dynamics

Material fire dynamics is the behaviour of a combustible material when it is directly exposed to fire conditions. Under sufficient heat and with available oxygen, materials exposed to fire conditions may ignite, begin to burn and release heat, and spread flame across the material surface. This section provides an overview of models to predict ignition, heat release rate, and flame spread of combustible materials.

### 4.3.1    IGNITION

A solid material exposed to sufficient heat will begin to decompose and off-gas or pyrolyze. If the gases released from the material are a flammable mixture and are exposed to a pilot ignition source, ignition will occur. The onset of ignition has been characterized in several ways including a critical surface temperature (ignition temperature), critical incident heat flux onto the material, rapid rate of rise in the surface temperature, critical mass loss rate, etc. To model the ignition of a material, a heat transfer model is usually invoked to predict the surface temperature. Most models will assume ignition occurs when the material surface reaches the ignition temperature of the material. All models will rely on experimental data to some extent to be able to adequately predict the ignition time.

The ignition properties of polymer laminates and sandwich composite materials are described in Chapter 3, and additional ignition data for two types of composite materials and plywood are provided in Table 4.5. These data were measured using the cone calorimeter test apparatus, which is described in ASTM E1354 'Standard Test Method for Heat and Visible Smoke Release Rate for Materials and Products Using an Oxygen Consumption Calorimeter' [21]. (A description of the cone calorimeter technique is given in Chapter 11). The data includes the time-to-ignition at different incident heat fluxes, minimum heat flux for ignition, and the estimated ignition temperature. In general, times-to-ignition decrease with an increase in incident heat flux. As demonstrated through some of the data, fire retardant additives may or may not have a significant impact on the time-to-ignition. Further information on the ignition times for flame retardant composite materials is provided in Chapter 8.

*Table 4.5. Ignition data for some solid materials. Data from Lattimer and Sorathia [24].*

| Material | Minimum Heat Flux $(kW/m^2)$ | Ignition Temperature (K) | Time to Ignition (sec) | | |
|---|---|---|---|---|---|
| | | | $25 \text{ kW/m}^2$ | $50 \text{ kW/m}^2$ | $75 \text{ kW/m}^2$ |
| Plywood | 13 | 622 | 304 | 22 | 8 |
| E-Glass FR Vinyl Ester | 17 | 670 | 387 | 80 | 34 |
| Sandwich Composite[1] | 15 | 650 | 306 | 70 | 28 |

1.    0.25 in. thick E-Glass FR vinyl ester skins with 3.0 in. thick balsa core.

The minimum heat flux is where a material is observed to ignite after being exposed to a heat flux for a specified time period (typically 10 – 20 minutes). Materials may ignite when exposed to heat fluxes less than the minimum heat flux, but they must be exposed for longer than the 10-20 minute period. Eventually, the heat flux could be reduced low enough that material ignition is not possible regardless of exposure duration. The lowest heat flux that ignition would be observed when exposed indefinitely is termed the critical heat flux. In the literature, reported measured critical heat fluxes are actually the minimum heat flux.

The ignition temperature can be determined from the minimum heat flux using an energy balance at the material surface:

$$q''_{min} = \varepsilon_s \sigma \left( T_{ig}^4 - T_\infty^4 \right) + h \left( T_{ig} - T_\infty \right) \qquad (4.22)$$

where $\varepsilon_s$ is the surface emissivity and $h$ is the convective heat transfer coefficient, which is assumed to be 0.010- 0.015 kW/m.K when the minimum heat flux is measured in the cone calorimeter.

The ignition of plastics and composite materials has been successfully modelled using semi-infinite solid heat transfer models, integral heat transfer models, and finite difference models. Except for the finite difference model, the application of these models requires that the material be thermally thick during the time it is being heated to the ignition temperature. A comparison of predicting the ignition times using the three different levels of heat transfer models is shown in Fig. 4.8 for a glass/polyester composite. In all models, the thermal properties of the material were selected to determine the best agreement between the predicted and measured ignition times over the range of incident heat fluxes. For predicting time-to-ignition, the finite difference model and integral model provided similar results. However, the solution of the integral model is faster and requires only the determination of the thermal inertia. The simple semi-infinite solid model provided similar results, which also compared well with the data.

The semi-infinite solid model is a simple equation that provides insight into some important aspects of predicting ignition. The surface temperature of a semi-infinite solid exposed to a constant heat flux is predicted using the following relation,

$$\frac{1}{\sqrt{t_{ig}}} = \frac{2}{\sqrt{\pi k \rho C}\left( T_{ig} - T_\infty \right)} q''_{net,s} \qquad (4.23)$$

where $q''_{net,s}$ is the net heat flux into the material surface, $k\rho C$ is the effective thermal inertia of the material, and $t_{ig}$ is the time-to-ignition. From this equation, a plot of the net heat flux into the material versus the inverse square root of the ignition time results in a straight line. The slope of this line is related to the effective thermal inertia of the material, $k\rho C$. The effective thermal inertia is a derived property that is model specific, and is not equal to the multiplication of the actual thermal properties of the material. This effective thermal inertia calibrates the model with the material data so that the heat transfer model being used can adequately predict the time-to-ignition. Therefore, the effective thermal inertia accounts for the changes in thermal properties with time and any solid decomposition or gas phase chemistry that may affect the ignition time of the material.

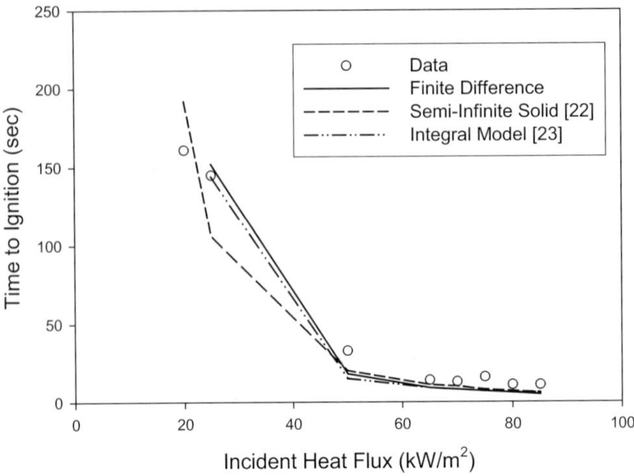


*Figure 4.8. Ignition times predicted for E-glass/polyester using different types of heat transfer models.*

## 4.3.2    HEAT RELEASE RATE

A material that has ignited will burn and release a certain amount of heat, termed heat release rate. Heat release rate is the energy generated by reacting the decomposition products from the fuel with oxygen. In an open environment, this heat release rate can be visually seen as a fire plume above the burning material. The heat release rate is dependent on the material chemical composition, the burning surface area, and the heat applied or heat flux onto the burning surface. The larger the burning surface area and the higher heat flux applied to the burning surface, the higher the heat release rate. As shown in Fig. 4.9, the heat flux onto the material surface is a combination of the heat from the flame at the material surface and heat from the surroundings, such as hot boundaries or objects, hot gases, or other nearby fires. As a result, a fire burning out in the open will have a lower heat release rate compared with a fire burning inside a compartment, where the hot walls and hot gas layer in the compartment will increase the heat flux onto the material surface.

Heat release rate for materials such as composites are typically measured using the cone calorimeter test apparatus described in ASTM E1354. Table 4.6 contains heat release rate values for some combustible materials when exposed to a range of heater incident heat flux levels, as measured by a water-cooled heat flux gauge. As seen in the table, the heat release rate generally increases with an increase in the incident heat flux from the heater.   Chapter 3 provides further information on the heat release rate properties of polymer composites.

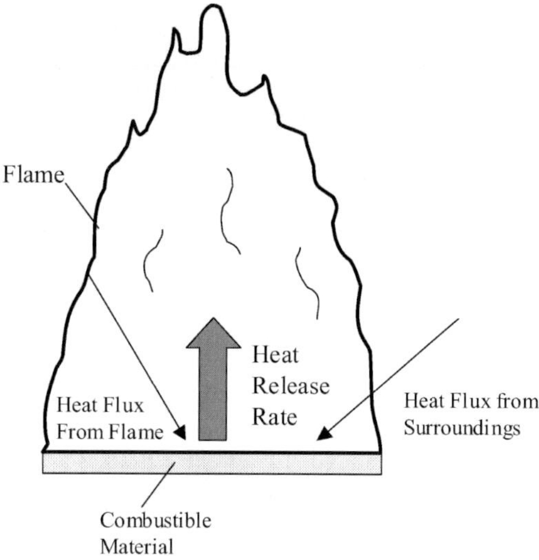


*Figure 4.9. Heat transferred to the material surface while burning.*

*Table 4.6. Heat release rate for some selected materials. Data from Lattimer and Sorathia [24].*

| Material | Test Average Heat Release Rate per Unit Area (kW/m$^2$) | | |
|---|---|---|---|
| | 25 kW/m$^2$ | 50 kW/m$^2$ | 75 kW/m$^2$ |
| Plywood | 87 | 135 | 210 |
| E-Glass FR Vinyl Ester | 60 | 80 | 110 |
| Sandwich Composite[1] | 80 | 100 | 120 |

1. 0.25 in. thick E-Glass FR vinyl ester skins with 3.0 in. thick balsa core.

Heat release rate curves for the materials in Table 4.6 are shown in Fig. 4.10. The simplest model for estimating heat release rate is using a constant heat of gasification. The heat of gasification is the amount of energy required to convert the solid material into gas and is calculated by dividing the net heat flux into the material by mass loss rate per unit area,

$$\Delta h_{g,eff} = \frac{q''_{net}}{\dot{m}''_f} = \frac{q''_{net}}{\left(\overline{Q}'' \big/ \Delta H_{c,eff}\right)} \tag{4.24}$$

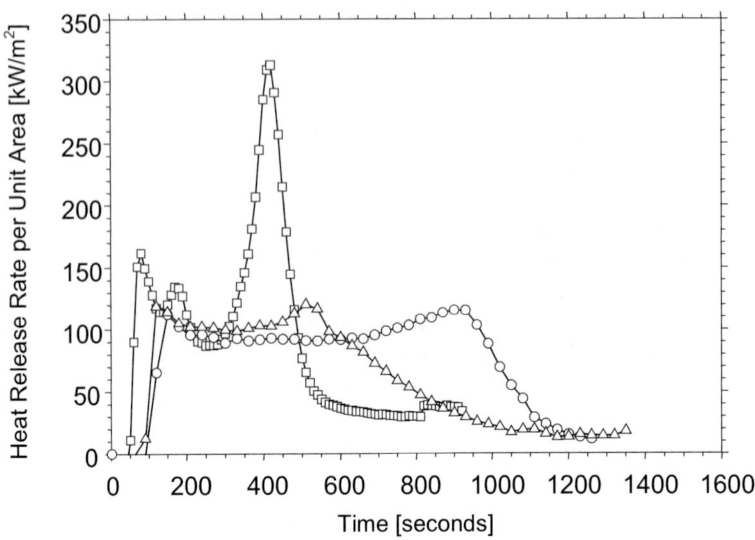

Figure 4.10.  *Heat release rate per unit area from plywood (□), E-Glass FR vinyl ester (O), and E-Glass FR vinyl ester sandwich composite with a balsa core (Δ) using the cone calorimeter with a heater incident heat flux of 50 kW/m². Data from Lattimer and Sorathia [24].*

Typically the cone calorimeter is used to determine the effective heat of gasification. The mass loss rate per unit area may be determined using the test average heat release rate divided by the effective heat of combustion. The effective heat of combustion is different to the heat of combustion that would be measured using a bomb calorimeter because it includes effects of incomplete combustion and other inefficiencies. Using this approach, the heat of gasification calculated would also be an effective heat of combustion. With the calculated effective heat of gasification, the heat release rate at any net heat flux onto the surface can be calculated by

$$Q'' = \frac{q''_{net}}{\Delta h_{g,eff}} \Delta H_{c,eff} \qquad (4.25)$$

The net heat flux in Eqns. (4.24) and (4.25) are the net heat flux applied at the front where material is decomposing or the pyrolysis front. For non-charring materials, this is simply the heat flux at the heat surface. Materials that char, like wood and composite materials, that have a non-combustible layer (glass and possibly char) between the heat surface and pyrolysis front need to be treated differently. The net heat flux used in Eqns. (4.24) and (4.25) should be the net heat flux at the pyrolysis front, which could change with time as the char or residual glass layer become thicker. This can be done by knowing the thermal properties of the char or glass and keeping track the location of the pyrolysis front using mass loss rate data. The thermal decomposition models discussed in Chapter 5 are capable of predicting this type of information.

### 4.3.3    FLAME SPREAD

Flame spread is the propagation of a flame along a material surface. As shown schematically in Fig. 4.11, flames may spread down, up or across a vertical surface or along a horizontal surface. In all types of flame spread, the flame is preheating unignited material. Figure 4.11 provides a comparison of different flame spread rates. The most rapid flame spread over a material is wind-aided flame spread. Wind-aided flame spread is where the flame extends out beyond the burning region and is in contact and preheats the unignited material. Examples of wind-aided flame spread include vertical flame spread on walls and flame spread along ceilings. In horizontal and downward flame spread, the flames preheat the material only by radiation causing the heat fluxes to be lower. This is sometimes referred to as opposed flow flame spread. As a result, horizontal and downward flame spread rates are typically ten times less than wind-aided flame spread.

Predicting flame spread over combustible surfaces is a complex phenomena that involves calculating nearly every aspect of fire dynamics discussed in this chapter thus far. Flame spread models need to predict heat fluxes from initiating fires and burning surfaces, the ignition of the material, heat release rate of the ignited material, material burning out, flame heights, and effects of gas temperatures on the heat flux to the material surface. Several flame spread models have been developed over the past fifteen years. Some of the first models are capable of predicting one-dimensional flame spread up a wall [25-27]. Based on the theory from the one-dimensional models, multi-dimensional flame spread models have been developed to predict flame spread in a corner configuration, see Fig. 4.12. The motivation for developing most of these models is to be able to predict the performance of combustible linings in a standard room corner fire test, like the ISO 9705 room corner test. As shown in Fig. 4.12, this test has the walls and ceiling of the room lined with a combustible material and a fire is placed in the back corner of the room. (Further details on the room corner test are provided in Chapter 11). The basic approach for modelling these types of fires was developed by Quintiere [28] and Karlsson [29]. These models represented the heat flux to the wall and ceiling as average heat fluxes over the region where flames existed. More recent models developed by Lattimer et al. [23,24] are capable of predicting heat flux distributions over the walls and ceiling, making the model applicable to a broader class of problems. Fig. 4.12 contains ISO 9705 test data on different composite materials compared with predictions using the model developed by Lattimer et al. [23].

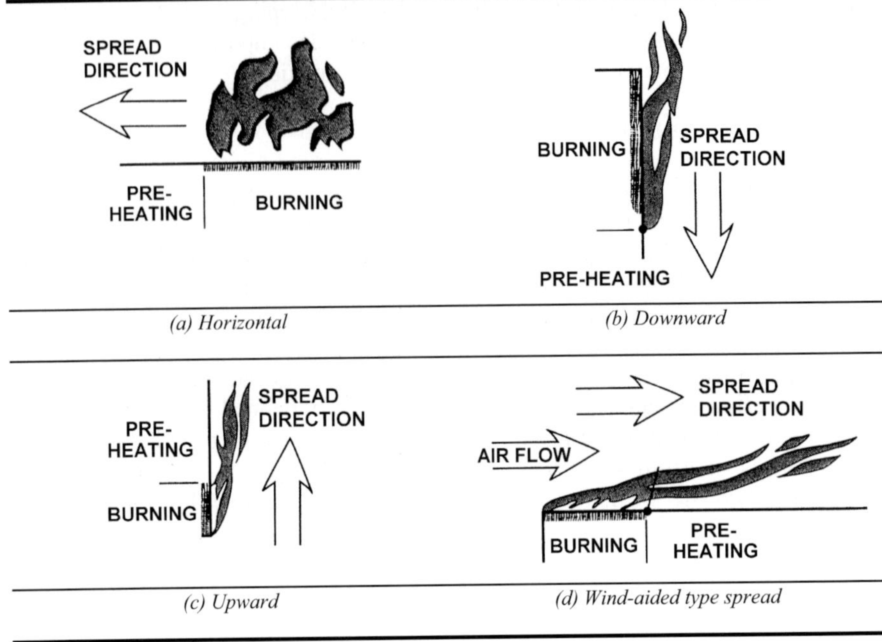


*Figure 4.11. Different types of flame spread. From Principles of Fire Behaviour 1ˢᵗ edition by J. Quintiere, © 1998. Reprinted with permission of Delmar Learngin, a division of Thomson Learning: www.thomsonrights.com. Fax 800 730-2215.*

## 4.4    Structural Modelling of Fire Response

This section addresses the structural response of composite structures to the threat of fire. Response is considered to include behaviours during a fire event, as well as under post-fire design loads. Analysis approaches are discussed that are currently available or potential areas of further investigation. Finite element analyses are the primary tools for assessment, complemented by application programs for specific details. Further information on modelling the mechanical response of composites during and after fire is provided in Chapters 6 and 7, respectively.

### 4.4.1    STRUCTURAL MODEL INTENT

It is important to establish the intent of the structural analysis to support a fire-qualified composite structure. This will dictate the extent of simulation required. In the structural design phase, conservative static analyses using 'worst case' structural degradation due to fire may be used. In test correlation or failure investigation analyses, more complex simulations including coupled thermodynamic structural solutions may be employed.

Analysis will include macro-level finite element analysis to yield internal load distributions, overall stability and frequency response.

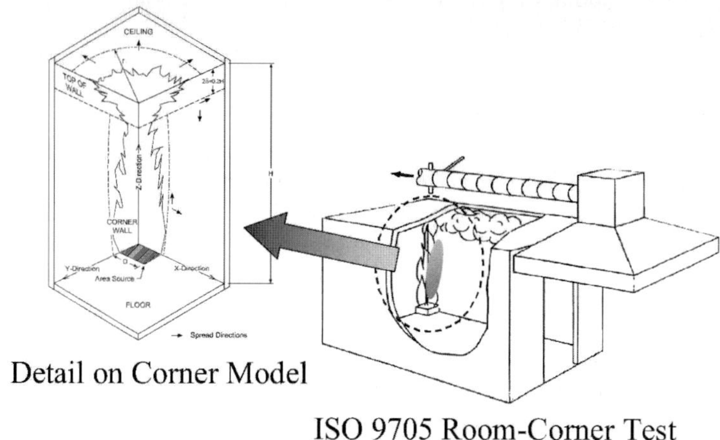



*Figure 4.12. Flame spread in a corner.*

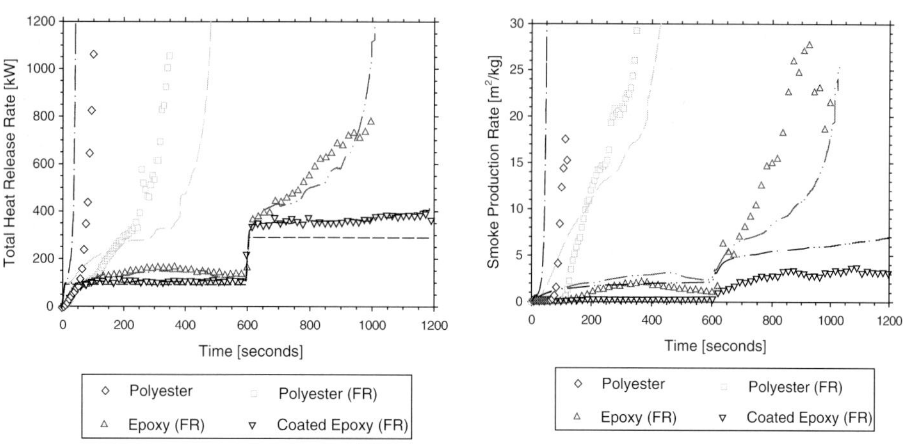



*Figure 4.13. ISO 9705 room corner fire test data (symbols) for various glass reinforced thermoset composites compared with predictions (lines) from the model of Lattimer et al. [23]. Dashed line in the heat release rate plot is the initiating fire.*

Detail finite element and other simulation models can be used for joint, attachment and material lamina level behaviour. Non-linear effects of contact, large displacement, material mechanical and thermal character as well as progressive damage must be considered. The limitations of software tools for the defined analysis goals must be recognized [30].

### 4.4.2    OVERALL LOAD PATH MODELS

An overall, sometimes called global, finite element model is required to predict structural behaviour at a fairly coarse level. This is necessary to establish the primary load paths, overall primary structural modes and frequencies, and homogenized material behaviour. Structural joints are typically represented by effective properties, ie. fasteners and bond lines are not discretely modelled. An example of this level of modelling is shown in Fig. 4.14.

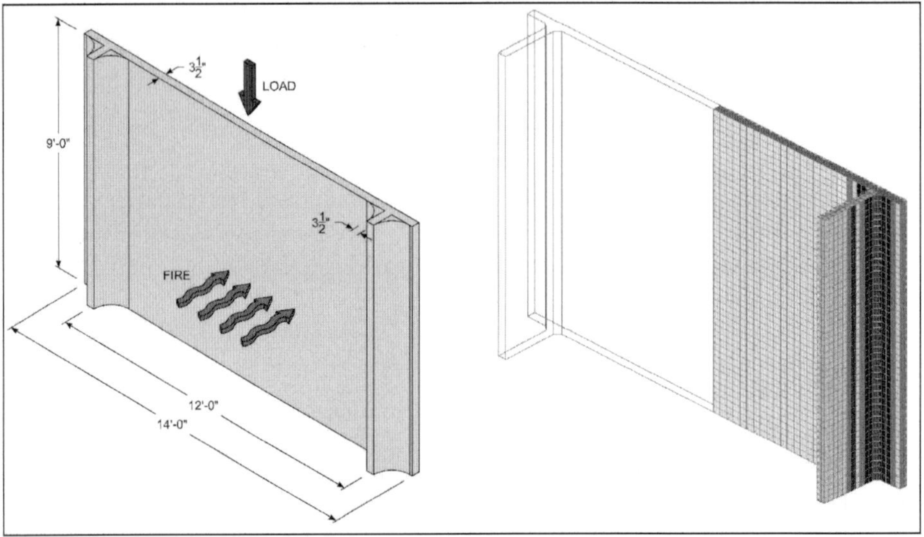


*Figure 4.14. Global model level of detail.*

### 4.4.3    LOAD INPUT FOR STRENGTH ANALYSIS

The loads required for this level of modelling are dictated by the extent of the geometric simulation. If an entire vehicle is modelled, for example a marine vessel, buoyant pressures balanced by inertial response of the structural mass will define critical seaway loading. If less than the entire vehicle is modelled, assumptions as to boundary loading must be made, typically based on coarse model or beam distributions. This established, the conditions and critical time points that meet the design criteria must be selected and defined.

*Frequency*
It is typically prescribed that primary structure have natural frequencies that are sufficiently isolated from external excitations. The level of modelling of the global model is usually sufficient to define frequency behaviour and its sensitivity to stiffness degradations due to fire response.

*Stability*
The primary concern of structural response to fire scenarios is collapse. The stability of a structural component typically requires local refinement and detail modelling. Sensitivity to the edge behaviour of a joint, ie. actual rotational stiffness as opposed to simple or fixed assumptions, is recognized. Free-body loads as boundary conditions from the overall model are employed. Analysis requires the inclusion of non-linear effects, precluding linear buckling solutions.

The analysis procedure for stability that includes all the non-linear effects of large deflection, material temperature, progressive material failure, etc requires a rigorous solution scheme. ABAQUS offers a 'modified Riks method' for the case where loading is proportional, providing solutions even in cases of complex, unstable response. User subroutines to prescribe material non-linearity and element progressive damage may be included. The concept of time is replaced by arc length in the Riks procedure, therefore time and strain rate effects such as viscous damping and rate dependent plasticity are not correctly simulated. Shown in Fig. 4.15 are analysis results for a stiffened panel under static compression, for an incrementally applied thermal load.

*Detail laminate Strength Analysis*
*Sublaminate analysis*
Typically, sublaminate analyses are performed to evaluate a laminate's behaviour under load, by examining the laminas and inter-lamina compliant strains. Again, the material system performance is assumed to be well characterized in both stiffness and strength. Strength data at the lamina level for temperature performance at and greater than the $T_g$ of the resin system is limited, if at all available.

A matrix of failure modes are examined including lamina failures in the form of transverse cracks in planes parallel to the fibres, fibre dominated failures in planes perpendicular to the fibres, and delaminations between layers. Analysis of a stacked laminate will employ one or more failure criteria for which the analyst is confident in results (based on experience and testing with the resin/fibre system). These include maximum stress, maximum strain, Tsai-Hill, tensor polynomial methods, well discussed in the noted references [31].

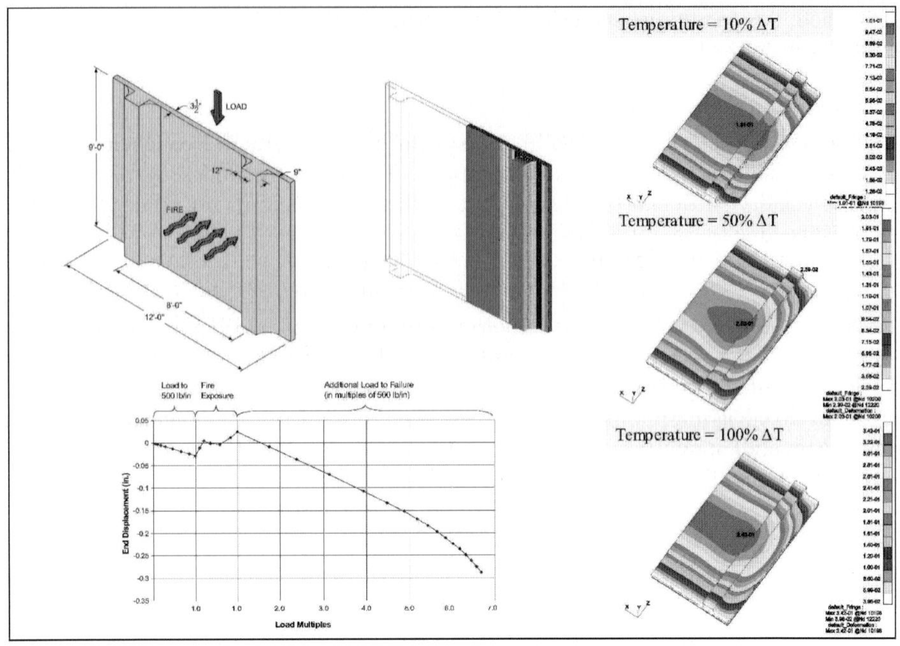


Figure 4.15. Non-linear collapse analysis under mechanical and thermal loading.

### RVE homogenization

At the macro-modelling level, equivalent "homogenized" laminate properties are used for plates in the simulation, post-processed to examine ply-by-ply response. Refined three-dimensional models employ repetitive volume element (RVE) properties. Analyses utilize damage material models for progressive failure simulations. The material response to elevated temperatures can be incorporated in these models.

### Material performance data

#### Stiffness

Lamina, i.e. ply, elastic properties are typically derived from a coupon test program or from documented material data, as in MIL-17 Handbook. Properties are measured at room temperature, as manufactured (dry) state, cold, dry and elevated temperature with moisture conditioning. These temperatures are established by maximum environment definitions and are typically well below the glass transition temperature ($T_g$) of specified resin systems. Classic lamination theory is used to synthesize the orthotropic properties of the stacked laminate. Recalling that Hooke's law relates the mechanical strains and stresses, the inclusion of temperature loads yields:

$$\{\sigma\} = [Q]\left(\{\varepsilon\} - \{\varepsilon^T\}\right) \qquad (4.26)$$

for stresses using $[Q]=[S]^{-1}$. It is recognized that the stress-strain relationship is highly non-linear in the fire scenario temperature ranges.

*Linear/non-linear elastic constants*
The discrete analysis of composite structure under high temperature-time histories presents a number of non-linear material behaviours. These include: (1) the initial non-linear modulus of a lamina/laminate as the resin system reaches $T_g$, (2) the degradation of the resin system as it reaches and exceeds combustion temperatures, (3) the non-linear behaviour of the fibre form as it approaches phase change temperatures, and (4) the degradation of fibre at high temperature. The initial strength and stiffness degrading effects are primarily important for the matrix dominated properties, $E_2$, $G_{12}$, $F_{tu2}$, $F_{cu2}$, $F_{su12}$ and interlaminar properties. A simple law of mixture approach for the degraded laminate is considered a good approximation of behaviour.

*Time/temperature dependence*
As noted, the initial time/temperature non-linearity is related to the $T_g$ of the resin system. An approximation of this behaviour for non-linear material input is derived from limited manufacturer provided data (Dow Derakane 510a shown as a reduction of base room temperature properties, Fig. 4.16).

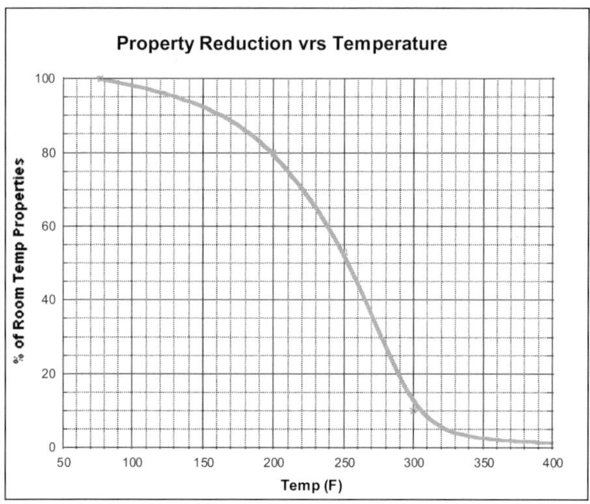


*Figure 4.16. Resin property reduction as function of temperature.*

*Strain rate effects*
The impact of strain rate effects must be considered if the mechanical or thermal loading is transient and/or short duration. Characterization data must be accumulated for the particular resin/reinforcement system to properly model.

*Law of mixtures approach*
Modelling of a laminate exposed to a fire condition will require behaviour data outside the typical characterization. The reduced volume fraction of resin as it pyrolizes is a mechanism for simulation of the non-linear structural response. To this end, coupling of the thermodynamic and structural models to quantify the extent of resin consumption through the time history is desirable.

*Repair strategies and simulations*
The post fire damage requires assessment and repair. Composite repair is typically a tailored "patch" with sufficient overlap with undamaged structure to return to full structure strength. Repair strategies must balance the cost and risk of repair with replacement. These decisions are supported by analysis.

*Efficiency of repair*
A doubler repair will require sufficient overlap area to unload the damaged structure. Load path eccentricities and bond-line flexibility often make full-strength repairs difficult. Mechanical fastening of patches have similar concerns compounded by the introduction of additional stress concentrations and failure modes. Simulations may be as simple as 1D flexible joint calculations or as complex as linear/non-linear FEA of bolted/bonded configurations.

*Performance of damaged material*
Post fire material performance must be understood to provide a comprehensive structural assessment. The extent of exposure, both time and temperature, impacts the strength and stiffness characteristics of the material. Test data to support these assessments must be generated particular to both the resin system and the reinforcement product form. These include post-fire exposure static and fatigue strengths, as well as fracture, thermal and moisture characteristics.

*Concluding Comments*
The analytical methods for the simulation of structural response before, during and after a fire scenario are available and have been demonstrated. The close coupling of the thermal and structural solutions is an area of potential growth and development.

There is a need for extensive testing to quantify stiffness and strength data for thermally exposed composite materials. High temperature, heat rate and thermal cycling effects as well as post-fire residual static and fatigue strength properties will be required for accurate simulations and proper assessment.

The adoption of standard methods and data reduction techniques for material characterization data at elevated temperatures above $T_g$ should be investigated. The accumulation and documentation of these properties for candidate materials, as in MIL-17, is strongly recommended.

## Nomenclature

| | |
|---|---|
| $A_T$ | surface area of compartment walls and ceiling excluding door [m$^2$] |
| $A_o$ | compartment door opening area [m$^2$] |
| $C$ | constant |
| $C_p$ | specific heat capacity compartment boundary [kJ/(kg K)] |
| $C_{p,air}$ | specific heat capacity of air at 300 K [1.0 kJ/(kg K)] |
| $d$ | burning width on a wall or burning width on one side of the corner [m] |
| $D$ | length of single side of square burner, diameter [m] |
| $E_2$ | Young's moduli in lamina normal to the fibre direction [GPa, ksi] |
| $F_{cu2}$ | ultimate compression stress in lamina normal to the fibre direction [MPa, ksi] |
| $F_{su2}$ | ultimate shear stress in lamina normal to the fibre direction [MPa, ksi] |
| $F_{tu2}$ | ultimate tension stress in lamina normal to the fibre direction [MPa, ksi] |
| $G_{12}$ | shear modulus in the plane of lamina [GPa, ksi] |
| $g$ | acceleration of gravity [9.81 m/s$^2$] |
| $h$ | convective heat transfer coefficient [kW/(m$^2$ K)] |
| $h_k$ | convective heat transfer coefficient for compartment boundaries [kW/(m$^2$ K)] |
| $H$ | distance between fuel source and ceiling or ceiling height [m] |
| $H_o$ | compartment door height [m] |
| $\Delta h_g$ | heat of gasification [kJ/kg] |
| $\Delta H_c$ | heat of combustion [kJ/kg] |
| $k$ | thermal conductivity of material or compartment boundary [kW/(m K)] |
| $l$ | flame optical path length [m] |
| $L_f$ | flame length [m] |
| $m_{comb}$ | combustible mass of fuel inside compartment [kg] |
| $m_t$ | total mass of fuel inside compartment [kg] |
| $\dot{m}_f$ | fuel mass loss rate [kg/s] |
| $\dot{m}_{ex}$ | mass flow rate of gas out of compartment [kg/s] |
| $\dot{m}_f''$ | fuel mass loss rate per unit area [kg/(s m$^2$)] |
| $Q$ | fire heat release rate [kW] |
| $\overline{Q}''$ | test average heat release rate per unit area [kW/m$^2$] |
| $Q^*$ | dimensionless parameter, $Q_D^* = \dfrac{Q}{\rho_\infty C_p T_\infty \sqrt{g} D^{5/2}}$ with $D$ being length scale |
| $r$ | distance from corner or stagnation point to measurement location [m] |
| $q''$ | heat flux [kW/m$^2$] |
| $t$ | time [s] |
| $t_{ig}$ | time to ignition [s] |
| $T$ | temperature [K] |
| $T_g$ | room gas temperature [K] |
| $T_s$ | material surface temperature [K] |

$T_\infty$     ambient temperature [300 K]

$x$     horizontal distance from corner or fire centerline or with distance into the material thickness [m]

$z$     vertical distance above base of fire [m]

$z'$     virtual source origin correction in tests with fires impinging on ceilings [m]

## GREEK

$\chi$     combustion efficiency [- -]

$\chi_r$     radiative fraction [- -]

$\delta$     compartment boundary thickness [m]

$\varepsilon$     emissivity [- -]

$\kappa$     extinction coefficient of fire gases [1/m]

$\eta$     combustible fraction of fuel [- -]

$\rho_\infty$     ambient density of air [1.2 kg/m$^3$]

$\rho$     compartment boundary density [kg/m$^3$]

$\pi$     constant [3.14159]

$\sigma$     Stefan-Boltzman constant [5.67 x 10$^{-11}$ kW/(m$^2$ K$^4$)]

$\tau_f$     transmissivity of flame [- -]

$\tau$     fire duration [s]


## SUBSCRIPTS

$conv$     convective

$D$     defined using $D$ as length scale

$eff$     effective

$f$     flame

$inc$     incident

$hfg$     heat flux gauge

$H$     defined using $H$ as length scale

$layer$     gas layer

$max$     max level

$min$     minimum

$net$     net

$peak$     peak

$rad$     radiative

$rr$     reradiated

$s$     material surface

## References


1. M.A. Delicatsios. Flame heights in turbulent wall fires with significant flame radiation. *Combustion Science and Technology*, 1984; 39:195-214.
2. B.Y.Lattimer, S.P. Hunt, U. Sorathia, M. Blum, T. Gracik, M. McFarland, A. Lee and G. Long. Development of a model for predicting fire growth in a combustible corner. NSWC Report, NSWCCD-TR-64-99/07, U.S. Navy, 1999.
3. G. Heskestad. Luminous height of turbulent diffusion flames. *Fire Safety Journal*, 1983; 5:103-108.
4. G. Back, C.L. Beyler, P. DiNenno and P. Tatem. Wall incident heat flux distributions resulting from an adjacent fire. In: *Proceedings of the 4th International Symposium on Fire Safety Science*, 1994, pp. 241-252.
5. Y. Hasemi, S. Yokobayashi, T. Wakamatsu and A. Ptchelintsev. Fire safety of building components exposed to a localized fire: Scope and experiments on ceiling/beam system exposed to a localized fire. In: *Proceedings of ASIAFLAM*, 1995, pp. 351-361.
6. B.Y. Lattimer. Heat fluxes from fires to surfaces. In: *The SFPE Handbook of Fire Protection Engineering*, 3rd edition, ed. P.J. DiNenno, SFPE, MA:Boston, 2001, Chapter 2-14.
7. B.Y. Lattimer and U. Sorathia. Thermal characteristics of fires in a non-combustible corner. *Fire Safety Journal*, 2003; 38:709-745.
8. J. Myllymaki and M. Kokkala. Thermal exposure to a high welded I-beam above a pool fire. In: *Proceedings of the 1st International Workshop on Structures in Fires*, Copenhagen, June 2000, pp. 211-226.
9. B. McCaffry. *NBSIR 79-1910*, National Bureau of Standards, Washington, D.C., 1979.
10. H.R. Baum and B.J. McCaffrey. Fire induced flow field: Theory and experiment. In: *Proceedings of the 2nd International Symposium on Fire Safety Science*, 1989, pp. 129-148.
11. L.T. Cowley. Behaviour of oil and gas fires in the presence of confinement and obstacles. Miscellaneous *Report TNMR.91.006*, Shell Research Limited, Thornton Research Center, Combustion and Fuels Department, Chester, UK, February, 1991.
12. J.J. Gregory, R. Mata and N.R. Keltner. Thermal measurements in a series of large pool fires. *Sandia Report Number SAND85-0196*, Sandia National Laboratories, Albuquerque, NM, 1987.
13. L.H. Russell and J.A. Canfield. Experimental measurements of heat transfer to a cylinder immersed in a large aviation fuel fire. *Journal of Heat Transfer*, August 1973.
14. B.J. McCaffrey, J. Quintiere and M. Harkelroad. Estimating room temperatures and the likelihood of flashover using fire test data correlations. *Fire Technology*, 1981; 17:98-119.
15. B. Karlsson and S. Magnusson. Combustible wall lining materials: Numerical simulation of room fire growth and the outline of a reliability based classification procedure. In: *Proceedings of the 3rd International Symposium in Fire Safety Science*, ed. G. Cox and B. Langford, London:Elsevier Applied Science, 1991.
16. S. Deal and C. Beyler. Correlating preflashover room fire temperatures. *Journal of Fire Protection Engineers*, 1990; 2:33-48.
17. M. Peatross and C. Beyler. Thermal environment prediction in steel-bounded preflashover compartment fires. In: *Proceedings of the 4th International Symposium on Fire Safety Science*, 1994, pp. 205-216.
18. T. Tanaka, I. Nakaya and M. Yoshida. Full scale experiments for determining the burning conditions to be applied to toxicity tests. In: *Proceedings of the First International Symposium on Fire Safety Science*, 1985, pp. 129-138.
19. V. Babrauskas and B. Williamson. Post-flashover compartment fires: Basis of a theoretical model. *Fire and Materials*, 1978; 2:39-53.
20. P.H. Thomas and J.M. Heselden. Fully-developed fires in single compartments: "A cooperative research programme of the Conseil International du Batiment," *Fire Research Note 923*, Fire Research Station, Borehamwood, 1972.
21. ASTM E1354, "Standard Test Method for Heat and Visible Smoke Release Rate for Materials and Products Using an Oxygen Consumption Calorimeter," American Society for Testing and Materials, *Vol. 4.07 Building Seals and Sealants; Fire Standards*; Dimension Stone, Conshohocken, PA, 2000.
22. J. Quintiere. *Fire and Materials*, 1981; 5:52-60.
23. B.Y. Lattimer, S.P. Hunt, M. Wright and C.L. Beyler. Corner fire growth in a room with a combustible lining. In: *Proceedings of the 7th International Symposium Fire Safety Science*, MA: Worcester, 2001.

24. B. Lattimer and U. Sorathia. Modelling fire growth in a combustible corner. *Fire Safety Journal*, 2003; 38:771-796.

25. H. Mitler. Predicting the spread rates on vertical surfaces. In: *Proceedings of the 23$^{th}$ International Symposium on Combustion*, The Combustion Institute, 1990, pp. 1715-1721.

26. A.K. Kulkarni, C.I. Kim and C.H. Kuo. Heat flux, mass loss rate and upward flame spread for burning vertical walls. *NIST-GCR-90-584*, U.S. Department of Commerce, 1990, p. 57.

27. C.L. Beyler, S.P. Hunt, N. Iqbal and F.W. Williams. A computer model of upward flame spread on vertical surfaces. In: *Proceedings of the 5$^{th}$ International Symposium Fire Safety Science*, ed. Y. Hasemi, 1997, pp. 297-308.

28. J. Quintiere. A simulation model for fire growth on materials subject to a room-corner test," *Fire Safety Journal*, 1993; 20:313-319.

29. B. Karlsson. Calculating flame spread and heat release rate in the room corner test, taking account of pre-heating by the hot gas layer. In: *Proceedings from Interflam '93*, 1993, pp. 25-37.

30. R.D. Cook. *Concepts and Applications of Finite Element Analysis*. New York:John Wiley and Sons, 1989.

31. J.C. Halpin. *Primer on Composite Materials*. Lancaster, PA:Technomic Publishing, 1984.

32. C. Herakovich. *Mechanics of Fibrous Composites*. New York:John Wiley and Sons, 1998.

33. E. Greene. Design guide for marine applications of composites", Ship Structure Committee, Washington, D.C. 1997.

34. Anon. *The Composites Materials Handbook MIL 17*, West Conshohocken,PA:ASTM International, 2002.

# Chapter 5

# Modelling the Thermal Response of Composites in Fire

## 5.1    Introduction

The thermal decomposition of fibre reinforced polymer composites in fire is a complex topic that involves the combined effects of thermal, chemical and physical processes. The thermal processes include heat conduction from the fire through the composite; heat generated or absorbed from the decomposition reactions of the polymer matrix, organic fibres and core material; heat generated by the ignition of flammable reaction gases; and convective heat loss from the egress of hot reaction gases and moisture vapours from the composite into the fire. The chemical processes include thermal softening, melting, pyrolysis and volatilisation of the polymer matrix, organic fibres and core material together with the formation, growth and oxidation of char. The physical processes can involve thermal expansion and contraction, internal pressure build-up due to the formation of volatile gases and vaporisation of moisture; thermally-induced strains; delamination damage; matrix cracking; surface ablation; and softening, melting and fusion of fibres. Many of these processes do not occur in isolation from each other, but usually influence other processes that add to the complexity of the behaviour of composites in fire. Understanding these processes and how they interact is essential to understanding the fire reaction and fire resistive properties of composite materials.

Analytical and finite element-based models have been developed to predict the response of polymer laminates and sandwich composite materials to high temperature and fire. The need for reliable thermal, thermal-physical and thermal-mechanical models was recognised in the 1970s when carbon fibre composites (eg. carbon/epoxy, carbon/phenolic) began to be used in high temperature aerospace applications, such as rocket nozzles, internal linings of solid rocket motors and ablative heat shields for re-entry spacecraft. For these applications, models were needed to predict the thermal response of composites exposed to very high heat fluxes (up to 300 kW/m$^2$) for short times (less than a few minutes). During the 1990s the growing use of glass/polyester and glass/vinyl ester laminates in large ship structures and offshore platforms required models for predicting the reaction of materials at lower heat fluxes (25-150 kW/m$^2$) and for longer times (more than 30 minutes).

The ability to accurately model the fire response of composites has several important benefits. Firstly, models can be used to rapidly assess the fire resistance of new design options for composite products. Secondly, models reduce the need to conduct expensive fire tests. Lastly, models can be used to further our understanding of the fire behaviour of traditional polymer composites and in the development of new fire-safe materials. These benefits are significant, however the need to combine theoretical analysis with experimental testing is essential to achieve a detailed understanding of the thermal response and fire performance of composites.

This chapter describes the thermal, thermal-chemical and thermal-chemical-physical models for predicting the behaviour of composite materials in fire. The models vary in complexity from simple analysis that only consider heat conduction when a composite is exposed to low heat flux through to complex models that consider a variety of processes that occur when a composite is exposed to high heat flux, including transient heat conduction, thermal expansion/contraction, pyrolysis, internal pressures and strains, flow of reaction volatiles and moisture vapour, char formation, delamination cracking and ablation.


## 5.2      Response of Composites to Fire

Before examining the models in detail, it is important to understand the sequence of events that occur when a composite material is exposed to high temperature fire. When a heat flux is applied to one-side of a polymer composite, then the first event is the conduction of heat into the material. The rate of heat conduction is governed by the incident heat flux and the thermal diffusivity of the virgin composite. The thermal diffusivity of most types of composite is low, particularly in the through-thickness direction, and therefore a steep temperature gradient can develop through the material. For example, when a thick composite is exposed to a medium-to-high heat flux (ie. above ~50 kW/m$^2$) the hot surface can heat-up at a rate approaching or exceeding ~1000°C/min whereas the back surface is heated by conduction at a much slower rate of typically 1-20°C/min. Heat conduction through composites is complicated by the highly anisotropic nature of their thermal properties. Most types of fibres have a higher thermal conductivity than the polymer matrix. For example, at room temperature the axial thermal conductivity of carbon and glass fibres is about 20 to 80 and 1 W/m.K respectively, whereas the conductivity of most polymers is only about 0.10 to 0.25 W/m.K. As a result, the rate of heat conduction along the lamina (ie. fibre direction) is much faster than in the through-thickness direction. Heat conduction is further complicated because the thermal conductivity and specific heat of composites vary with temperature.

The conduction of heat through a composite causes it to expand or contract depending on the temperature. At temperatures below the glass transition temperature of the polymer matrix, and the amount of expansion is determined by the linear coefficient of thermal expansion of the virgin material. However, due to the thermal gradient through the material the expansion will be non-uniform in the through-thickness direction; being greatest at the hot surface and decreasing with distance below the surface. The thermal conductivity of some types of carbon fibre are anisotropic, and when heated will expand in the transverse direction and contract slightly in the axial (or fibre) direction. Therefore, it is possible for a carbon fibre composite to expand in the through-thickness direction and simultaneously contract in the planar direction.

At temperatures below the decomposition temperature of the polymer matrix, the transfer of heat energy occurs mainly by conduction with a small amount of energy being absorbed in the thermal expansion. When the surface of a composite material reaches a sufficiently high temperature, the polymer matrix and organic fibres (eg. aramid, UHMW polyethylene) begin to decompose. The temperature at which decomposition commences depends on the composition and chemical stability of the organic material, heating rate and fire atmosphere, although typically it is within the range of 250 to 400°C. As the temperature rises the organic matrix and fibres are degraded in the sequence of endothermic reactions that usually occur by random chain scission and possibly end-chain scission and chain stripping. These reactions yield low molecular weight gaseous products, and eventually the organic material is completely degraded to a porous carbonaceous char. The reaction volatiles flow through the char layer towards the hot surface of the composite. As the temperature rises towards the surface the volatiles may be decomposed by secondary reactions into smaller gas species. Heating above ~100-150°C also causes the vaporisation of moisture presence in the polymer matrix. Vaporisation of moisture absorbed by aramid fibres will also occur in this temperature range. The decomposition reactions of most polymer matrices and organic fibres are endothermic and therefore temporarily delay the conduction of heat through the reaction zone. The volatilisation of any water present in the laminate will have a similar 'cooling' effect on heat conduction.

The reaction volatiles and vapourised moisture are initially trapped due to the low gas permeability of composites, and this leads to a rapid rise in internal pressure and a large expansion of the material. The gas pressure within decomposing laminates has been measured up to ~10 atm [1], although it is speculated the pressure can be as high as 200 atm [2]. Because the polymer matrix is heated to well above the glass transition temperature, the pressure exerted by the trapped gases onto the soft, complaint matrix can lead to the formation of gas-filled pores, delaminations and matrix cracks.

Eventually the matrix becomes sufficiently porous and cracked that the reaction volatiles and water vapour can flow through the degraded region of the composite into the fire environment. This outflow of hot gases has a cooling effect by convection, thereby attenuating (albeit slightly) the conduction of heat to the reaction zone. The extent that the pyrolysis gases will cool the composite depends on the heat capacity of the gases. The higher the heat capacity of the gas the greater is the cooling effect. In addition, when the gases reach the hot surface of the composite they may form a protective thermal boundary layer (in the absence of any external convection processes, such as airflow). The gases can also diffuse into the virgin composite, although due to the low gas permeability this process is considerably slower than the outward flow.

Endothermic decomposition of the matrix and organic fibres continues until the reaction zone reaches the rear-face of the laminate, where the last of the combustible material is degraded to volatiles and char. At this stage the decomposition process ceases unless the temperature is high enough to induce pyrolysis reactions between the fibres and char. When the temperature exceeds ~1000°C then the char can react with the silica network in glass fibres, resulting in considerable mass loss [3]. In the case of carbon fibre composites, the carbon fibres and char may oxidise when exposed to fire in an oxygen-rich environment. Ablation can also occur at high temperatures (generally above 1000°C), which is accelerated by high velocity airflow over the composite surface that can have an erosive effect.

The processes that occur when a polymer laminate is exposed to high heat flux are summarised in Table 5.1. Figure 5.1 shows the approximate temperatures over which the different processes occur in a glass/polymer composite. The situation is even more complex with sandwich composites because the core material has a large influence on the fire response. An accurate prediction of the thermal behaviour of both laminates and sandwich materials requires the solution of a complex numerical model which should include all of the processes listed in Table 5.1.

*Table 5.1. Summary of the main processes when a composite is exposed to fire.*

| |
| --- |
| Anisotropic heat conduction through virgin material and char |
| Thermal expansion/contraction |
| Decomposition of polymer matrix and organic fibres |
| Pressure rise due to formation of combustion gases and vapourisation of moisture |
| Flow of gases from the reaction zone through the char zone |
| Flow of gases into the virgin composite |
| Thermally-induced strains. |
| Formation of delamination and matrix cracks |
| Reactions between char and fibre reinforcement |
| Ablation |

Mathematical modelling of the fire performance of composite materials is based largely on theoretical studies performed since the mid-1940s on the fire behaviour of wood [4-9]. Many of the processes that occur in burning wood are similar to those described above for composites. Burning wood is essentially modelled as a two-phase material consisting the residual char and virgin (unpyrolysed) material, as shown schematically in Fig. 5.2. Numerous studies – most notably by Kung [7], Kansa et al. [8] and Fredlund [9] – have developed thermal-chemical models that consider the processes of transient heat conduction, endothermic decomposition reactions of wood, convection flow of volatile gases, and combustion of volatiles at or near the solid surface.

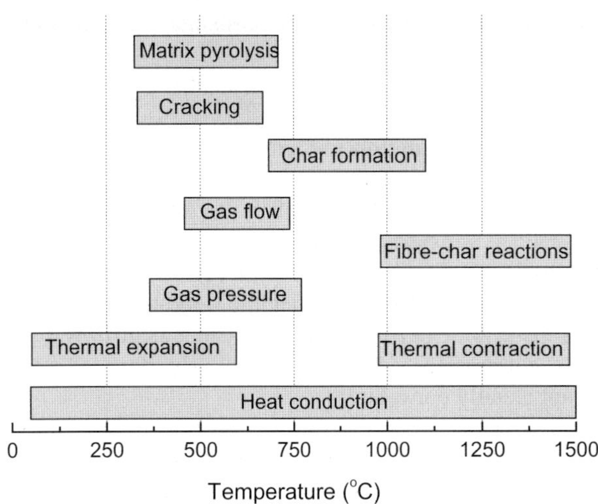


*Figure 5.1. Effect of temperature on the various responses of a fibreglass composite. (The temperatures are approximate, and will vary with the composition of the material and fire conditions).*

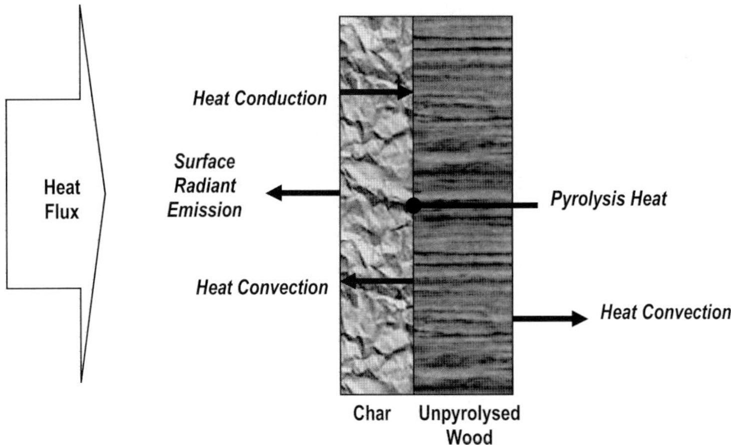


*Figure 5.2. Schematic of the thermal decomposition of wood.*

The current state-of-the-art for modelling the fire response of composites is defined by work published between the mid-1980s and mid-1990s by research groups lead by Henderson (from the University of Rhode Island) [10-16], Sullivan (Marshall Space Flight Center) [17-20], Springer (Stanford University) [21-23], Dimitrienko (NPO Mashinostroeniya) [2,24,25] and Gibson (University of Newcastle-upon-Tyne) [26-28]. The next section provides a review of the main mathematical models for composite materials, beginning with the simplest models that only consider the effect of heat conduction to more sophisticated models that consider many of the thermal, chemical and physical processes listed in Table 5.1.

## 5.3    Modelling Heat Conduction in Composites

In the study of heat transfer in solid materials it is customary to consider three distinct modes of thermal energy transfer: conduction, convection and radiation. However, for simplicity in the analysis, almost all the mathematical models for composites only consider the effect of heat conduction under the condition of one-sided heating. The influence of heat transfer by external convection, such as airflow across the hot surface of a composite, is not usually considered. Similarly, the radiation of heat from a composite is also not usually considered.

The simplest model is a one-dimensional analysis that considers heat conduction in the through-thickness (x-) direction of a composite material that is heated from one side, as shown schematically in Fig. 5.3. This model assumes that the composite is a thermally-thick slab with a uniform in-plane temperature distribution. The cool (rear) surface of the slab is assumed to be adiabatic. The 1D heat conduction model is expressed as [29]:

$$\rho C_p \frac{\partial T}{\partial t} = \frac{\partial}{\partial x}\left[ k_x \frac{\partial T}{\partial x} \right]$$

(5.1)

where $T$ is the temperature, $t$ is time, and $x$ is the distance below the hot surface in the through-thickness direction. $\rho$ and $C_p$ are the density and specific heat of the composite, respectively, and $k_x$ is the thermal conductivity of the composite in the through-thickness direction. The values for $\rho$, $C_p$ and $k_x$ are assumed to be independent of temperature, although as described later, this is not true in reality. The left-hand side of the equation represents the change in thermal energy per unit volume and the right-hand side represents the energy flux due to conduction. With appropriate boundary conditions, this equation provides a starting point for predicting the temperature distribution in a flat composite plate subjected to one-sided heating.

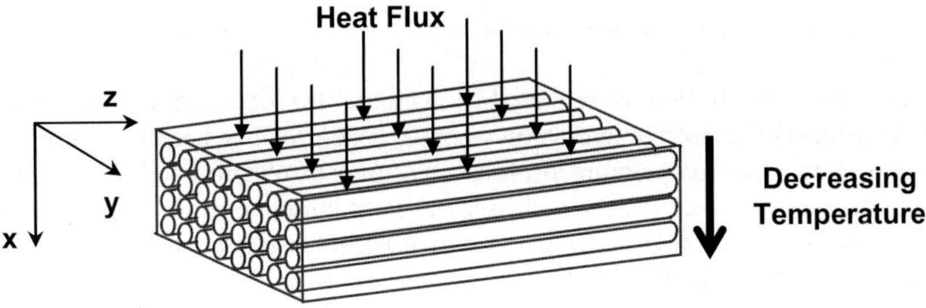


*Figure 5.3. Schematic of one-dimensional heat conduction through a composite material exposed to one-sided, uniformly distributed heating.*

One-dimensional heat transfer analysis is used in many of the mathematical models described shortly that also consider other thermal, chemical and physical processes when composites are exposed to fire. In fact, in the description of the thermo-chemical and thermo-physical models presented later in this chapter it is assumed that the composite is subjected to a one-sided heat flux that is applied uniformly over the surface, and therefore heat conduction can be modelled using 1D heat transfer theory. However, multi-dimensional heat transfer analysis is required to model the temperature distribution in composites exposed to localised surface heating. Griffis et al. [30] and others have used a two-dimensional heat conduction model to determine the radial and through-thickness temperature distribution in a anisotropic composite material exposed to localised, short-duration heating.

Charles and Wilson [31], Milke and Vizzini [32] and Asaro et al. [33] have used a three-dimensional heat transfer model to predict heat conduction in the x, y and z-directions of orthotropic composites. The x-direction is taken to be the through-thickness direction whereas the y- and z-directions define the planar directions. The heat conduction model is expressed as:

$$\rho C_p \frac{\partial T}{\partial t} = \frac{\partial}{\partial x}\left[k_x(T)\frac{\partial T}{\partial x}\right] + \frac{\partial}{\partial y}\left[k_y(T)\frac{\partial T}{\partial y}\right] + \frac{\partial}{\partial z}\left[k_z(T)\frac{\partial T}{\partial z}\right] \qquad (5.2)$$

where $k_x(T)$, $k_y(T)$ and $k_z(T)$ are the thermal conductivities in the x, y and z directions, respectively. As with Eqn. 5.1, the 3D heat conduction model assumes that the thermal conductivity of a composite does not vary with temperature. It is also assumed that other thermally activated processes, such as resin decomposition, convective flow of volatiles, etc., do not affect the heat conduction process. For these reasons, Eqns. 5.1 and 5.2 are only accurate for predicting the temperature of composites exposed to low heat flux fires that do not induce combustion or degradation of the organic matrix and fibres. For most types of polymer composites, this heat flux limit below which pyrolysis does not occur is in the range of 10 to 20 kW/m².

Heat conduction modelling is remarkably accurate for determining the temperature distribution through polymer laminates when exposed to low heat flux. For example, Figure 5.4 shows the temperature profiles measured by Asaro et al. [33] at the front face, mid-section and back face of a glass/vinyl ester laminate irradiated with a low heat flux for over one hour. This heat flux was insufficient to degrade the polymer matrix, and therefore heat conduction is the dominant thermal process. The curves in Fig. 5.4 show the theoretical temperature rise calculated using Eqn. 5.2, and excellent agreement is observed with the experimental temperature values.

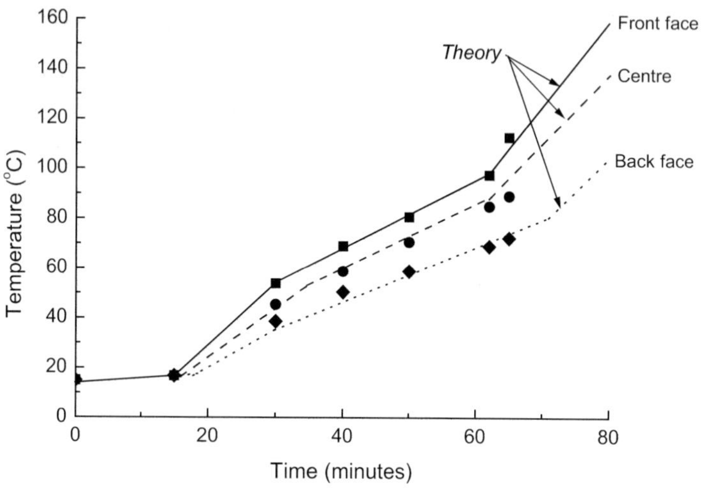


*Figure 5.4. Comparison of theoretical and measured temperature profiles in a glass/vinyl ester laminate exposed to a low heat flux. The theoretical curves were calculated using a 3D heat conduction model. Reproduced from Asaro et al. [33].*

As mentioned, the influences of convection and radiation on heat transfer are usually ignored when modelling the thermal response of composite materials in fire. However, Griffis et al. [30,34] propose a heat transfer model that considers both conduction and convection on the temperature rise of composites. Milke and Vizzini [32] have also proposed a model that analyses the influences of conduction and radiation on the thermal response of composites. In this model, the temperature rise is calculated using 3D heat conduction theory (ie. Eqn. 5.2). The effect of radiation loss to the environment due to the increase in temperature ($T$) is calculated using:

$$Q_r = \varepsilon\sigma\left(T^4 - T_\infty^4\right) \tag{5.3}$$

where $\varepsilon$ is the emissivity of the surface, $T_\infty$ is the ambient temperature and $\sigma$ is the Stefan-Boltzmann constant. Milke and Vizzini [32] have shown that this modelling approach of analysing heat conduction and radiation can accurately predict the temperature rise in carbon/epoxy laminates subjected to low heat fluxes (~8 and 19 kW/m$^2$).

## 5.4     Modelling the Fire Response of Composites

Pering, Farrell and Springer [21] developed the first model to predict the thermal response of composite laminates that involved thermal decomposition of the polymer matrix. The model analyses the increase in thermal energy in a material due to the combined effects of heat conduction and pyrolysis of the matrix. Heat conduction is calculated using the 1D heat transfer equation (ie. Eqn 5.1) while the heat of pyrolysis is determined from the theoretical mass loss rate. The one-dimensional equation proposed by Pering et al. [21] is:

$$\rho C_p \frac{\partial T}{\partial t} = \frac{\partial}{\partial x}\left[k_x \frac{\partial T}{\partial x}\right] + \frac{\partial m}{\partial t}Q_p \tag{5.4}$$

where $\partial m/\partial t$ is the mass rate of vapour generated per unit volume and $Q_p$ is the heat of pyrolysis, which must be experimentally determined. In this analysis, energy transfer by convection is assumed to be negligible and the volatile gases produced by the pyrolysis reaction are assumed to be immediately removed from the composite and therefore do not affect the temperature.

Pering et al. [21] was able to obtain an accurate estimation of the mass loss for composite laminates in a fire using their model. Figure 5.5 shows the effect of increasing heating time on the normalised mass loss of a carbon/epoxy composite during exposure to a gas flame with a temperature of 540°C. The normalised mass loss represents the mass loss of polymer matrix caused by pyrolysis divided by the original mass of the matrix. After a very short induction period during which the composite did not loss any weight, the normalised mass loss increased rapidly with time as the matrix was thermally decomposed to volatiles and a small amount of residual char. The mass loss curve quickly stabilised at unity which indicates the matrix was completely consumed by the fire. The data points indicate the measured mass loss while the curve shows the theoretical mass loss calculated using Eqn. 5.4, and excellent agreement is observed.

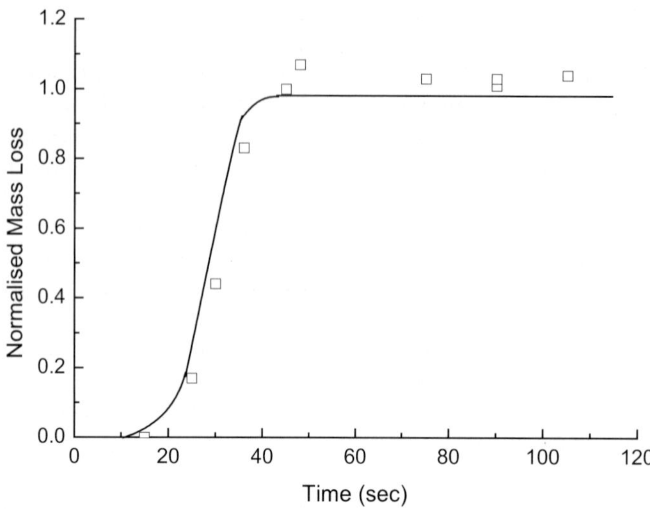


*Figure 5.5. Comparison of the measured and calculated mass loss of a carbon/epoxy laminate exposed to a fire with a flame temperature of 540°C. The mass loss curve was calculated using Eqn. 5.4. Reproduced from Pering et al. [21].*

Shortly after the work by Pering et al. [21], Henderson and colleagues [10] developed a more sophisticated model that considers the processes of heat conduction, pyrolysis, and diffusion of decomposition gases. The process of heat conduction is modelled using 1D heat transfer theory, although a unique feature of the analysis is that changes to the heat conduction rate caused by variations in the transverse thermal conductivity of the laminate with increasing temperature are considered. The diffusion of decomposition gases from the reaction zone in the laminate through the char structure to the char surface/fire interface is analysed using convective mass transfer theory. Lastly, the decomposition reactions are modelled using single or multiple-order kinetic rate theory. Another unique feature of the model is that the decomposition reaction of the polymer matrix as well as the carbon-silica reaction that can occur between char and glass fibres at high temperatures are both analysed in the model.

Henderson et al. [10] based their model on theoretical work into the fire response and decomposition of wood, particularly the models by Kung [7] and Kansa et al. [8]. The 1D equation derived by Henderson et al. to predict the thermal response of glass-reinforced polymer laminates is expressed as:

$$\rho C_p \frac{\partial T}{\partial t} = k \frac{\partial^2 T}{\partial x^2} + \frac{\partial k}{\partial x}\frac{\partial T}{\partial x} - \dot{m}_g C_{pg} \frac{\partial T}{\partial x} - \frac{\partial \rho}{\partial t}\left(Q_i + h - h_g\right) \tag{5.5}$$

where $i$ equals 1 and 2 for the matrix decomposition and carbon-silica reactions, respectively. The value for $k$ is taken to be the through-thickness thermal conductivity of the material. The first term on the right-hand side of the equation considers the effect of heat conduction. The second term also accounts for heat conduction, although it considers the influence of changing transverse thermal conductivity on the rate of heat conduction. The thermal conductivity is a function of both temperature and the stage of the decomposition reaction. However, it is not possible to theoretically calculate the change in thermal conductivity with temperature, and therefore this term must be measured experimentally over the temperature range of interest. Section 5.5 provides further information on the effect of temperature on thermal conductivity, and how it can be determined. The third term in Eqn. 5.5 considers the internal convection of thermal energy due to the flow of hot decomposition reaction gases through the char structure. This process has a cooling effect on a laminate, and therefore this is a negative term. The last term is the rate of heat generation or consumption resulting from matrix decomposition and the char-glass fibre reactions, where $Q_i$, $h$ and $h_g$ are the heat of decomposition, enthalpy of the solid phase, and enthalpy of the volatile gas, respectively. The term is expressed as a negative term for endothermic reactions that absorb heat and a positive term for exothermic reactions. In the final term, the decomposition reaction rates are determined from the mass loss rate using the Arrhenius kinetic rate equation:

$$\frac{\partial m}{\partial t} = -A_i m_o \left[ \left( m - m_f \right) / m_o \right]^{n_i} e^{\left( -E/RT \right)} \tag{5.6}$$

where $A_i$, $E$ and $n_i$ are the pre-exponential factor, activation energy and order of the reaction, respectively, and these must be determined experimentally by thermogravimetric analysis. $R$ is the universal gas constant. The parameters $m_o$, $m_f$ and $m$ are the initial, final and instantaneous mass of the material[1].

In modelling the thermal response of a composite material, Henderson et al. [10] also considered the radiant and convective heat exchange at the hot surface. This was analysed using:

---

[1] Most models that consider the effect of pyrolysis on the thermal response of composites in fire assume that decomposition of the polymer matrix can be analysed using the conventional Arrhenius equation, in which the reaction rate is only dependent on the type of polymer and temperature. Dimitrienko et al. [25] proposed a modified form of the Arrhenius equation to consider the dependence of the reaction rate on both temperature and volatile gas pressure in the pores of a decomposing composite. The modified Arrhenius equation is expressed as:

$$\frac{\partial m}{\partial t} = A \left[ 1 - \frac{p}{p_o} e^{-(2E/RT)} \right]^{1/2} . e^{-E/RT}$$

where $p$ is the local extrema of gas pressure in the pores and $p_o$ is the ambient pressure, which can be assumed to equal 0.1 MPa.

$$-k\frac{\partial T}{\partial x} = \bar{h}(T_\infty - T) + \sigma(\varepsilon_r \alpha_m T_r^4 - \varepsilon_m T_k^4) \qquad (5.7)$$

where $\varepsilon_r$ and $\varepsilon_m$ is the emissivity of the heat source and composite material, respectively whereas $\alpha_m$ is the absorptivity of the composite. A $\varepsilon_m$ value for of 0.9 is commonly used for composites.

Henderson et al. [10] evaluated the accuracy of their model by comparing the theoretical temperature against measured temperature profiles for a glass/phenolic composite exposed to a high heat flux (279.9 kW/m$^2$) for up to 800 seconds. A large heat flux was used to generate a sufficiently high temperature to cause significant decomposition of the polymer matrix and induce reactions between the glass fibres and char. Figure 5.6 compares the calculated and measured temperature profiles at different depths through a 3 cm thick specimen, and good agreement is observed. Henderson and Wiecek [14] refined the model further to include the combined effects of thermal expansion of the composite and storage of decomposition gases within the composite. This had the affect of improving slightly the reliability of the model in predicting the temperature rise.

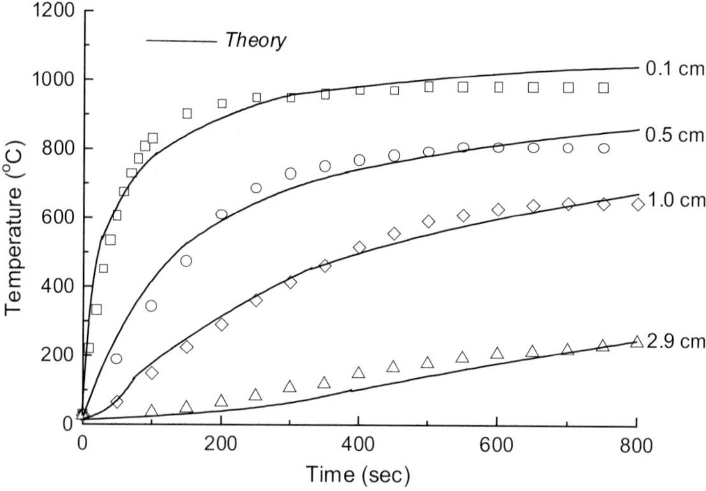


*Figure 5.6. Comparison of theoretical and experimental temperature profiles determined at different locations through a glass/phenolic composite exposed to a heat flux of about 280 kW/m$^2$. The theoretical temperatures were calculated using the Henderson et al. [10] model. Reproduced from Henderson et al. [10].*

A simplified version of the model developed by Henderson et al. [10] was proposed by Gibson and colleagues [26] to predict the fire performance of GRP laminates. The model was simplified by assuming that glass-char reactions do not occur, and therefore is applicable to fires with heat fluxes below ~125 kW/m$^2$ (or ~1000°C). The model was originally developed to predict the fire performance of GRP panels when exposed to hydrocarbon fires in order to aid in the development of fire-resistant wall and floor structures for offshore drilling platforms. More recently it has been used to model the fire behaviour of marine-grade GRP laminates exposed to cellulosic fires and low to intermediate heat flux conditions (25-100 kW/m$^2$) [28].

Gibson et al. [26] also simplified the model by Henderson et al. [10] by assuming the thermal and gas transport properties are constant during the decomposition of the polymer matrix. The thermal conductivity and specific heat properties of the composite are assumed to remain constant with increasing temperature, and the ambient temperature values are generally used in the analysis. The thermal properties are calculated using rule-of-mixture analysis presented in section 5.5. It is also assumed that the decomposition reaction rate can be predicted using a first-order Arrhenius equation, which is valid for many of the polymer systems used in offshore composite structures, such as polyesters, vinyl esters and epoxies. However, the model is also able to model higher-order decomposition reactions, such as phenolic resins that degrade in a multiple-stage process.

The model by Gibson et al. [26] analyses the three most important energy transfer processes that occur in a composite exposed to fire; namely conductive heat transfer through the material, endothermic decomposition of the polymer matrix, and convective mass transfer of volatile products from the reaction zone to the hot composite surface. The model is expressed as a one-dimensional non-linear equation that incorporates these processes, and is similar in form to Eqn. 5.5:

$$\rho C_P \frac{\partial T}{\partial t} = \frac{\partial}{\partial x}\left(k \frac{\partial T}{\partial x}\right) - \dot{M}_G \frac{\partial}{\partial x} h_G - \rho A \left[\frac{(m - m_f)}{m_o}\right]^n e^{-\frac{E}{RT}} \left(Q_P + h_c - h_g\right) \quad (5.8)$$

The first term on the right-hand side of the equation relates to unsteady-state heat conduction through the composite. The second term defines the magnitude of the mass flux of volatile decomposition products through the composite towards the fire. The last term is the endothermic decomposition term that defines the pyrolysis reaction rate of the polymer matrix. By iteratively solving the equation for increasing temperature as the composite is heated by fire it is possible to calculate the following fire properties: time-to-ignition, mass loss rate and char formation. The model can also be solved by finite element analysis [35,36].

Like the model by Henderson et al. [10], Eqn. 5.8 takes no special account of char formation, which is believed to be beneficial in prolonging integrity, or of the fact that the fibres on the hot face of the laminate may progressively fall away after a period of exposure in a fire. Nevertheless, the model has proven accurate in the determination of the fire performance of many types of GRP laminates [26-28,35,36]. Figure 5.7 gives a comparison of the theoretical and measured temperatures plotted as a function of time for a glass/polyester composite exposed to a hydrocarbon fire. The term $x/L$ represents that distance below the hot surface ($x$) divided by the specimen thickness ($L$), which in this case is 10.9 mm. The temperatures were determined at the hot face, a distance 1/10th through the composite, half-way ($x/L$ = 5/10), and at the cold face. There is excellent agreement between the predicted and experimental temperatures. Mouritz et al. [28] recently showed that the model can accurately predict the time-to-ignition, mass loss, mass loss rate and char growth of GRP laminates.

Another significant model for predicting the fire performance of a decomposing composite was proposed Florio, Henderson, Test and Hariharan [16]. This model was built upon the original work by Henderson et al. [10,14], and considers not only the processes of heat conduction, convection heat flow of decomposition gases, and pyrolysis, but in addition includes the effects of thermal expansion and the internal pressure rise due to the formation of volatile gases. Florio et al. [16] provide a full description of the derivation of the model, and the full energy-balance equation is expressed as:

$$m_g C_p \frac{\partial T}{\partial t} = \dot{m}_g \, \Delta A \Delta x C_p \frac{\partial T}{\partial t} + \Delta A \Delta x \frac{\partial}{\partial x}\left( k\Phi \frac{\partial T}{\partial x} \right) + h_r \Delta A \Delta x (T_s - T_g)$$

$$+ \Delta A \Delta x \frac{D(\Phi P)}{Dt} - \frac{\partial m}{\partial t}\left[ -h_g + h_g(T_s) + v^2/2 \right] \tag{5.9}$$

where $\Delta A$ is the cross-sectional area, $\Delta x$ is the thermal expansion of the control system, $\Phi$ is porosity, $v$ is gas velocity and $P$ is the internal gas pressure. The first term on the right-hand side of Eqn. 5.9 accounts for the net rate of volatile gas flow, the second term considers the effect of heat conduction, the third term presents the net rate of energy transfer by convection of gases, and fourth term accounts for the rate of work of expansion, and the last term accounts for energy consumption/accumulation due to the decomposition reactions.

Using the model, it is possible to predict the temperature, mass loss, porosity and volumetric expansion of the composite material as well as the temperature, pressure, mass flux and mass storage of the gases. The model was validated by Florio et al. [16] using a 3 cm thick glass/phenolic composite exposed to a heat flux of ~280 kW/m$^2$. (This is the same material and heat flux used by Henderson et al. [10] to validate their model). Figure 5.8 compares the analytical and experimental temperature profiles as a function of time for different depths through the laminate. It is seen that the agreement is good, although it is not any better than that found by Henderson et al. [10] using their simpler model. It appears that considering the influences of thermal expansion and gas pressure on the internal energy does not significantly improve the prediction of temperature in a decomposing composite material.

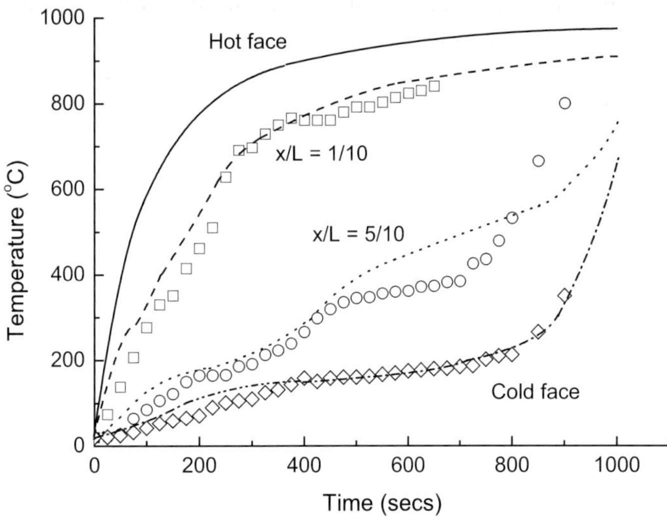


*Figure 5.7. Comparison of theoretical and experimental temperature profiles determined at different locations through a glass/phenolic composite exposed to a heat flux of about 280 kW/m². The theoretical temperatures were calculated using the Florio et al. [16] model. Reproduced from Florio et al. [16].*

A unique feature of the model by Florio et al. [16] is the ability to calculate the internal pressure rise caused by the accumulation of volatile gases. This is important in understanding the diffusion properties of hot gases through a composite and the formation of delamination cracks, which occur when the pressure exceeds the interply failure stress. Rumamurthy et al. [1] measured the pressure rise within a burning glass/phenolic composite to assess the accuracy of the model. The pressure was measured using thin pressure-sensitive tubes placed at different locations within the composite. Figure 5.8 compares the theoretical and measured pressure-time responses at depths of 0.6 cm and 2.55 cm below the hot surface of the 3 cm thick glass/phenolic specimen. The pressure due to volatilisation (P) is normalised to the ambient pressure inside the composite before fire testing ($P_o$). The agreement between the calculated and measured pressures is poor, with neither the pressure values nor the change in pressure with time being accurately predicted. However, Florio et al. [16] do not attribute this discrepancy to the model, but instead believe that problems encountered when measuring the gas pressure caused the experimental values to be lower than the actual pressure.

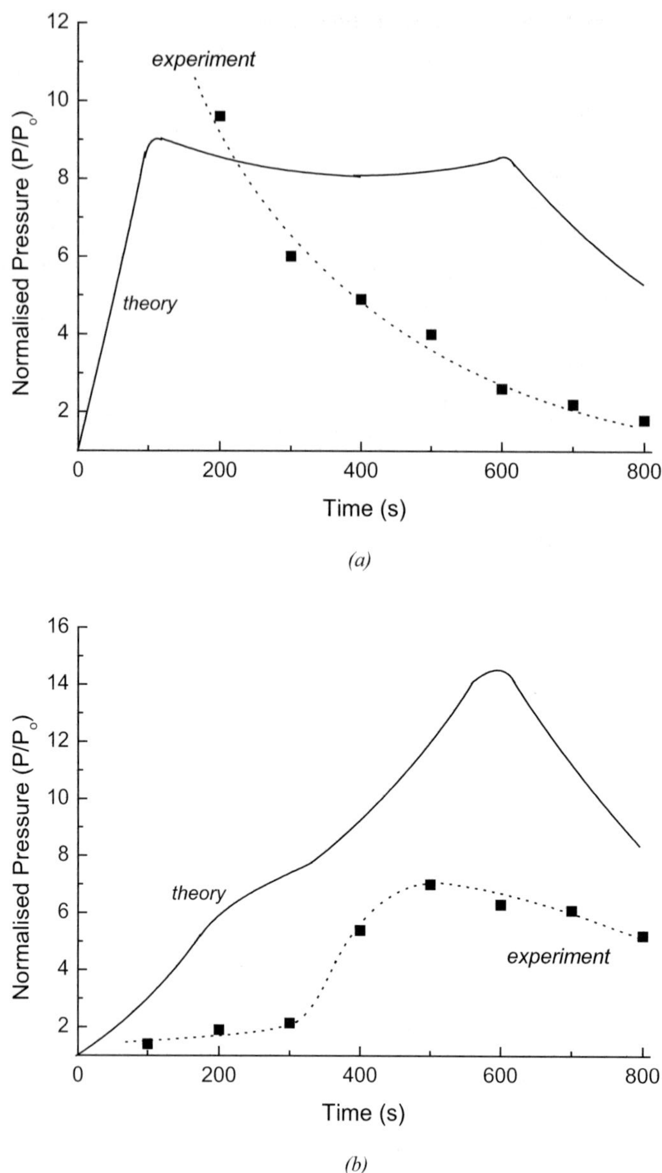



Figure 5.8. Comparison of theoretical and experimental pressure-time profiles determined at depths of (a) 0.6 cm and (b) 2.25 cm below the hot surface of through a 3 mm thick glass/phenolic composite exposed to a heat flux of about 280 kW/m$^2$. The theoretical pressures were calculated using the Florio et al. [16] model. Reproduced from Florio et al. [16].

Several models have been developed to predict the thermal expansion of a composite material at high temperatures [12,15,37]. Predicting the thermal expansion is a major challenge because a composite material can expand and contract at different rates as the temperature rises. For example, Fig. 5.9 shows the volumetric expansion – defined by the fractional length change - of a glass/phenolic composite when heated at the rates of 5 and 20°C/min to a maximum temperature of 2000°C. When heated from room temperature, the composite initially expands at a very slow linear rate with increasing temperature. The magnitude of this increase is barely discernible in Fig. 5.9. This is due to the expansion of the glass fibres and polymer matrix. At about 300°C, when pyrolysis of the matrix commences, the expansion rate rises rapidly due to the pressure build-up from the formation of volatile gases. As the temperature increases the composite continues to expand as a greater amount of the polymer matrix is degraded to volatiles. Figure 5.9 shows that the expansion is greater at the higher heating rate, and this is because the pyrolysis reactions are dependent on the heating rate. As the heating rate is increased, the rate of gas generation at a particular temperature is also increased, resulting in greater expansion. The composite begins to contract when the temperature rises above ~500°C due to formation of char, and the rate of contraction increases rapidly above 1600°C as a result of glass fibre-carbon (char) reactions. This complex relationship between expansion and temperature is observed in many types of composite materials [11-13,38,39].

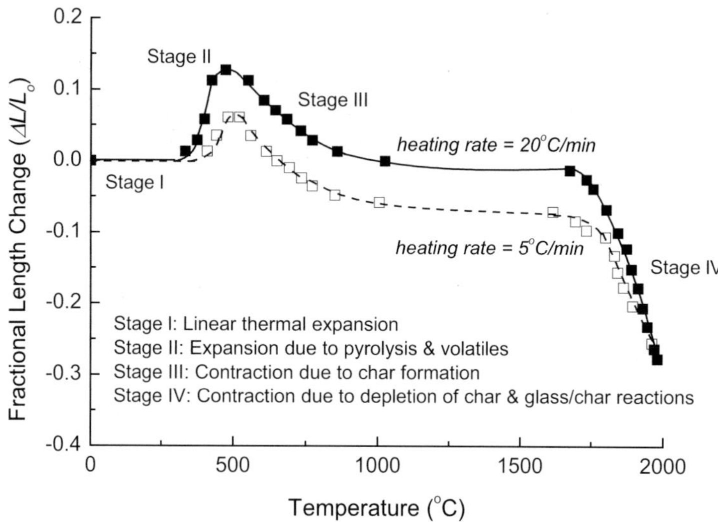


*Figure 5.9. Fractional length change with increasing temperature for a glass/phenolic composite heated at 5 and 20°C/min. The different stages of expansion and contraction are indicated. Data from Florio et al. [15].*

Florio et al. [15] developed a thermal-physical model to predict the expansion of composites when exposed to fire. The expression is calculated using the non-linear partial differential equation:

$$\frac{1}{L_o}\frac{\partial L}{\partial t} = \alpha_v F \frac{\partial T}{\partial t} + \alpha_c (1-F)\frac{\partial T}{\partial t} + \frac{\partial}{\partial t}\left(\eta T + \xi \frac{m}{m_o}\right) \tag{5.10}$$

where $L$ and $L_o$ are the instantaneous and initial lengths, respectively, $m$ and $m_o$ are the instantaneous and initial mass, respectively, and $\alpha_v$ and $\alpha_c$ are the linear expansion coefficients of the virgin material and char, respectively. The parameters $\eta$ and $\xi$ are the heating rate and decomposition coefficients, and these must be experimentally determined over the temperature range of interest. As yet, the reliability of this model to predict the thermal expansion of decomposing composite materials in a fire has not been evaluated.

Another important model to analyse the fire response of polymer laminates was developed in the early 1990s by McManus and Springer [22]. The model is the first analytical method that can simultaneously predict the thermal and stress response of a composite exposed to elevated temperature. The thermal response component of the model follows a similar approach to the work by Henderson and Wiecek [14], and analyses the thermal processes of heat conduction, pyrolysis and volatilisation so that theoretical predictions can be made of the temperature, pressure distribution, vapour and volatile formation rates, and amount of char. A key feature of the model is that the stress within a decomposing composite can also be calculated. The model analyses the strain arising from an externally applied stress as well as from internally generated strains from thermal expansion, pressure exerted by volatile gases and moisture vapour, and volumetric changes due to the formation of char. The total strain within a composite is determined using the governing equation:

$$\varepsilon_{ij} = S_{ijkl}\sigma_{kl}^{m} + \Lambda_{ij}\Delta P + \alpha_{ij}\Delta T + \beta_{ij}\Delta(MC) + \chi_{ij}\Delta\upsilon_c \tag{5.11}$$

where $\alpha$, $\beta$ and $\chi$ are the temperature, moisture and charring expansion coefficients, and $\Delta P$, $\Delta T$, $\Delta(MC)$ and $\Delta\upsilon_c$ are the changes from a reference value of the pressure, temperature, moisture content and char volume, respectively. The first-term on the right-hand side of Eqn. 5.11 accounts for the strain generated by an externally applied stress ($\sigma$). The tensor $S$ is the compliance of the laminate. In the absence of an externally applied stress, the first-term equals zero. The second term considers the internal strain arising from pressure exerted from the reaction volatiles, in which $\Lambda$ represents the compliance of the laminate when subjected to the internal gas pressure. The third, fourth and last terms of the equation account for the strain arising from the thermal expansion, vaporisation of moisture, and formation of char. All of the parameters – $S$, $\Lambda$, $\alpha$, $\beta$, $\chi$ - must be determined experimentally before the total strain can be calculated.

McManus and Springer [23] have shown that their model can predict the formation of delamination cracks in a laminate exposed to fire. Delamination is assumed to occur when the magnitude of the strain calculated using Eqn. 5.11 exceeds the intraply failure strain, which can be predicted using the Tsai-Wu, maximum stress or some other failure criteria. Figure 5.10 shows the effect of heating time on the maximum depth of delamination cracks in a carbon/phenolic composite exposed to a propane flame. The data points and step-shaped curve show the measured and calculated delamination depths, respectively, and excellent agreement is observed. Since this model was developed by McManus and Springer over ten years ago, several other models for predicting the strain within a decomposing composite material in fire have been proposed [18-20,40,41]. Thermal-mechanical models capable of predicting the reduction to the stiffness, strength and creep resistance of composites exposed to elevated temperature and fire are described in greater detail in Chapter 6.

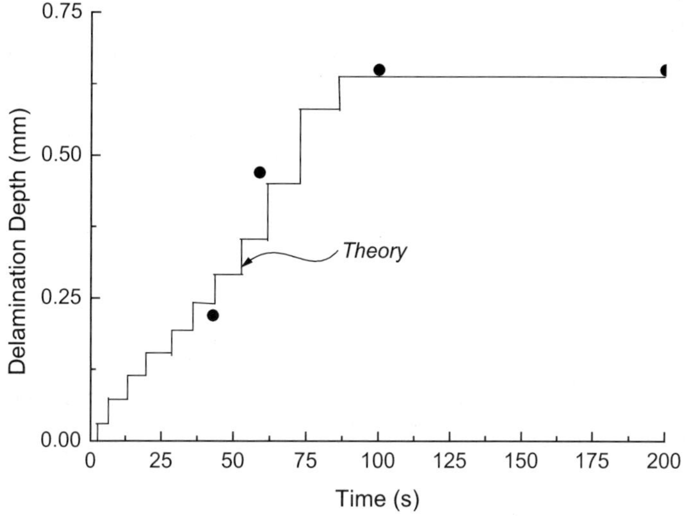


*Figure 5.10. Effect of heating time on the depth of delamination cracking in a carbon/phenolic laminate exposed to a gas flame. The theoretical increase in delamination depth was calculated using the thermal-mechanical model by Springer and McManus [22]. Reproduced from Springer and McManus [23].*

In addition to the models described, several other models have been proposed to predict the fire performance of polymer laminates [eg. 17-19,37,40-43]. Models have also been proposed for sandwich composite materials with combustible skins and non-combustible cores [44,45]. However, a model for calculating the fire behaviour of sandwich composites with a combustible core has not been developed.

Table 5.2 presents a summary of the various thermal, chemical and physical processes that can be analysed using the fire performance models for composites. It is also worth noting that the reliability of several of the models have not been assessed against experimental data, and therefore should be used with caution. Another point to note is that while most of the models were developed for composite materials with aerospace or ship applications (eg. epoxy, phenolic, polyester, vinyl ester-based laminates), they are applicable to most types of thermoset matrix composites. The applicability of the models to thermoplastic laminates or composites reinforced with combustible fibres (eg. aramid, UHMW polyethylene) is not well understood, although the models can be modified to analyse such materials.

*Table 5.2. Summary of processes that can be predicted using fire models for composites.*

| Fire Process | Reference |
| --- | --- |
| Heat conduction/temperature | 10,14,16,17,18,19,21,22,25,26,29,31,32,33,37,40,41 |
| Pyrolysis | 10,14,16,17,18,19,21,22,25,26,32,37,40,41 |
| Volatile convective flow | 10,14,16,17,21,22,25,37 |
| Char formation/mass loss | 10,14,16,17,21,22,26 |
| Internal pressure | 16,17,18,19,22,25 |
| Thermal expansion | 16,17,18,19,22,25 |
| Thermal stresses | 2,17,18,19,40,41 |

## 5.5    Modelling the Thermal Properties of Composites

Accurate prediction of the fire response of a polymer composite using the thermal models described above requires knowledge of the thermal and physical properties of the material over the temperature range of interest. The properties include specific density, thermal conductivity, gas permeability and specific heat of both the virgin material and char, and in many composites these properties change considerably when heated to high temperature in a fire. In such situations, it is necessary to include in the analysis the variation in these properties with temperature.

An immense amount of theoretical and experimental work has been done on the thermal conductivity of composites [46-55]. A large number of models have been proposed for calculating the equivalent thermal conductivity of laminates from the thermal properties and volume content of the fibres and polymer matrix. The models vary in mathematical complexity depending on the type of laminate being analysed. Simple models are able to accurately determine the equivalent thermal conductivity of unidirectional and cross-ply laminates whereas more complex models that consider the influence of fibre architecture have been developed for woven textile and three-dimensional orthogonally reinforced composites. The number of models is too large to cover them all here, and so a few of the more commonly used equations are described.

The simplest equations for determining the equivalent thermal conductivity of a laminate under isothermal conditions are the 'Geometric Mean Model' [56] and the 'Series or Stacked Plate Model' [47]. The models assume that all the fibres are straight, aligned and evenly dispersed through the laminate, perfect bonding exists between the fibres and matrix, and the material is free of porosity and other defects. It is also assumed that the material does not thermally decompose. The Geometric Mean Model can be used to estimate the thermal conductivity in the fibre direction, and is expressed using:

$$k_{II} = k_f V_f + k_m (1 - V_f) \qquad (5.12)$$

while the Series/Stacked Plate Model is used to calculate the through-thickness thermal conductivity:

$$k_{\perp} = k_m \frac{k_f (1 + V_f) + k_m (1 - V_f)}{k_f (1 - V_f) + k_m (1 + V_f)} \qquad (5.13)$$

where $k_f$ and $k_m$ are the thermal conductivities of the fibre and matrix, and $V_f$ is the volume fraction of fibres. Despite the simplicity, an accurate predict of the thermal conductivity of many types of composite materials can be achieved using these expressions. Ott [49] reports that the difference between the theoretical thermal conductivity calculated using the equations and the measured thermal conductivity can be less than 10% for many materials, and their simplicity and accuracy makes them popular when analysing the thermal behaviour of composite materials in fire. However, the geometric and series models are not suited for all types of composite materials, and more sophisticated models are needed to consider the influences of fibre architecture and porosity on the equivalent thermal conductivity. Ott [49] gives a comprehensive review of models for calculating thermal conductivity of polymer laminates.

Henderson et al. [10] have proposed modified versions of the Geometric and Series models to calculate the thermal conductivity of a decomposing composite from the thermal properties of the virgin material and char. The in-plane and through-thickness equivalent thermal conductivities are approximated using:

$$k_{II} = k_v V_v + k_c (1 - V_v) \qquad (5.14)$$

and

$$k_{\perp} = \frac{k_v k_c}{k_v (1 - V_v) + k_c V_v} \qquad (5.15)$$

where $k_v$ and $k_c$ are the thermal conductivities of the virgin laminate and char, respectively, and $V_v$ is the volume fraction of virgin material remaining in the decomposing composite, which is determined using:

$$V_v = \frac{m - m_f}{m_o - m_f} \qquad (5.16)$$

Equations 5.14 and 5.15 are commonly used for calculating the equivalent thermal conductivities of composites when predicting the fire performance using the models described earlier.

While it is possible to calculate the thermal conductivity of a decomposing composite under isothermal conditions using Eqns. 5.14 and 5.15, a theoretical model for predicting the change in conductivity with temperature is not available. Although, analytical models have been proposed to calculate the change in thermal conductivity of materials with temperature [rg.57], they are only applicable to materials that experience a continuous linear increase in thermal conductivity with temperature. This is not the case with polymer composites and char, which show non-linear behaviour due to changes in the heat conduction mechanism with temperature. For example, Fig. 5.11 shows the variation in the through-thickness thermal conductivity of a glass/epoxy composite with temperature [25]. When fibreglass composites are initially heated above room temperature their thermal conductivity rises due to increases in the conductivities of both the fibres and polymer matrix. When the pyrolysis temperature of the polymer matrix is reached, the conductivity drops substantially due to the formation of a porous char network. After completion of the pyrolysis reactions, the thermal conductivity rises again due to the increasing conductivity of the char with temperature. Due to the complex relationship between thermal conductivity and temperature, it is necessary to experimentally measure the thermal properties over the temperature range of interest using techniques such as infra-red flash thermography. More basic techniques, such as non-uniform heating of a composite and measuring the rate of heat conduction over a fixed distance by thermocouples, can also be used.

It is possible to predict the change in thermal conductivity with temperature for some composite materials using empirical curve-fit equations [14,25,36] propose that the thermal conductivities of virgin laminate ($k_v$) and char ($k_c$) are empirically related to temperature by the polynomial equations:

$$k_v = k_{v1} + k_{v2}T \qquad (5.17)$$

and

$$k_c = k_{c1} + k_{c2}T + k_{c3}T^2 + k_{c4}T^3 \qquad (5.18)$$

where $k_{vj}$ is the curve fit coefficients for the virgin laminate (j = 1,2) and $k_{cj}$ is the curve fit coefficients for the char material (j = 1,2,3,4). Figure 5.12 shows the change in thermal conductivity of a virgin glass/phenolic composite and its char with temperature. The data points show the experimentally measured values, and the curves show the predicted in change in thermal conductivity using Eqns. 5.17 and 5.18, and good agreement is observed.

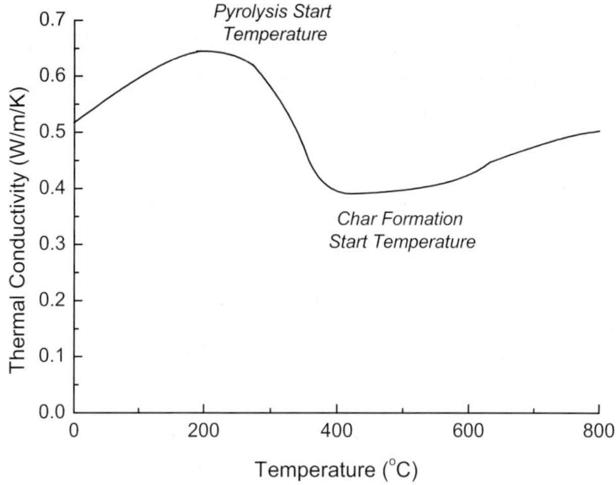



*Figure 5.11. Effect of temperature on the through-thickness thermal conductivity of a glass/epoxy composite. Reproduced from Dimitrienko [25].*

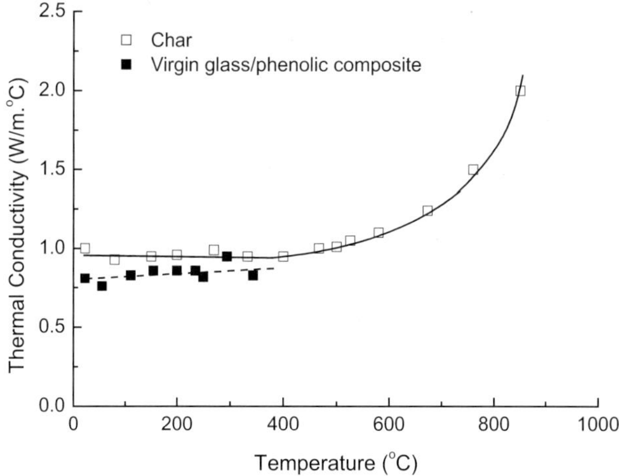



*Figure 5.12. Effect of temperature on the thermal conductivity of a glass/phenolic composite and its char. The curves show the predicted change in conductivity using empirically-based polynomial equations. Reproduced from Henderson et al. [10].*

Dimitrienko et al. [24] has proposed an alternate approach to calculate the change in thermal conductivity of a composite material with increasing temperature:

$$k = \frac{k_o}{(1-V_m^o)}\left(\frac{T}{T_o}\right)(V_m + n_\lambda V_c)$$

(5.19)

where $k_o$ is the coefficient of thermal expansion at room temperature, $T_o$, and $n_\lambda$ is an empirical constant. $V_m$ and $V_c$ are the volume fractions of polymer and char in the composite. Although, the reliability of this equation to predict the thermal conductivity of a decomposing laminate has not yet been validated against experimental conductivity data.

The specific heat of a composite is another important thermal property that influences the fire response of composite materials [58]. The specific heat of a polymer laminate under isothermal conditions can be calculated using [59]:

$$C_p = \frac{1}{\rho_c}[C_{pf}\rho_f V_f + C_{pm}\rho_m(1-V_f)]$$

(5.20)

where $\rho_c$ is the density of the composite, $\rho_f$ and $\rho_m$ are the densities of the fibre and matrix, and $C_{pf}$ and $C_{pm}$ are the specific heats of the fibre and matrix, respectively. Based on rule-of-mixtures analysis, Henderson et al. [10] proposes that the specific heat of a decomposing laminate in a fire can be determined using:

$$C_p = C_{p_v}V_v + C_{p_c}(1-V_v)$$

(5.21)

where $C_{pv}$ and $C_{pc}$ is the specific heats of the virgin material and char, respectively. Henderson et al. [10], Kalogiannakis et al. [55] and others report that the specific heat of composites fluctuate with increasing temperature as the polymer matrix undergoes decomposition whereas the specific heat of the char rises steadily with temperature, as shown in Fig. 5.13. Henderson and Wiecek [14] report that the specific heats for the virgin laminate and char are functions of temperature, and are calculated by:

$$C_{pv} = C_{pv1} + C_{pv2}T$$

(5.22)

and

$$C_{pc} = C_{pc1} + C_{pc2}T$$

(5.23)

As with thermal conductivity, the specific heats for the laminate and char must be measured experimentally over the temperature range of interest, and then these expressions curve-fitted to the data.

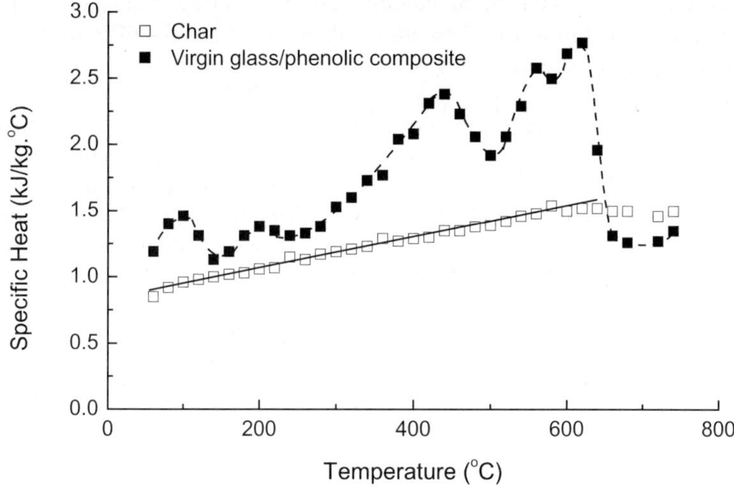


*Figure 5.13. Effect of temperature on the specific heat of a glass/phenolic composite and its char. The curve through the data for the char is calculated using an empirical polynomial function. Adapted from Henderson et al. [10].*


## 5.6      Concluding Remarks

Modelling the response of composite materials to fire is a complex scientific problem because of the many thermal, chemical and physical events that must be analysed. A model needs to consider the thermal processes of heat conduction, convection and radiation; the chemical events of pyrolysis, fibre-char reactions and char oxidation; and the physical processes of thermal expansion, thermal contraction, internal pressure due to reaction volatiles and moisture, out-gassing of volatiles, and delamination and matrix cracking.

Considerable progress has been made since the 1980s in the development of mathematical models for predicting the fire response of composite laminates. Simple thermal models that only consider the influence of heat conduction can accurately predict the temperature rise in composites when the radiant heat flux is below the level that induces decomposition of the polymer matrix or organic fibres. More sophisticated models have been developed to predict the thermal response of a decomposing laminate exposed to high temperature. The models consider the effects of heat conduction, pyrolysis, internal pressure and strain, volatile and moisture flow, and char formation. Such models have the ability to compute such parameters as the temperature rise, extent of char formation, thermal expansion, internal pressure and delamination cracking. However, the accuracy of many models has only been assessed for one or two types of composite materials exposed to a single fire scenario. Most models have not been rigorously tested against experimental fire data for a wide variety of composite materials. It is necessary that the models be more thoroughly evaluated against a large body of experimental fire test data to ensure they have the versatility to the applied to any type of composite material.

Other important issues remain in the development of more accurate and robust models. Firstly, the ability to model the thermal response of sandwich composite materials is presently underdeveloped. Most models are only applicable to laminated composites, and while a model exists for sandwich composites the material must have a non-combustible core. Most sandwich composites used in structural applications have an organic core such as PVC foam, phenolic foam, polyurethane, Nomex, balsa or some other combustible material. A model is required that can predict the fire response of sandwich composites constructed of combustible face skins and a flammable core.

A need exists for a model that can consider the influence of non-uniform rates of heating through laminates and sandwich composites exposed to fire. It is well known that the heating rate at the hot surface can be above $1000^{\circ}$C/min while at the cold face it can be under $20^{\circ}$C/min. Several thermally-controlled processes are sensitive to the heating rate, including the pyrolysis temperature, pyrolysis rate, and formation rate of volatiles. It is important that models consider the influence of heating rate to more accurately predict the thermal response of composites.

A major limitation with modelling decomposing composites in fire is inadequate data on the change in thermal conductivity, specific heat and density for polymers, fibres and char with increasing temperature. A large database of information exists over a moderately low temperature range ($\sim 20$-$200^{\circ}$C), however limited high temperature data is available. It is essential that this data be acquired for a diverse variety of polymers and fibres as well as for the different types of chars produced during decomposition. There is also a need to develop a theoretical (rather than the existing empirical) model for calculating the change to the thermal properties of composites and char with increasing temperature. Such a model needs to be incorporated into the existing thermal-chemical-physical models to more accurately predict the thermal response of laminates and sandwich composite materials in fire.


## References


1.    H. Ramamurthy, F.L. Test, J. Florio and J.B. Henderson. Internal pressure and temperature distribution in decomposing polymer composites. In: *Proceedings of the Ninth Heat Transfer Conference*, Jerusalem, Israel, August 1990.
2.    Y.I. Dimitrienko. Thermomechanical behaviour of composite materials and structures under high temperatures: 2. Structures. *Composites*, 1997; 28A:463-471.
3.    M. Ladacki. Silicon carbide in ablative chars. *Journal of the American Institute of Aeronautics & Astronautics*, 1966; 4:1445-1447.
4.    C.H. Bamford and D.H. Malan. The combustion of wood, Part 1. *Proceedings of the Cambridge Philosophical Society*, 1946; 42:166-182.

5.   T.R. Munson and R.J. Spindler. Transient thermal behaviour of decomposing materials: Part 1 General theory and application to convective heating. *RAD-TR-61-10*, AVCO Corporation, May 1961.

6.   K.A. Murty. Thermal decomposition kinetics of wood pyrolysis. *Combustion & Flame*, 1977; 29: 311-324.

7.   H.C. Kung. A mathematical model of wood pyrolysis. *Combustion & Flame*, 1972; 18:185-195.

8.   E.J. Kansa, H.E. Perlee and R.F. Chaiken. Mathematical model of wood pyrolysis including internal forced convection. *Combustion & Flame*, 1977; 29:311-324.

9.   B. Fredlund. Modelling of heat and mass transfer in wood structures during fire. *Fire Safety Journal*, 1993; 20:39-69.

10.  J.B. Henderson, J.A. Wiebelt and M.R. Tant. A model for the thermal response of polymer composite materials with experimental verification. *Journal of Composite Materials*, 1985; 19:579-595.

11.  MR Tant, JB Henderson and CT Boyer. Measurement and modelling of the thermochemical expansion of polymer composites. *Composites*, 1985; 16:121-126.

12.  J.B. Henderson and M.R. Tant. Measurement of thermal and kinetic properties of a glass-filled polymer composite to high temperatures. *High Temperatures-High Pressures*, 1996; 18:17-28.

13.  J.B. Henderson and M.P. Doherty. Measurement of selected properties of a glass-filled polymer composite. *High Temperatures-High Pressures*, 1987; 19:95-102.

14.  J.B. Henderson and T.E. Wiecek. A mathematical model to predict the thermal response of decomposing, expanding polymer composites. *Journal of Composite Materials*, 1987; 21:373-393.

15.  J. Florio, J.B. Henderson and F.L. Test. Measurement of the thermochemical expansion of porous composite materials. *High Temperatures – High Pressures*, 1989; 21:157-165.

16.  J. Florio, J.B. Henderson, F.L. Test and R. Hariharan. A study of the effects of the assumption of local-thermal equilibrium on the overall thermally-induced response of a decomposition, glass-filled polymer composite. *International Journal of Heat & Mass Transfer*, 1991; 34:135-147.

17.  R.M. Sullivan. A finite element method for thermochemically decomposing polymers, *PhD dissertation*, The Pennsylvania State University, 1990.

18.  R.M. Sullivan and N.J. Salamon. A finite element method for the thermochemical decomposition of polymeric materials – I. Theory. *International Journal of Engineering Science*, 1992; 30:431-441.

19.  R.M. Sullivan and N.J. Salamon. A finite element method for the thermochemical decomposition of polymeric materials – II. Carbon phenolic composites. *International Journal of Engineering Science*, 1992; 30:939-951.

20.  R.M. Sullivan. A coupled solution method for predicting the thermostructural response of decomposing, expanding polymeric composites. *Journal of Composite Materials*, 1993; 27:408-434.

21.  G.A. Pering, P.V. Farrell and G.S. Springer. Degradation of tensile and shear properties of composites exposed to fire or high temperature. *Journal of Composite Materials*, 1980; 14:54-66.

22.  H.L. McManus and G.S. Springer. High temperature behaviour of thermomechanical behaviour of carbon-phenolic and carbon-carbon composites, I. Analysis. *Journal of Composite Materials*, 1992; 26:206-229.

23.  H.L. McManus and G.S. Springer. High temperature behaviour of thermomechanical behaviour of carbon-phenolic and carbon-carbon composites, II. Results. *Journal of Composite Materials*, 1992; 26:230-255.

24.  Y.I. Dimitrienko. Thermal stresses and heat-mass-transfer in ablating composite materials. *International Journal of Heat and Mass Transfer*, 1995; 38:139-146.

25.  Y.I. Dimitrienko. Thermomechanical behaviour of composite materials and structures under high temperatures: 1. Materials. *Composites*, 1997; 28A:453-461.

26.  A.G. Gibson, Y-S. Wu, H.W. Chandler, J.A.D. Wilcox and P. Bettess. A model for the thermal performance of thick composite laminates in hydrocarbon fires. *Revue de L'Institut Francais du Petrole*, 1995; 50:69-74.

27.  N. Dodds, A.G. Gibson, D. Dewhurst and J.M. Davies. Fire behaviour of composite laminates. *Composites*, 2000; 31A:689-702.

28.  A.P. Mouritz, A.G. Gibson, Y. Wu, C.P. Gardiner and Z. Mathys. Validation of the Gibson model for the fire reaction properties of fibre-polymer composites., *Fire & Materials*, (in press).

29.  C.I. Chang. Thermal effects on polymer composite structures. *Theoretical & Applied Fracture Mechanics*, 1986; 6:113-120.

30. C.A. Griffis, J.A. Nemes, F.R. Stonesifer and C.I. Chang. Degradation in strength of laminated composites subjected to intense heating and mechanical loading. *Journal of Composite Materials*, 1986; 20:216-235.

31. J.A. Charles and D.W. Wilson. A model for passive thermal nondestructive evaluation of composite laminates. *Polymer Composites*, 1981; 2:105-111.

32. J. Milke and A.J. Vizzini. Thermal response of fire-exposed composites. *Journal of Composites Technology & Research*, 1991; 13:145-151.

33. R.J. Asaro, M. Dao and N. Schultz. Fire protection techniques for commerical vessels: structural fire protection modelling. *Flame Retardant Polymers*, 1998; 113-127.

34. C.A. Griffis, R.A. Masumura and C.I. Chang. Thermal response of graphite epoxy composites subjected to rapid heating. *Journal of Composite Materials*, 1981; 15:427-442.

35. M.R.E. Looyeh, P. Bettess and A.G. Gibson. A one-dimensional finite element simulation for the fire-performance of GRP panels for offshore structures. *International Journal of Numerical Methods for Heat & Fluid Flow*, 1997; 7:609-625.

36. M.R.E. Looyeh and P. Bettess. A finite element model for the fire-performance of GRP panels including variable thermal properties. *Finite Elements in Analysis & Design*, 1998; 30:313-324.

37. G.T. Boyer and W.C. Thomas. An analytical investigation of charring composites undergoing thermochemical expansion. In: *Proceedings of the ASME National Heat Transfer Conference*, Denver, Colorado, August 1985, Paper 85-HT-54.

38. J.D. Buch. Thermal expansion behavior of a thermally degrading organic matrix composite. In: *Thermomechanical Behavior of High-Temperature Composites*, ASME Publication AD-04, 1982, ASME, New York, pp. 35-49.

39. C.G. Goetzel. High-temperature properties of some reinforced phenolic composites. *High Temperatures-High Pressures*, 1980; 12:131-146.

40. M.E. Tuttle, A.M. Mescher and M.L. Potocki. Mechanics of polymeric composites exposed to a constant heat flux. *Journal of the American Society of Mechanical Engineers*, 1997; 80:157-164.

41. M.L. Potocki, M.E. Tuttle and A.M. Mescher. Behaviour of polymeric composites exposed to a heat flux simulating fire. In: *Proceedings of the Conference on the Mechanical Behaviour of Advanced Materials*, ASME, 1998, pp. 325-332.

42. C.S. Vatikiotis and D. Salinas. Heat transfer in a fibrous composite with combustion. In: *Proceedings of the ASME/AIChE National Heat Transfer Conference*, Orlando, Florida, 27-30 July 1980.

43. D. Salinas, Y.W. Kwon and E.A. Faxlanger. Failure of unidirectional composites exposed to fires. In: *Proceedings of Non-Classical Problems of the Theory and Behavior of Structures Exposed to Complex Environmental Conditions*, AMD Vol 164, American Society of Mechanical Engineers, 1993, pp. 115-128.

44. M.R.E. Looyeh, K. Rados and P. Bettess. Thermomechanical responses of sandwich panels to fire. *Finite Elements in Analysis & Design*, 2001; 37:913-927.

45. P. Krysl, W. Ramroth and R.J. Asaro. FE modelling of FRP sandwich panels exposed to heat: uncertainty analysis. In: *Proceedings of the SAMPE Technical Conference*, 16-20 May 2004, Long Beach CA.

46. G. Springer and S. Tsai. Thermal conductivity of unidirectional materials. *Journal of Composite Materials*, 1967; 1:166-173.

47. R.C. Progelhof, J.L. Throne and R.R. Ruetsch. Methods for predicting the thermal conductivity of composite systems: A review. *Polymer Engineering & Science*, 1976; 16:9.

48. L. Han and A. Cosner. Effective thermal conductivities of fibrous composites. *Journal of Heat Transfer*, 1981; 103:387-392.

49. H.-J. Ott. Thermal conductivity of composite materials. *Plastics, Rubber Processing and Applications*, 1981; 1:9-24.

50. J.P. Fanucci. Thermal response of radiantly heated kevlar and graphite/epoxy composites. *Journal of Composite Materials*, 1987; 21:129-139.

51. B. James, G. Wostenholm, G. Keen and S. McIvor. Prediction and measurement of the thermal conductivity of composite materials. *Journal of Physics D: Applied Physics*, 1987; 261-268.

52. C. Havis, G. Peterson and L. Flectcher. Predicting the thermal conductivity and temperature distribution in aligned fiber composites. *Journal of Thermophysics*, 1989; 3:416-422.

53. Y. Gowayed, J.-C. Hwang and D. Chapman. Thermal conductivity of textile composites with arbitrary preform structures. *Journal of Composites Technology & Research*, 1995; 17:56-62.

54. H. Tai. Equivalent thermal conductivity of two- and three-dimensional orthogonally fiber-reinforced composites in one-dimensional heat flow. *Journal of Composites Technology & Research*, 1996; 18: 221-227.
55. G. Kalagiannakis, D. Van Hemelrijck and G. Van Assche. Measurements of thermal properties of carbon/epoxy and glass/epoxy using modulated temperature differential scanning calorimetry. *Journal of Composite Materials*, 2004; 38:163-174.
56. W. Knappe, H.-J. Ott and G. Wagner. Calculation and measurement of the thermal conductivity of GRP. *Kunststoffe*, 1978; 68:13.
57. M.N. Ozisik. *Basic Heat Transfer*, Tokyo:McGraw-Hill Kogakusha Ltd, 1977.
58. J.N. Zalameda. Measured through-the-thickness thermal diffusivity of carbon fiber reinforced composite materials. *Journal of Composites Technology & Research*, 1995; 21:98-102.
59. K.K. Chawla. *Composite Materials*, New York, Springer, 1987.

# Chapter 6

# Structural Properties of Composites in Fire

## 6.1    Introduction

A large quantity of information is available on the fire reaction properties of polymer composites (as described in chapters 2 to 5), and the level of fire hazard associated with their use is known for a large number of materials.  We also have a good understanding of the chemical, thermal and physical mechanisms that control reaction properties such as time-to-ignition, heat release rate, flame spread rate, smoke production and toxicity. In short, we have quite a good quantitative understanding of the fire reaction behaviour of composites. Unfortunately, less is known about the fire resisting properties, such as burn-through resistance, dimensional stability and structural integrity, especially when the structure is under load.  Moreover it is not usually possible to estimate the fire resistive behaviour based solely on the known fire reaction properties.  A composite that has good fire reaction properties, such as low heat release rate and smoke yield, may not necessarily have good fire resistive properties.  Composites with a polymer matrix having high thermal stability, decomposition temperature and char yield may not necessarily have better fire resistive properties than more flammable materials.  For example, phenolic laminates generally show better fire reaction properties than unsaturated polyester laminates, including longer ignition time, lower heat release rate, slower flame spread and less smoke, but their mechanical properties can often degrade more rapidly in fire. Until recently, little was known about the structural properties of composites in fire. Understanding the structural performance in fire is a critical safety issue because the loss in stiffness, strength and creep resistance can cause composite structures to distort and collapse; possibly resulting in injury and death. Structural properties of composites in fire are therefore arguably as important to safety as the fire reaction properties that have generally been more widely studied.

This chapter describes the effect of elevated temperature and fire on the mechanical properties of composites. Several models for predicting the loss in the mechanical properties are described, and data on the losses in elastic modulus and strength at high temperatures and in fire are presented. The models are described in terms of the properties of the constituent materials: the fibre and matrix. In doing this we will refer the reader to some standard texts on the mechanical properties of composites [1-3], although a short description of micromechanical models is presented here. We discuss how conventional micromechanical modelling approaches can be adapted to consider both thermal effects and ultimately the fire behaviour of composites under load. In some cases, when all the necessary properties of a particular composite material are known as a function of temperature, this approach may not be required. In the majority of cases, however, the full range of information will not be available and it will be necessary to use models for the temperature dependence of composite properties [4-7] to calculate and model behaviour. Some small consideration will also be given to the time dependence of properties, which has been described in greater detail elsewhere [8,9].

There are relatively few references in the literature to the fire behaviour of moderately-sized composite structures under load. Two interesting examples are the work of Massot [10] which describes the behaviour of deep pultruded beams, and of Greene [11] which relates to research carried out for the US Navy on large panels under in-plane and out-of-plane loading. Several studies, including [12-26] and going back to early work on the behaviour of wood [12] and ablative materials [13], have been used to model and interpret the physical response of composites to fire. The most relevant achievement in this area, that was discussed in chapter 5, was the development of the Henderson equation [16,17], which accurately describes the thermal and decomposition effects during exposure to heat flux. The improved quantitative understanding that derived from these modelling studies has led to several useful publications on the measurement and modelling of the behaviour of composite structures under load [27-36].

## 6.2     Laminate Properties

### 6.2.1     PLY STRUCTURE AND PLY ARCHITECTURE

The majority of composite components take the form of laminates, which are plate or shell-like structures, built up from individual layers or plies [1-3], as shown in Fig. 6.1. It is this form of construction, and the inherent strength of many curved shell-like structures, that gives composite parts some of their advantageous properties. In laminates the reinforcement is handled and processed not as individual fibres, but as bundles containing many hundreds of fibres, which are known as fibre tows or rovings. These are the basic unit of a composite laminate. Strength and stiffness can be enhanced by using high performance fibres in the plane of the laminate and, if desired, these properties can be further enhanced by concentrating fibres in one or more particular location or directions.

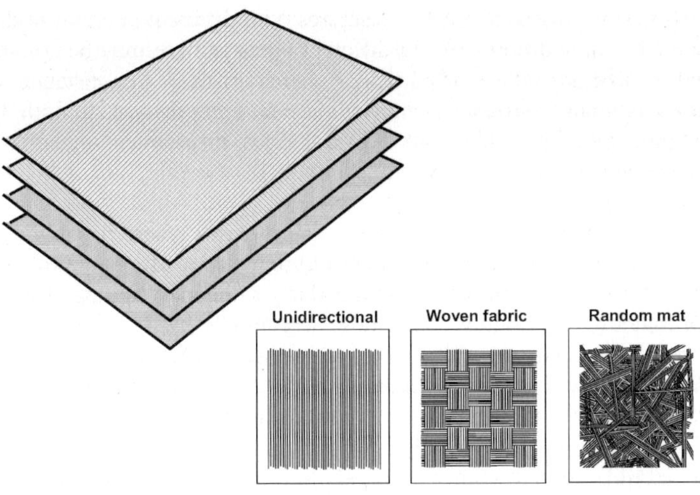


*Figure 6.1. Ply structure of laminated composites and common examples of ply architecture.*

As shown in Fig. 6.1, there are various ways of arranging the tows in each ply. Furthermore, plies of different types are often combined in a single laminate. The angle of the plies may be varied through-the-thickness of the laminate. Laminate theory [1-3] is the well-established procedure used to predict the properties of a laminate from the properties of the individual plies. The ply properties can be calculated from those of the constituents using a set of relationships known as micromechanics models [1-3].

The term, 'fibre architecture' refers to the way the fibre rovings are arranged within each ply. The simplest is the unidirectional (U/D) ply that consists simply of closely spaced parallel fibres. U/D plies can be achieved in a number of ways. In the processes of pultrusion or filament winding, for instance, unidirectional tows of reinforcement are incorporated directly into the structure of the product. In other processes the U/D plies may be pre-impregnated with resin, or they may comprise parallel strands of reinforcement, stitched or woven together with just a sufficient number of transverse fibres to hold them together.

A common method of achieving strength in two orthogonal directions is to use reinforcement in the form of a woven fabric. Woven glass fibre fabrics are often referred to as woven rovings. Many types of weave patterns are possible (plain, satin, twill, etc.); the choice being determined by factors such as drapeability of the fabric, i.e. its ability to conform to complex, doubly-curved shapes without wrinkling and with a minimum of waste.  The number of fibre tows in the warp and weft directions of a fabric may sometimes be varied, to enable structures with different strength in different directions to be made.  In addition, tows of different types of fibre may be combined in a 'hybrid' fabric to take advantage of particular characteristics.  For instance, aramid fibre, which has exceptional resistance to impact but poor compressive strength, is often combined with glass or carbon fibres in hybrid fabrics to produce impact resistant laminates with adequate compressive strength.

For general ease of coverage of complex shapes, and in cases where roughly equal in-plane strength is required in all directions, random fibre mats are often employed.  Here, the fibre orientation is roughly random within the plane of the mat.  Chopped strand mat (CSM) reinforcement is probably the most commonly-used glass fibre-based reinforcement for general purpose and marine mouldings.  CSM is manufactured by chopping rovings, depositing them on a moving belt, and applying a binder to ensure the integrity of the mat prior to processing.  A lower cost alternative to CSM is 'spray lay-up', in which rovings are fed to a chopper gun and then sprayed *in-situ* onto the laminate in a mould, along with catalysed liquid resin.  Although chopped reinforcement provides similar properties in all in-plane directions, the strength and stiffness values are reduced because of this and also because the strands are discontinuous.

A further type of random-in-plane reinforcement is continuous strand mat or 'swirl' mat. This is similar to CSM, except that the strands, instead of being chopped, are laid down in a swirling motion to achieve an overall random in-plane fibre architecture.  Swirl mat is often used in automated processes when random fibre reinforcement is needed, for instance in pultruded products.

'Quasi-isotropic' laminates provide a higher strength alternative to random mats when uniform in-plane properties are required and when reinforcements other than glass are used.  The term refers to any continuous fibre laminate in which plies are equally distributed between '*m*' equally-spaced directions, where $m$ must be 3 or more. Such laminates have in-plane elastic properties that are invariant with respect to angle, and the strength is also approximately constant in all directions.  The most common examples of quasi-isotropic laminate construction are (i) laminates with equal numbers of plies at 0°, +120° and -120° (the $m$ = 3 case) and (ii) laminates with plies at 0°, 90°, +45° and   -45° ($m$ = 4).  The latter is the most widely-used.

'Multiaxial' or 'non-crimp' reinforcing mat is manufactured by a special process in which several unidirectional plies of reinforcement are assembled at the required angles without the presence of undulating warp and weft fibres.  The fibre layers are held together by through-stitching with a small number of fine polyester fibres.  The use of multiaxial reinforcement is expanding because the absence of fibre undulations gives improved strength and stiffness.  The multiaxial architecture also enables complex sequences of ply angle to be achieved without the need for special alignment of each ply. Stitching several plies together increases the laminate thickness deposited at each stage, resulting in improved productivity. The limiting factor in this type of product is that thicker layers of reinforcement become progressively more difficult to impregnate with resin.


## 6.2.2    FIBRE CONTENT

The composition of each ply of a composite may be defined by either the weight fractions or the volume fractions of the components.  If the densities of the phases are $\rho_1$ and $\rho_2$ then the density of the composite is given, in terms of the volume fractions, $V_1$ and $V_2$, by:

$$\rho = V_1\rho_1 + V_2\rho_2 \tag{6.1}$$

Alternatively, using weight fractions, $W_1$ and $W_2$ :

$$\frac{1}{\rho} = \frac{W_1}{\rho_1} + \frac{W_2}{\rho_2} \tag{6.2}$$

Of course, volume fractions can be expressed in terms of weight fractions by:

$$V_1 = \frac{\dfrac{W_1}{\rho_1}}{\dfrac{W_1}{\rho_1} + \dfrac{W_2}{\rho_2}} \quad \text{and} \quad V_2 = \frac{\dfrac{W_2}{\rho_2}}{\dfrac{W_1}{\rho_1} + \dfrac{W_2}{\rho_2}} \tag{6.3}$$

and vice-versa:

$$W_1 = \frac{V_1\rho_1}{V_1\rho_1 + V_2\rho_2} \quad \text{and} \quad W_2 = \frac{V_2\rho_2}{V_1\rho_1 + V_2\rho_2} \tag{6.4}$$

The above expressions assume only two phases to be present, but can be readily extended to cover the case of three or more components.


## 6.2.3    MICROMECHANICS RELATIONSHIPS

These relationships are used to calculate the ply properties from those of the constituents, and they are often required when engineering calculations are needed and measured values of properties are not available.

The familiar relationships for the Young's moduli parallel ($E_1$) and perpendicular ($E_2$) to the fibres of a unidirectional laminate are:

$$E_1 = E_f V_f + E_m V_m \tag{6.5}$$

and

$$E_2 = \left( \frac{V_f}{E_f} + \frac{V_m}{E_m} \right)^{-1} \tag{6.6}$$

These well-known 'law of mixtures' expressions are obtained from assumptions of uniform strain and uniform stress, respectively. They represent approximate upper and lower bounds to the modulus of the composite, regardless of the shape or topology of its phases. More accurately defined bounds are to be found in the literature [1-3].

For unidirectional fibre composites, Eqn. (6.5) usually corresponds closely to the modulus parallel to the fibre direction, although it may be necessary to apply a correction for the effect of fibre undulation or finite fibre length. Although often used to obtain an estimate of the transverse modulus, Eqn. (6.6) significantly underestimates this quantity. Its accuracy can be greatly improved by making allowance for the constraining effects of the fibres by replacing $E_m$ with its equivalent 'plane strain' value, equal to $E_m \left( 1 - v_m^2 \right)$. It should also be borne in mind, when using Eqn. (6.5) for anisotropic fibres such as carbon or aramid, the correct value of $E_f$ is the transverse value for the particular fibre.

The more sophisticated micromechanics models given in the literature [1-3] may be used when more accurate estimates are needed for elastic constants. These, however, are not very flexible when account needs to be taken of different types of fibre architecture. The alternative approach is to use the semi-empirical Halpin-Tsai equations [1-3], which will be discussed in the next section.

The principal Poisson's ratio of a unidirectional ply is also given by a law of mixtures expression:

$$v_{12} = v_f V_f + v_m V_m \tag{6.7}$$

The other Poisson's ratio can be found from:

$$v_{21} = \frac{E_2 v_{12}}{E_1} \tag{6.8}$$

### 6.2.4    HALPIN-TSAI EQUATIONS

The Halpin-Tsai equations are a set of semi-empirical relationships that enable a particular composite property to be expressed or interpolated in terms of the values of the equivalent fibre and matrix properties, taking account of the fibre volume fraction and the arrangement or topology of the phases. Although other micromechanics relationships are available, some of which are acknowledged to have a more rigorous theoretical basis, the Halpin-Tsai equations are the most widely used because of their relative simplicity of form and their ability to describe experimental data with a reasonable degree of accuracy.  The general form is:

$$P = HS\left(P_f, P_m, V_f, \xi_P\right) = P_m \frac{1 + \xi_P \eta_P V_f}{1 - \eta_P V_f} \quad \text{where } \eta_P = \frac{P_f - P_m}{P_f + \xi_P P_m} \tag{6.9}$$

Here, the function, $HS$, is used to represent the Halpin-Tsai relationship. $P_f$ and $P_m$ are the equivalent fibre and matrix values, respectively, of the property in question. $\xi_P$ is the 'contiguity' factor that relates to the influence of the phase arrangement for the particular property, identified by the sub-script.

Although the Halpin-Tsai equations have no formal derivation, they resemble certain elasticity-based expressions for elastic properties of mixtures of materials. They also reduce to some well-known analytical expressions in special cases.  The most extreme special cases are $\xi_P = 0$ and $\xi_P = \infty$. Here the Halpin-Tsai equations for modulus reduce to the lower and upper bound expressions, respectively, Eqns. (6.5) & (6.6).

Table 6.1 shows some recommended values of $\xi_P$ for particular examples of ply architecture.   Those familiar with the literature will be aware of inconsistencies regarding the most appropriate values of $\xi_P$. In fact, given the empirical nature of the Halpin-Tsai equations it is permissible to adjust $\xi_P$ to enable the most accurate fit to be obtained to each particular set of experimental data. With fibres that are themselves anisotropic, i.e. carbon and aramid, it is necessary to use the appropriate $P_f$ value for the property in question (i.e. the fibre transverse modulus should be used when calculating the ply transverse modulus).  The same applies for the shear moduli.

When the elastic constants of the polymer matrix are known, the Halpin-Tsai relationships can be used to derive all the elastic constants for laminates possessing a particular reinforcement content and architecture.  When doing this it is useful to bear in mind the relationships between the elastic constants for an isotropic material, namely

$$G = \frac{E}{2(1+v)}; \quad K = \frac{E}{3(1-2v)}; \text{ and } G = \frac{3K(1-2v)}{2(1+v)} \tag{6.10}$$

These relationships apply for all isotropic matrix and fibre materials (i.e. all resins and glass fibres, but not carbon or aramid). Frequently the value of the Young's modulus only is known. In this case the shear modulus may be found by assuming a particular value for the Poisson's ratio; 0.35 is a typical value for resins and 0.28 for glass fibre. When the resin is above its glass transition temperature, $T_g$, the Poisson's ratio becomes larger, eventually approaching the value of 0.5, often assumed for rubbers. Fortunately the effect of the value chosen for Poisson's ratio is not usually very large.

*Table 6.1. Values of the contiguity factor for the Halpin-Tsai equations.*

| Case | $\xi_P$ |
|---|---|
| Perfectly aligned continuous fibres.  Young's modulus, $E_{11}$, parallel to fibres. | $\xi_{11} = \infty$ (equivalent to Eqn. (6.6)) |
| Imperfectly aligned (or wavy) continuous fibres.  Young's modulus, $E_{11}$, parallel to fibres. | $\xi_{11} < \infty$ (often in the range, 100-1,000, adjusted to allow for lack of perfect fibre alignment) |
| Aligned discontinuous fibres, length, $L$ and diameter, $D$.  Young's modulus, $E_{11}$, parallel to fibres. | $\xi_{11} = 2\,L/D$ |
| Aligned continuous or discontinuous fibres.  Young's modulus, $E_{22}$, perpendicular to fibres. | $\xi_{22} = 2$ or value to achieve best fit to data |
| Aligned continuous or discontinuous fibres.  Shear modulus, $G_{12}$, for shear parallel to fibres. The same expression also applies for $G_{13}$. | $\xi_{12}$ (or $\xi_{13}$) = 1 or value to achieve best fit to data. |
| All, the above, principal Poisson's ratio, $v_{12}$ | $\xi_{v_{12}} = \infty$  (equivalent to Eqn. (6.7)) |

When characterising the high temperature behaviour of composites or their mechanical behaviour in fire it is necessary to measure or to model elastic constants and other mechanical properties as a function of temperature. In the case of the elastic constants the information available is often limited.  For instance only one of the elastic constants may be known as a function of temperature.  In this case it is possible to derive the resin modulus vs. temperature relationship by using the inverse of the Halpin-Tsai equations. The other elastic constants can then be calculated once the resin modulus is known. The inverse of the Halpin-Tsai relation can be found by manipulating Eqn. (6.9), which gives the following positive solution of the quadratic expression for resin modulus as a function of composite and fibre moduli:

$$P_m = \frac{P\left(\xi_P + V_f\right) - P_f\left(1 + \xi_P V_f\right) + \sqrt{\left(P\left(\xi_P + V_f\right) - P_f\left(1 + \xi_P V_f\right)\right)^2 + 4PP_f\xi_P\left(1 - V_f\right)^2}}{2\xi_P\left(1 - V_f\right)} \tag{6.11}$$

Alternatively, when pure resin samples are available, the resin modulus can be measured directly.

## 6.2.5   EFFECT OF VOID CONTENT

Most composites, other than those manufactured by the prepreg/autoclave route, contain an appreciable void content, often up to 3-5% by volume.  In the case of phenolic-based composites this is even higher (up to 25%) due to the evolution of water during the condensation reaction involved in cure. When modelling elastic properties it is convenient to assume the voids to be a component of the resin phase and reduce the modulus of this phase accordingly. Using the 'self-consistent' approach the shear modulus and bulk modulus of an isotropic solid containing spherical voids are given by [37]:

$$G = \frac{3}{16}(\phi_V - 3)K_0 - \frac{1}{4}(5\phi_V - 2)G_0 + \frac{1}{2}\sqrt{\left(\frac{3}{8}(3 - \phi_V)K_0 + (5\phi_V - 2)G_0\right)^2 - 9\left(\phi_V - \frac{1}{2}\right)K_0 G_0} \qquad (6.12a)$$

and

$$K = \frac{1 - \phi_V}{\dfrac{1}{K_0} + \dfrac{5}{4}\dfrac{\phi_V}{G}} \qquad (6.12b)$$

where $G$ and $K$ are the overall shear modulus and bulk modulus, respectively, $\phi_V$ is the void volume fraction in the phase, and $G_0$ and $K_0$ are the shear and bulk moduli in the absence of voids. The above relationships should be used in conjunction with the relationships between the elastic moduli, Eqn. (6.10). Alternatively, the almost linear relationship between the elastic constants and void content is approximated by:

$$\frac{E}{E_0} \cong \frac{G}{G_0} \cong 1 - 1.93\phi_V \qquad (6.13)$$

The constant in this expression depends slightly on the value of the Poisson's ratio, but the expression is sufficiently accurate for most purposes up to void fractions of about 0.2. An alternative experimentally-based relationship has been proposed for ceramic materials containing voids [38].  This is also sufficiently accurate for most purposes for a void-containing resin phase:

$$\frac{E}{E_0} = 1 - 1.9\phi_V + 0.9\phi_V^2 \qquad (6.14)$$

## 6.3      Measurement of Elastic Constants

The elastic moduli can be measured by a variety of methods. One of the simplest means of measuring the Young's modulus in a particular direction, within laminates or plies, is the tensile test, supplemented by the use of an accurate strain-measuring device, such as a clip-on extensometer or strain gauge. This method can also be used to determine Poisson's ratio if the transverse strain is measured simultaneously with the longitudinal strain. Variations of the method can also be used to determine elastic constants in compression; although it is frequently justifiable with composites to make the assumption that, at low stress levels, the constants are the same in tension and compression. In addition to Young's moduli, in-plane shear moduli can be deduced from off-axis measurements with strain gauge rosettes, by means of laminate theory.

Both strain gauges and extensometers pose limitations for high temperature use. In the case of laminate samples where the architecture is the same through-the-thickness, flexural measurements can provide a very accurate method of measuring Young's moduli at elevated temperature. In this case the displacement can be measured with sufficient accuracy by the machine displacement in the case of testing machines, or by an externally mounted displacement transducer in the case of dead weight loading. Flexure thus enables higher measurement temperatures to be accessed. Either three-point or four-point flexure may be used. The four-point bending configuration has the advantage that no correction for shear deflection is required. However, it has the disadvantage that measurement of the curvature of the central section during the test can present experimental difficulties.

In the other bending cases, care must be taken to compensate for the shear deflection, which can be very significant in the case of aligned fibre or woven composites: for these materials the ratio of longitudinal modulus to through-thickness shear modulus is large, even when the resin phase is below its $T_g$. Above $T_g$, the shear modulus relaxes much more than the extensional modulus, so the ratio becomes even less favourable.

For a monolithic rectangular beam the flexural deflection is given by the sum of the components due to bending and shear. For a rectangular beam, length, $L$, under constant shear force along its length, $F$, a rough estimate of the shear defection can be obtained by assuming the shear stress to be constant through the depth (which is not true in practice). This gives $\tau = \dfrac{F}{bt}$, so we may deduce that the shear strain is $\gamma = \dfrac{F}{Gbt}$ and a rough estimate of the shear deflection is $\delta_s = \dfrac{FL}{Gbt}$. An accurate estimate needs to account for the fact that the shear stress is not constant through the depth, but varies parabolically from zero at the surfaces to a maximum of $1.5F/bt$. For a rectangular section the shear deflection is:

$$\delta_s = \frac{6}{5}\frac{FL}{Gbt} \tag{6.15}$$

which is just a little higher than the rough estimate obtained above.

It is useful, using this estimate, to give expressions for the total deflection for two bending cases of practical interest for measurement. For three-point loading, simply supported:

$$\delta = \frac{WL^3}{4EBt^3} + \frac{3WL}{10btG} = \frac{WL^3}{4EBt^3}\left(1 + \frac{6}{5}\frac{E}{G}\left(\frac{t}{L}\right)^2\right)$$

(6.16)

and for a single cantilever beam, with one fixed end:

$$\delta = \frac{4WL^3}{EBt^3} + \frac{6WL}{5btG} = \frac{WL^3}{4EBt^3}\left(1 + \frac{3}{10}\frac{E}{G}\left(\frac{t}{L}\right)^2\right)$$

(6.17)

It is interesting to note that the shear component of deflection is four times less in the single cantilever case.  For unidirectional glass fibre samples, below the $T_g$ of the resin, $E/G$, can be of the order of 10, so for a beam of $L/t = 10$, the shear component would be 12% in the former case and 3% in the latter.  For a unidirectional carbon beam, however, $E/G$ may be six times higher, giving 72% and 18%, respectively.  However, these figures will change substantially when the resin approaches $T_g$ because the shear modulus, being more resin-dependent, will fall by a greater proportion than the axial modulus, in which case the shear deflection can easily eclipse that due to bending.  This effect can seriously influence the high temperature measurement of elastic properties, and compensation always needs to be made.

The specialised thermal technique of DMTA (dynamic mechanical and thermal analysis) uses the single cantilever beam configuration to measure the elastic properties at elevated temperature.  The three-point bend method is more accurate than the single cantilever beam because a 'fixed' or 'encastré' support is always difficult to achieve in practice.  Nevertheless, for experimental convenience the single cantilever method is often employed in DMTA.

Different loading methods may be used when measuring elastic properties; these include:
- constant deformation rate, as with a mechanical testing machine,
- constant load creep, measuring the modulus corresponding to different times (the isochronous modulus), and
- sinusoidally varying loading, as in DMTA.

Although these methods will give similar values for moduli at room temperature, where no relaxation takes place, they may produce appreciable differences in values at elevated temperature.  The creep and DMTA cases are most amenable to rigorous theoretical treatment, taking into formal consideration the full relaxation characteristics of the material.  A full formal treatment of the time-dependent behaviour of polymers and composites is outside the scope of this book, but the reader concerned with a rigorous and correct approach is advised to consult one of the specialist texts [e.g. 8].

The DMTA method in which the load and deflection of the sample vary sinusoidally with time, has some experimental advantages, the main one being that a broad range of temperatures can be encompassed in a single experimental scan. A range of frequencies may also be scanned automatically during a single measurement sequence, giving additional useful information on the activation energy of relaxation processes. A disadvantage of DMTA is that the raw modulus values may not be as accurate as in other techniques, for a range of experimental reasons; certainly it is necessary, as outlined above, to make corrections for the shear deflection. A good means of improving the accuracy of DMTA is to measure the room temperature or 'relaxed' values of the moduli using a more accurate method such a tension or three-point flexure. The DMTA measurements can then be corrected accordingly.

Creep or isochronous measurements, using a dead weight to apply a constant load, are a very accurate means of measuring modulus in flexure, but again the shear deflection needs to be taken into account. It should also be borne in mind that isochronous creep measurements are not exactly equivalent to the constant frequency measurements obtained in DMTA because each technique averages the relaxation times of the material in different ways. Fortunately, for most of the polymers employed in composites the peaks in the relaxation spectra are broad, so that in practice isochronous modulus results obtained at a time of say 10 s in creep are very similar to those obtained at a frequency of 0.1 Hz in DMTA. Measurements at constant displacement rate are the most difficult to relate rigorously to the fundamental relaxation behaviour of the material. However, in practice the data obtained over a particular loading time will be broadly comparable with the isochronous value corresponding to the same period, although the latter will always give slightly lower modulus values in relaxation regions. Finally, it is necessary with all measurements of elastic constants to ensure that the strain level is low enough for the material to be considered to be in the linear region, where stress is accurately proportional to strain.

## 6.4      Mechanical Properties as a Function of Temperature

This section discusses the effect of elevated temperature, up to decomposition of the polymer matrix, on the mechanical properties of composites. With thermally stable, post-cured laminates, changes in properties with temperature can be considered reversible up to the point where decomposition of one of the phases, usually the polymer matrix, begins. With fire it is necessary to consider changes beyond this point and a method of achieving this will also be discussed.

### 6.4.1   MECHANICAL PROPERTY CHANGES UP TO THE ONSET OF DECOMPOSITION

There is broad interest in this topic in relation to both fire behaviour and the high temperature applications of composites.   Figure 6.2 shows the typical relationship between the mechanical property of a laminate and temperature under isothermal conditions (i.e. with constant temperature through the material). Properties that show this type of relationship include Young's modulus, shear modulus and compressive strength.

Ideally, for a particular composite system, each elastic constant or strength value would be measured and expressed as a function of temperature. However, there are few composite systems where all the required data is available in this form. It will be assumed here, mainly for convenience, that all the mechanical properties, including strength, can be fitted to relationships similar in form to that for the modulus, as in Fig. 6.2. This assumption appears to work well for all thermosetting resin systems, which happen to be amorphous. On heating an amorphous polymer from room temperature only one transformation - the glass transition - occurs before decomposition, so the problem is really one of fitting a suitable relationship to the property vs. temperature relationship in this region.

A polynomial in temperature is one way of describing the variation of, say Young's modulus in the transition region, and this assumption has been used by other workers [4,33] over a relatively narrow thermal region. The approach can sometimes be inconvenient because a polynomial of order at least six is generally needed if a simple polynomial is used to describe the property-temperature relationship right across the $T_g$ region with the required accuracy. Moreover, such a relationship only behaves reliably within the range of the fitted data.  This problem was overcome by Kulkarni and Gibson [4] who fitted a polynomial in normalised terms, to give:

$$\frac{P(T)}{P_0} = 1 - a_1\left(\frac{T - T_0}{T_g - T}\right) - a_2\left(\frac{T - T_0}{T_g - T}\right)^2 - a_3\left(\frac{T - T_0}{T_g - T}\right)^3 \tag{6.18}$$

where $P(T)$ is a particular property, and $P_0$ is the value of that property at some low temperature, say room temperature $T_0$. $a_1$, $a_2$ and $a_3$ are fitting constants.

Another model, due to Mahieux and Reifsnider [5,6], assumes that increasing temperature has the effect of increasing the number of intermolecular bonds in the resin.

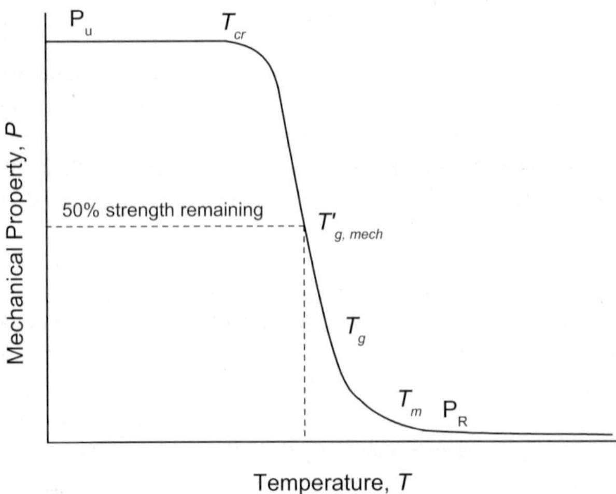



*Figure 6.2. Schematic of the effect of iso-thermal heating on the mechanical property of a laminate.*

The bond 'strength' is assumed to follow a cumulative Weibull distribution as a function of temperature, giving rise to a simple functional relationship of the form:

$$P(T) = P_R + (P_U - P_R)exp\left(-\left(\frac{T}{T_0}\right)^m\right)$$

(6.19)

where $P_U$ and $P_R$ are the unrelaxed (low temperature) and relaxed (high temperature) values of that property, respectively, $T_0$ is the relaxation temperature, and $m$ is the Weibull exponent. For this model the temperatures are in Kelvin. Some success was achieved in fitting modulus vs. temperature data for several polymers. Acceptable fitting of property data can be achieved for most thermosets with $m$ values in the range 9-21.

A number of empirical functions showing the required antisymmetric behaviour were examined by Gibson et al. [7] for fitting mechanical properties to thermoset laminates. Particular success was achieved with functions based on the hyperbolic tanh function, which leads to a relationship of the following form, which describes behaviour as a function of both temperature and time-scale:

$$P(T) = P_U - \frac{P_U - P_R}{2}\left(1 + \tanh\left(k'\left(\ln(a_T) + \frac{H}{R}\left(\frac{1}{T_g} - \frac{1}{T}\right)\right)\right)\right) \qquad (6.20)$$

where $k'$ is a constant describing the breadth of the relaxation and $T_g$ (in Kelvin) is the temperature of the mechanically observed glass transition, corresponding to a 50% reduction in the property value (see Fig. 6.2).

Equation 6.20 assumes Arrhenius time-temperature dependence to apply. This is convenient because a shift in time-scale or frequency is related to an equivalent change in temperature by:

$$\ln(a_T) = \frac{H}{R}\left(\frac{1}{T_{ref}} - \frac{1}{T}\right) \qquad (6.21)$$

where $T_{ref}$ is the original or reference temperature. Provided that the relaxation process is governed by Arrhenius kinetics, this relationship applies independently of the profile of the relaxation process. $a_T = t/t_0$ for creep measurements, where $t$ is time and $t_0$ is a reference time. Alternatively $a_T = \omega_0/\omega$ for DMTA measurements, where $\omega$ is frequency and $\omega_0$ is the reference frequency. $H$ and $R$ are the activation energy and gas constant, respectively.

There is ongoing discussion in the literature regarding the applicability or otherwise of Arrhenius kinetics to the modelling of transition behaviour [8, 9]. Recent experience suggests that the approach works fairly well with most thermoset-based composites.

In many cases, the results may all be generated on the same time-scale, so $a_T$ is zero and the expression simplifies to:

$$P(T) = P_U - \frac{P_U - P_R}{2}\left(1 + \tanh\left(k'\frac{H}{R}\left(\frac{1}{T_g} - \frac{1}{T}\right)\right)\right) \qquad (6.22)$$

Like the Mahieux and Reifsneider expression (Eqn. (6.19)), this involves only four fitting constants, in this case: $P_U$, $P_R$, $k'\frac{H}{R}$ and $T_g$. Equation (6.22) predicts property-temperature curves that are anti-symmetric when plotted against the reciprocal of temperature. In many cases the curves are also close to being anti-symmetric against linear temperature so some simplification can be achieved:

$$P(T) = P_U - \frac{P_U - P_R}{2}\left(1 + tanh\left(k\left(T - T_g\right)\right)\right) \tag{6.23}$$

Finally, the unrelaxed and relaxed quantities, $P_U$ and $P_R$, may themselves vary with temperature. Treating one or both as linear functions of temperature requires one or two additional fitting constants.

There are no theoretical restrictions on the shape of the property-temperature relationship so, in practice, the choice may be made between Eqns. (6.18), (6.19), (6.22) or (6.23) on the basis of which best fits the data. Figure 6.3 shows examples of the Young's moduli of a polyester and vinyl ester resin as a function of temperature, fitted by Eqn. (6.22). Here it was necessary to assume the unrelaxed moduli to be temperature-dependent. The two resin types show very similar behaviour. Indeed, for many purposes the data could probably be described by the same relationship, changing only the value of $T_g$. This opens up the possibility of modelling thermosetting composites as 'generic' systems, defined only by their mechanical $T_g$, a useful procedure when the full modulus vs. temperature data are not available. The majority of thermosetting resins show unrelaxed Young's moduli in the fairly narrow range of 3.3-4.5 GPa. A 'generic' procedure would provide sufficiently accurate data for most design and modelling purposes.

It should, however, be borne in mind that the mechanical $T_g$ referred to here is generally different from the $T_g$ value measured by thermal techniques such as DSC, the latter being commonly about 15-20°C higher. The heat deflection temperature (HDT), which is often about 20°C higher still than the DSC $T_g$, provides a very rough relative comparison of the thermal capabilities of different resin or composite systems. Often the HDT is the only thermal parameter available to characterise the system. In some cases, it may be practical to assume the mechanical $T_g$ to be 20-30°C below the HDT.

In some composite structures the state of cure may not be complete at the start of life. Here, it is common for the cure state to advance during the course of any elevated temperature testing, producing increases in $T_g$ and resin strength with rising temperature. To achieve consistent results when this happens it may be necessary to post-cure the test samples, but this procedure of course changes the state of the material. In some cases it is possible to model the progression of cure during testing or fire simulation simply by shifting the $T_g$ value in Eqns. (6.18), (6.19), (6.22) or (6.23). When high test temperatures are used the cure state probably advances significantly during testing, even when the material has been previously well-cured. Fortunately the effect is masked to some extent by the magnitude of other effects during actual or simulated fire testing.

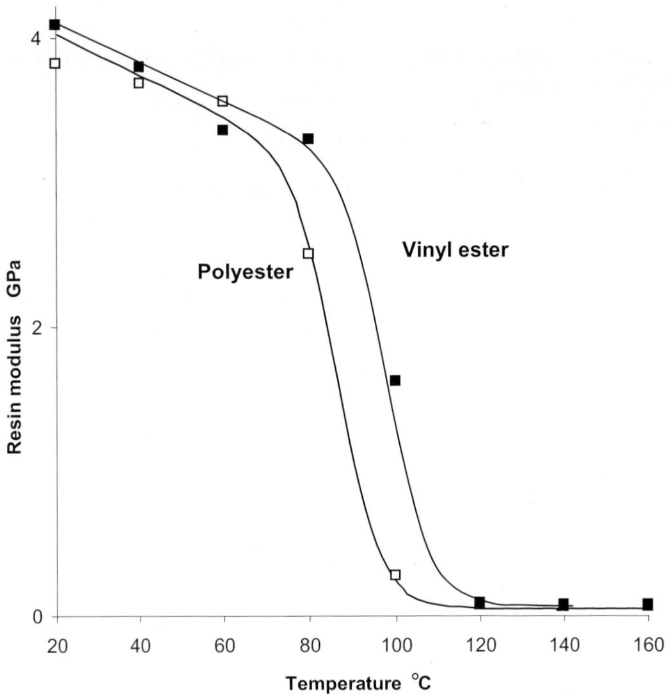


*Figure 6.3.  Young's moduli of polyester and vinyl ester resins as a function of temperature, fitted by Eqn. (6.22) to DMTA data at 1Hz.*

## 6.4.2    EFFECTS OF DECOMPOSITION ON MECHANICAL PROPERTIES

In addition to the effect of temperature it is also necessary to account for resin decomposition.  Little is currently known about the effect of decomposition on the mechanical properties. Until a more appropriate relationship can be found, it is proposed that each mechanical property be modified by a power law factor, $R^n$, in residual resin content.  Equation (6.23) thus becomes:

$$P(T) = \left( \frac{P_U + P_R}{2} - \frac{P_U - P_R}{2} tanh(k(T - T')) \right) R^n \qquad (6.24)$$

This has the characteristics that, for $R = 1$, Eqn. (6.24) reduces to Eqn. (6.23), whereas for $R = 0$, a zero property value is predicted.  It is reasonable to assume that some properties, including the elastic constants and compressive strength depend strongly on resin content while the tensile strength, which more closely reflects the strength of the reinforcement, is much less dependent. In a recent investigation it was found that, for woven fabric laminates, the decline in the resin-dependent properties was described well by $n = 1$, while for the tensile strength, $n = 0$ [7]. Future work will probably provide more accurate values.


## 6.5      Modelling of Properties

### 6.5.1    UNIDIRECTIONAL PLIES

For perfectly aligned unidirectional plies of continuous fibres the longitudinal modulus and the principal Poisson's ratio are given by the law of mixtures expressions, Eqns. (6.5) & (6.7). The transverse and shear moduli can be modelled most conveniently using the Halpin-Tsai equations. The assumption of perfect alignment is realistic in the case of unidirectional prepreg-based materials and in composites manufactured by filament winding. In other cases, however, including woven plies, where the rovings undulate, and products where manual operations play a role in manufacture, the fibre alignment in nominally unidirectional materials is often less than perfect. In these situations it is possible to allow for imperfect fibre alignment by employing the Halpin-Tsai equations to model the fibre direction modulus, as well as the other constants, adjusting the continuity factor accordingly, as in Table 6.1. In the general unidirectional case, therefore, the four ply elastic constants are given by:

$$E_{11} = HS\left(E_f, E_m, V_f, \xi_{11}\right)$$
$$v_{12} = v_f V_f + v_m V_m \qquad (6.25)$$
$$E_{22} = HS\left(E_f, E_m, V_f, \xi_{22}\right)$$
$$G_{12} = HS\left(E_f, E_m, V_f, \xi_{12}\right)$$

where the Halpin-Tsai expressions are as described in equation 6.9. The principal Poisson's ratio relates to the strain in the 2-direction resulting from a stress in the 1-direction. The other Poisson's ratio can be found, as discussed previously from equation 6.8.

As an example of the use of the above relationships, Fig. 6.4 shows the transverse modulus of unidirectional pultruded samples. This data was obtained by DMTA as a function of temperature and frequency in the range 0.01-100 Hz. Figure 6.5 shows the master-curve obtained from these results, assuming Arrhenius temperature dependence using Eqn. (6.22).

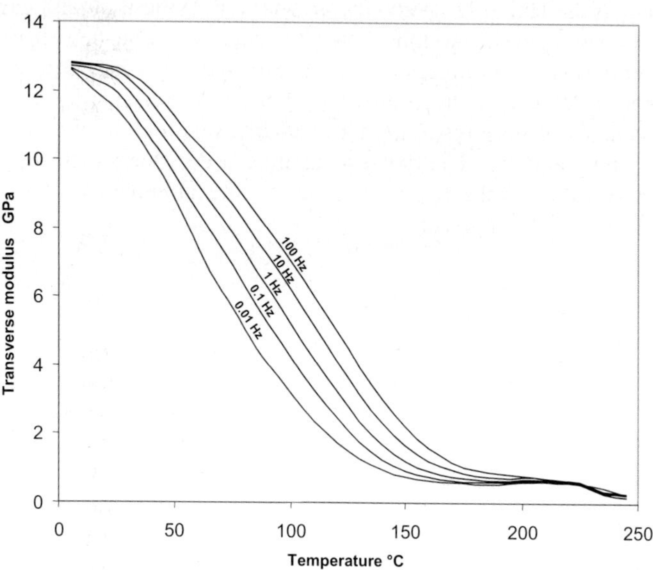


*Figure 6.4.  Transverse storage modulus of unidirectional glass/polyester pultrusion at frequencies of 0.01, 0.1, 1, 10 and 100 Hz. Material courtesy of Fiberline Composites a.s. References [39, 40].*

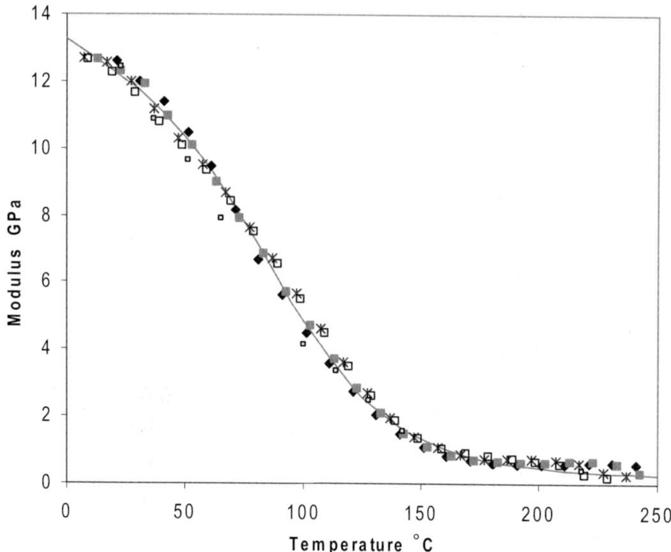


*Figure 6.5.  Transverse modulus of glass/polyester pultrusions: master curve calculated from the data of Fig. 6.4 using Eqn. (6.22). Creep moduli at 1, 10 and 100 s have also been added to the data, and can be seen to follow the same relationship. Material courtesy of Fiberline Composites a.s, References [39, 40].*

Using the inverse of the Halpin-Tsai equations (with $\xi = 2$) these moduli can be used to obtain the corresponding resin moduli, which are shown in Fig. 6.6.  This shows the advantage of using DMTA measurements in a resin-sensitive direction to obtain data that can be directly related to the properties of the base resin.  In situations where representative samples of pure resin are not available, this may be the only method of obtaining resin modulus data.  The data can then be used, along with the Halpin-Tsai equations, to derive values of the elastic constants for the laminate.

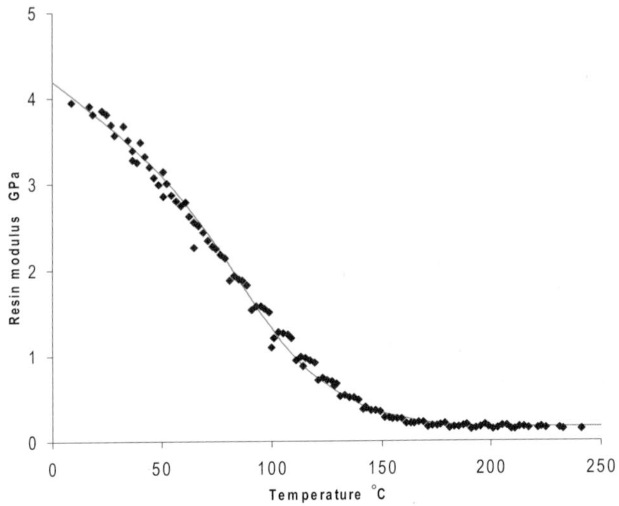

*Figure 6.6.   Predicted Young's modulus of polyester pultrusion resin (1 Hz DMTA or 1 sec creep), derived from the transverse modulus data of Fig. 6.5. Material courtesy of Fiberline Composites a.s, references [39, 40].*

Figure 6.7 shows the stiffness properties of a glass/polyester pultrusion, calculated from the transverse DMTA moduli as described above. Pultrusions take many sectional forms, including box sections, 'I' beams and channels.  The basic structure within the laminate, however, is most often a sandwich comprising a core of unidirectional material, that forms the main structural element, surrounded by skins of random swirl mat.  The role of the skins is to protect the unidirectional core from physical and environmental damage.  Figure 6.7 shows the predicted axial Young's modulus of the core material, compared to measured values, which agree fairly well.  The modulus of the swirl mat skins was also calculated, using a $\xi$ value of 200 in the procedure outlined in the next section.  The overall or 'effective' Young's modulus of the sandwich section was also calculated and is shown.  It is interesting to note that the transverse modulus of unidirectional material declines more rapidly with temperature than the modulus of the mat material.  This is because the decline in resin modulus has the greatest effect on this particular property.

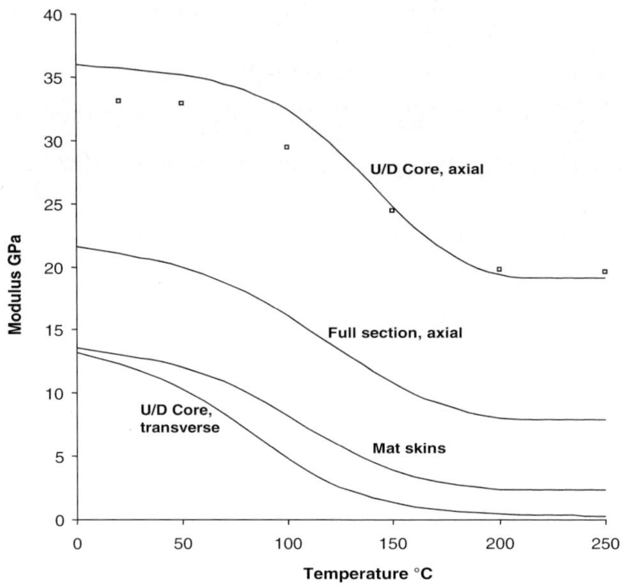


*Figure 6.7.  Predicted Young's moduli of a glass/polyester pultrusion and its components, determined from the resin modulus data of Fig. 6.6.  The full section comprises 60% by volume of unidirectional core, along with skins of random swirl mat. The experimental points shown are one second axial creep moduli for the core. Material courtesy of Fiberline Composites a.s, references [39, 40].*

## 6.5.2    ELASTIC CONSTANTS IN OFF-AXIS DIRECTIONS

When the elastic constants are needed for directions other than the ply direction, the well-known transformation equations can be used [1-3].  First, the ply elastic constants are converted to reduced stiffness constants:

$$Q_{11} = \frac{E_{11}}{1 - v_{12} v_{21}}$$

$$Q_{22} = \frac{E_{22}}{1 - v_{12} v_{21}} = Q_{11} \frac{E_{22}}{E_{11}} \tag{6.26}$$

$$Q_{12} = v_{12} Q_{22}$$

$$Q_{66} = G_{12}$$

These can then be transformed to obtain the corresponding values in the new coordinate direction, assumed to be at an angle, $\theta$, to the ply axis:

$$\overline{Q}_{11} = Q_{11} \cos^4 \theta + 2(Q_{12} + 2Q_{66}) \sin^2 \theta \cos^2 \theta + Q_{22} \sin^4 \theta$$

$$\overline{Q}_{12} = (Q_{11} + Q_{22} - 4Q_{66}) \sin^2 \theta \cos^2 \theta + Q_{12}\left(\sin^4 \theta + \cos^4 \theta\right)$$

$$\overline{Q}_{22} = Q_{22} \cos^4 \theta + 2(Q_{12} + 2Q_{66}) \sin^2 \theta \cos^2 \theta + Q_{11} \sin^4 \theta$$

$$\overline{Q}_{66} = (Q_{11} + Q_{22} - 2Q_{12} - 2Q_{66}) \sin^2 \theta \cos^2 \theta + Q_{66}\left(\sin^4 \theta + \cos^4 \theta\right)$$

$$(6.27)$$

The $\overline{Q}$ values are the terms in the matrix of transformed stiffness constants.

## 6.5.3    CROSS-PLY LAMINATES AND LAMINATES WITH WOVEN PLIES

Cross-ply lamination is a common form of construction. It can be achieved by stacking unidirectional plies manually with a pattern of 0° and 90° orientations. A similar effect is also achievable through the use of multiaxial non-crimp reinforcement. Woven fabric (i.e. woven roving) reinforcement can also be readily modelled by assuming it to be analogous to a cross-ply laminate. The properties of a single ply are given, as above, by:

$$E_{11,ply} = HS\left(E_f, E_m, V_f, \xi_{11}\right)$$

$$\nu_{12,ply} = \nu_f V_f + \nu_m V_m$$

$$E_{22,ply} = HS\left(E_f, E_m, V_f, \xi_{22}\right)$$

$$G_{12,ply} = HS\left(E_f, E_m, V_f, \xi_{12}\right)$$

$$\nu_{21,ply} = \nu_{12,ply} \frac{E_{22,ply}}{E_{11,ply}}$$

$$(6.28)$$

Here the axial Young's modulus is modelled using Halpin-Tsai, instead of the law of mixtures, because it enables a small allowance to be made for the 'waviness' of the woven yarns, as discussed above. It is assumed that a woven ply is equivalent to a combination of 0° and 90° plies in which $X$ is the volume fraction of 0° plies. Assuming the strain at any point to be equal in both plies the overall properties in each direction can be obtained by a weighted average, so the engineering constants are simply:

$$E_{11} = E_{0,ply} X + E_{90,ply} (1 - X)$$

$$E_{22} = E_{90,ply} X + E_{0,ply} (1 - X)$$

$$(6.29)$$

$$\nu_{12} = \nu_{0,90,ply} X + \nu_{90,0,ply} (1 - X) = \nu_{0,90,ply}\left(X + \frac{E_{0,ply}}{E_{90,ply}} (1 - X)\right)$$

$$G_{12} = G_{0,90,ply}$$

Often the assumption can be made that $X = 0.5$. Sometimes, however, woven fabrics are designed with a deliberate bias in properties in one of the orthogonal directions, so that $X \neq 0.5$. Even with fabrics having the same nominal fibre content in both directions there may be small manufacturing-related differences between properties in the warp $(0^\circ)$ and weft $(90^\circ)$ directions, which must be allowed for in design. Figure 6.8 shows the Young's modulus of a woven glass/vinyl ester laminate calculated from the resin data of Fig. 6.3, using this procedure. Measurements of creep moduli, also shown, confirm the accuracy.

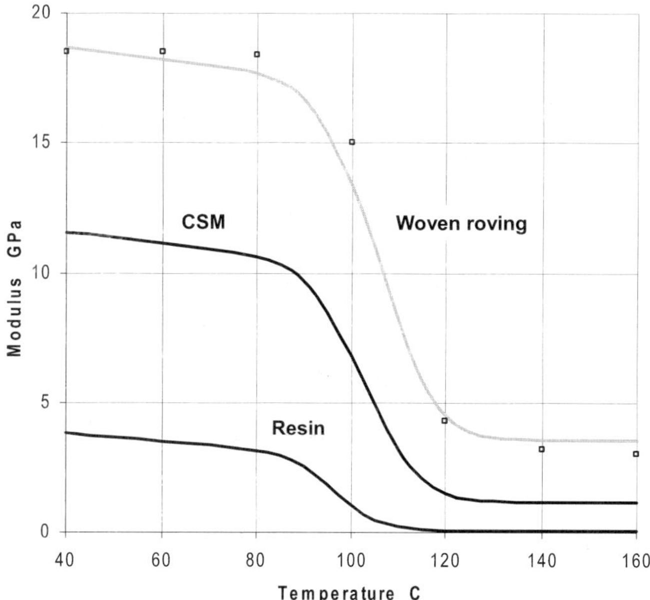


*Figure 6.8. Young's moduli of a woven glass/vinyl ester laminate calculated from the resin data of Fig. 6.3, also shown here. The experimental points are the results of measurements of 1 second creep moduli [41].*

### 6.5.4    QUASI-ISOTROPIC & RANDOM-IN-PLANE MAT REINFORCEMENTS

This section applies both to 'quasi-isotropic' laminates with continuous plies and to laminates where the strands have random orientation in the plane. The latter case would include chopped strand and 'swirl' mat laminates. In both these laminates the material is regarded as equivalent to a randomised array of idealised unidirectional tows or plies, so it is necessary to look first at the properties of these 'plies'.

In this case the unidirectional stiffness constants of each 'ply' can be modelled as before (Eqn. (6.25)). The transformation equations (Eqns. (6.27)) can then be used, working on the assumptions that equal numbers of plies or strands are oriented in every direction in-plane. Substituting the in-plane average values for $\left\langle \cos^4 \theta \right\rangle = \left\langle \sin^4 \theta \right\rangle = 3/8$ and $\left\langle \sin^2 \theta \cos^2 \theta \right\rangle = 1/8$ into Eqn. (6.27) gives:

$$Q_{11} = Q_{22} = \frac{3}{8}Q_{11,ply} + \frac{3}{8}Q_{22,ply} + \frac{1}{2}Q_{66,ply} + \frac{1}{4}Q_{12,ply}$$

$$Q_{12} = \frac{1}{8}Q_{11,ply} + \frac{1}{8}Q_{22,ply} - \frac{1}{2}Q_{66,ply} + \frac{3}{4}Q_{12,ply} \qquad (6.30)$$

$$Q_{66} = \frac{1}{8}Q_{11,ply} + \frac{1}{8}Q_{22,ply} + \frac{1}{2}Q_{66,ply} - \frac{1}{4}Q_{12,ply}$$

Finally, the engineering constants can be found from:

$$v_{12} = v_{21} = Q_{12}Q_{22}$$
$$E_{11} = E_{22} = Q_{11}\left(1 - v_{12}^2\right) \qquad (6.31)$$
$$G_{12} = Q_{66}$$

An alternative, less complex procedure involves the approximate relationships:

$$E_{11} = E_{22} = \frac{3}{8}E_{11,ply} + \frac{5}{8}E_{22,ply} \qquad (6.32)$$

and

$$G_{12} = \frac{1}{8}E_{11,ply} + \frac{1}{4}E_{22,ply} \qquad (6.33)$$

These give fairly similar predictions to Eqns. (6.30) & (6.31). The chopped strand mat data shown in Fig. 6.8 was calculated using the former procedure from the resin data and from one room temperature measurement of the Young's modulus.


### 6.5.5     HIGH TEMPERATURE STRENGTH VALUES

As with elastic properties there is a scarcity of high temperature strength data suitable for use in simulations of fire behaviour. It is desirable to obtain data at temperatures a little higher than the range normally studied for design purposes, even though decomposition may be beginning to take place. For tensile strength measurements this involves some difficulties because the temperatures are often beyond the range of testing ovens. There are also problems with grip slippage and failure in the grips due to the softening of the resin. One method of overcoming these problems is to use the heated gauge length set-up shown in Fig. 6.9. Here a conventional 'dog bone' sample is used for the tensile test. Instead of using a temperature-controlled enclosure, however, metal blocks with cartridge heaters are lightly clamped around the surface of the gauge length. The end tabs are then gripped in the conventional manner. This has the advantage of allowing the grip region of the sample to be kept cool, so as to avoid slippage or deformation. Additional insulation is generally required around the upper grip, to prevent it from being warmed by air convection from the hot blocks.

*Figure 6.9. Tensile test sample with heated gauge length to permit strength measurements at elevated temperature [40, 41].*

Figure 6.10 shows the results of tensile strength measurements up to 400°C on three types of woven fabric laminate: glass/vinyl ester, glass/polyester and glass/polypropylene. With continuous fibre laminates there is significant strength retention at elevated temperatures due to the strength of the reinforcement. The drop in strength due to the initial softening of the resin is greater than would have been anticipated on the basis of the resin contribution to strength. If resin softening were the only phenomenon occurring then a relatively small drop would be anticipated. The cause of this larger fall in strength is still under discussion, but a major element is probably the loss of the 'composite action' that ensures that the fibres are all subjected to the same strain level, despite minor waviness and misalignment. This ensures that the fibres in a composite below $T_g$ all tend fail at around the same strain. When the matrix ceases to make a contribution the effects of path differences, or differences in waviness may cause fibres to fail at different strain levels, with a consequent loss of effectiveness.

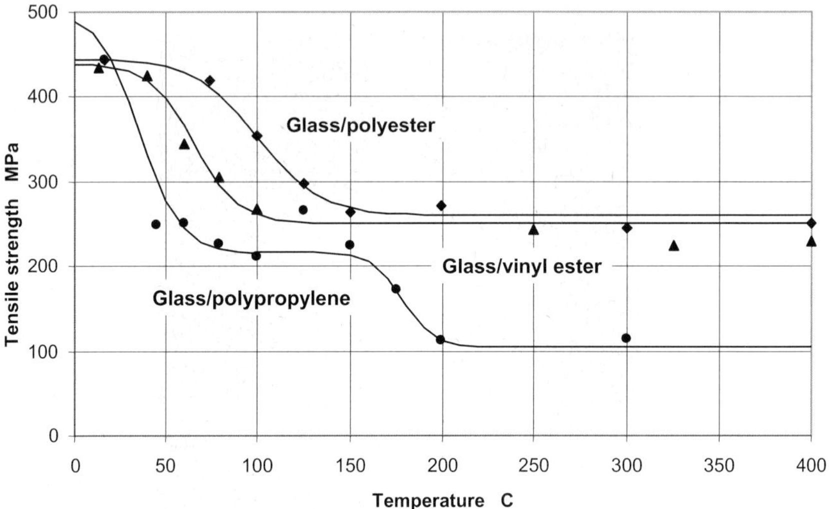


*Figure 6.10.   Elevated temperature tensile strength of glass/vinyl ester, glass/polyester and glass/polypropylene laminates [41].*

As with the elastic properties it is desirable with strength to be able to fit a functional relationship with temperature. The use of Eqn. (6.23) has been found useful in this respect and generally fits the strength data well, and it is used to fit the curves shown in Fig. 6.10. The exception is the glass/polypropylene laminate. Here the resin passes through not one, but two transitions; namely the glass transition (which is just below room temperature for polypropylene) and the crystalline transition, or $\alpha$-relaxation that occurs prior to melting. In the glass/polypropylene case, therefore, an additional relaxation process is added to the curve-fitting model, so that Eqn. (6.22) was generalised to include a further term:

$$P(T) = P_1 - \frac{P_1 - P_2}{2}\left(1 + tanh(k_1(T - T_1))\right) - \frac{P_2 - P_3}{2}\left(1 + tanh(k_2(T - T_2))\right) \qquad (6.34)$$

where the subscripts 1, 2 and 3 refer to the low, intermediate and high temperature states.  Mahieux and Reifsnieder [5,6] proposed a similar method of extending their model (Eqn. (6.19)) to describe the multiple transitions in thermoplastics.

Determination of compressive strength at elevated temperature also entails experimental difficulties.  To achieve this, the jig shown in Fig. 6.11 can be used.  This again involves a set of heated close-fitting blocks that fulfilled the dual role of anti-buckling jig and sample heater.

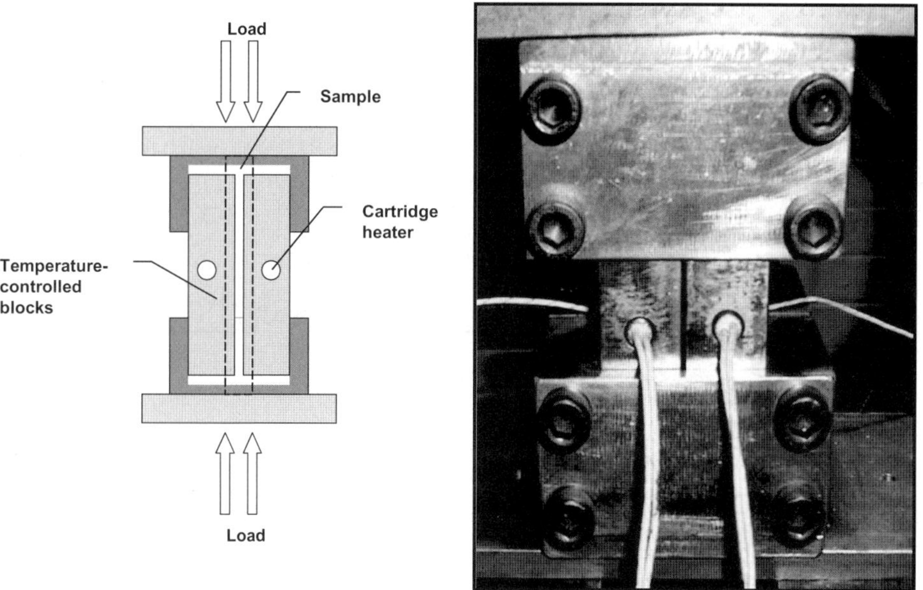


*Figure 6.11.   Combined heating and anti-buckling jig used for the measurement of compressive strength at elevated temperature. (a) schematic of jig and (b) assembly with heated blocks in place [40, 41].*

Figure 6.12 shows high temperature compressive strength results for several glass laminates.  The mechanism of compressive failure is very different from that in tension, and involves the formation of a band of kinked material, probably according to the mechanism described by Budiansky and Fleck [42].  Compressive failure is initiated via a shear deformation process in a region where the fibres are less than perfectly aligned in the plane of loading. In woven laminates there are plenty of examples of such regions due to the natural undulation of the weave. Such imperfectly aligned regions experience high levels of shear loading between fibres, that eventually triggers failure by local shear deformation. The compressive failure mode of the laminate therefore reflects the shear stress-strain behaviour of planes containing the fibres and can be modelled on this basis [32]. The key factor in compressive failure is the reduction in the shear properties of the resin that occurs around $T_g$ in the case of the thermosets, or the melting point, $T_m$, in the case of polypropylene.

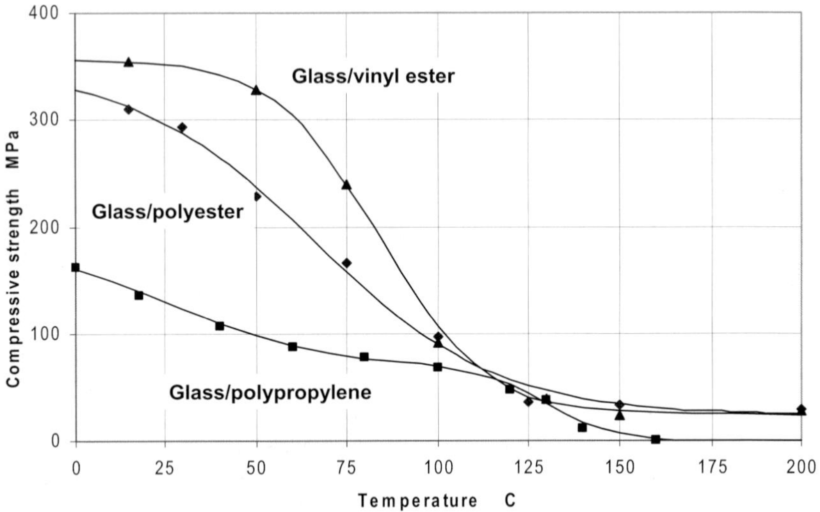


*Figure 6.12. Elevated temperature compressive strength of glass/vinyl ester, glass/polyester and glass/polypropylene laminates [41].*

While the room temperature compressive strength of the thermosets shown in Fig. 6.12 is fairly similar to the tensile strength, that of the glass/polypropylene samples is substantially lower. This reflects the lower intralaminar matrix shear strength of the thermoplastic system. The other key feature of the compressive behaviour is that it occurs at a generally lower temperature than the tensile failure, being essentially a resin-dependent property. The level of compressive strength retained above $T_g$ for the thermosets is quite low compared to the tensile case. With glass/polypropylene composites the compressive strength above the melting point of 165°C is effectively zero. Figure 6.13 compares the tensile and compressive strength of pultruded glass/polyester as an example of the greater sensitivity of compressive strength to temperature.

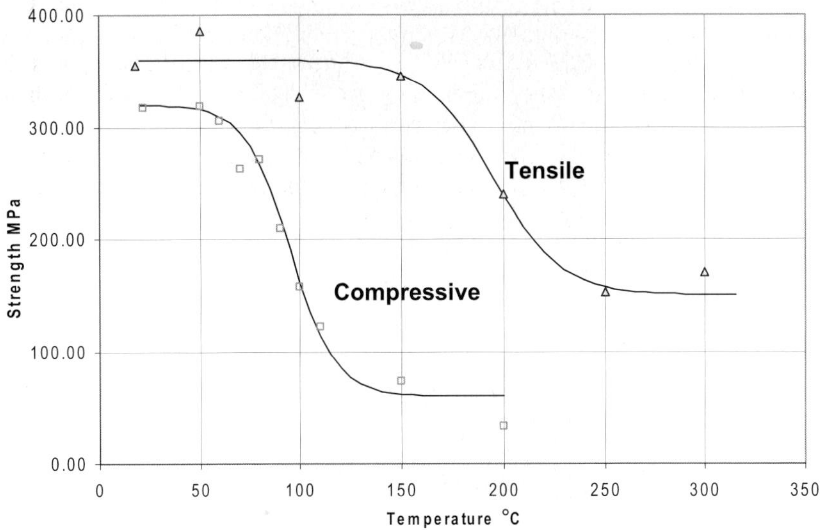


*Figure 6.13.   Elevated temperature tensile and compressive strength of glass/polyester pultrusions [40].
Material supplied courtesy of Fiberline Composites a.s.*

## 6.6     Fire Resistance of Laminates under Load

The simplest method of assessing the fire resistance of a composite system under load is to carry out small scale tests that are, in effect, stress-rupture tests [7,27-29,30,32-36]. Of course, for the qualification of full-scale composite structures much larger scale testing is generally needed. Nevertheless, small scale tests give very useful indications of the mode of failure and whether or not some form of fire protection will be needed. Small scale tests are most conveniently carried out under conditions of constant heat flux, rather than one of the standard time-temperature profiles, for reasons of scale, and because times-to-failure of unprotected samples are often quite short. Gas burners [7,33,35] as well as electrical radiant heaters [28-30,32-35] have been used as the thermal radiation source for this type of measurement. A propane burner has been used successfully for this purpose (Fig. 6.14), a procedure which involves low cost, but offers the possibility of constant heat flux values well in excess of 100 kW/m$^2$. Testing can be carried out with a conventional mechanical testing machine, albeit with some additional safety and fume extraction features. Figure 6.15 shows examples of the results obtained for woven glass laminates: it can be seen that the tensile load-bearing capability falls off rapidly under the effect of one-sided heat flux. Nevertheless, in the case of thermosetting materials containing continuous fibre reinforcement a reasonable residual level of strength is maintained.

*Figure 6.14. Propane burner test on a polymer laminate under tension [40, 41].*

For measurements under compressive load, some researchers have employed simple compression on panels or plates [28-30,32,34,36]. One significant experimental problem is to ensure that true compressive failure occurs, as opposed to Euler buckling. This requires samples of sufficiently low aspect ratio to be employed. One method of suppressing local buckling is to restrain the edges of the specimen, as shown in Fig. 6.16. This type of restraint is similar in principle to that in the well-known Boeing compression test [43]. The edge restraint raises the load required for buckling failure, enabling laminate specimens of reasonable size to be placed under significant compressive stress during fire testing.

Figure 6.17 shows compressive failure results for woven glass/thermoset laminates under a heat flux of 75 kW/m$^2$, obtained in constrained compression. The rate of decline under compressive load is similar to that for tension (Fig. 6.15). However, in contrast to the tensile case, the residual load-bearing capability at times longer than ~100 seconds is very low. Behaviour in compression is clearly the Achilles' heel of composite behaviour under load in fire.

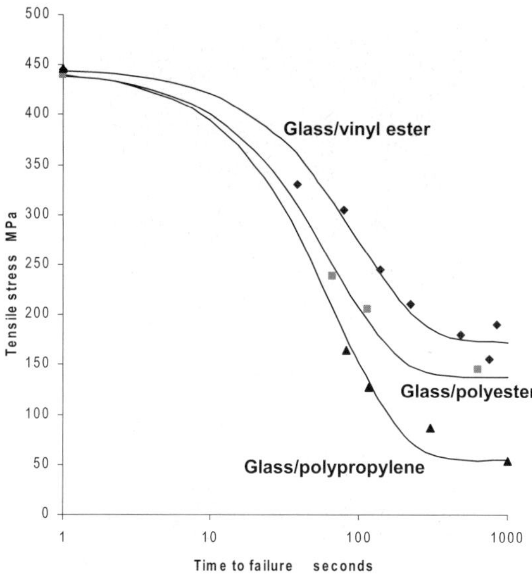



*Figure 6.15.  Fire test results under tensile load at a constant propane burner heat flux of 75 kW/m² on glass/vinyl ester, glass/polyester and glass/polypropylene composites [41].*

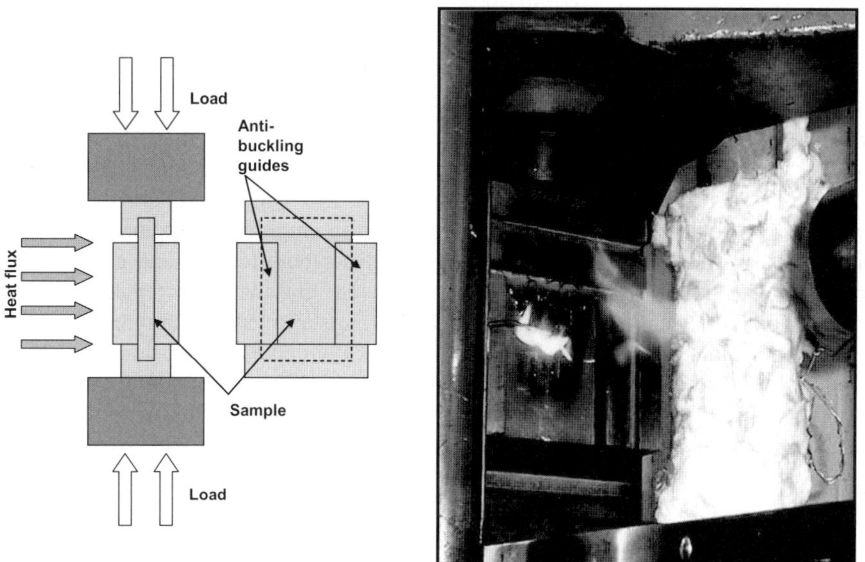



*Figure 6.16.  Restrained compression test. (a) Schematic of picture frame restraint jig  and (b) propane burner test on a polymer laminate under compression.*

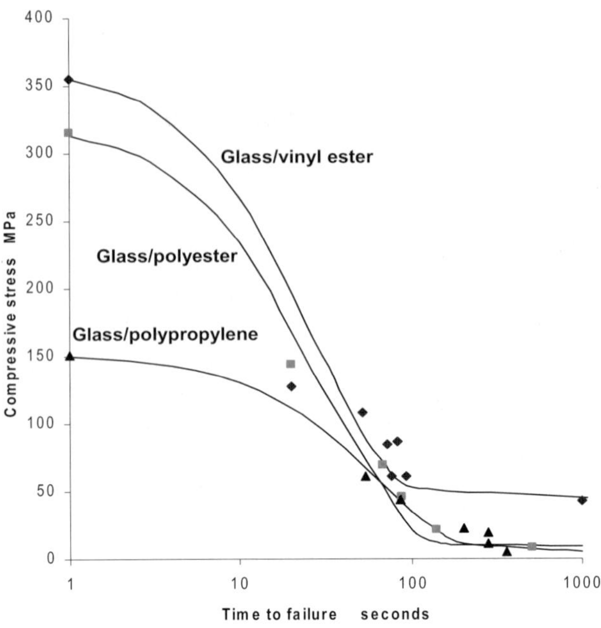


*Figure 6.17.   Fire test results under compressive load, at a constant propane burner heat flux of 75 kW/m$^2$ on glass/vinyl ester, glass/polyester and glass/polypropylene composites [41].*

Electrical radiant heat sources are a convenient and easily calibrated means of applying a known heat flux. Mouritz et al. [44] employed the radiant element from a cone calorimeter in the vertical configuration to heat vertical laminate specimens when under constant load. The benefit of this test is that the heat flux is easily adjusted and easily calibrated using the heat flux meter procedure normally employed in calibrating a cone calorimeter. Results are shown in Figs. 6.18 & 6.19 for both tension and compression covering a wide range of heat fluxes for woven glass/vinyl ester and woven glass/phenolic laminates. The normalised stress is the static tensile/compressive stress applied to the laminate when exposed to the heat flux divided by its tensile/compressive strength at room temperature. The failure times increase with a reduction in the applied stress and heat flux. The failure times of the phenolic laminate are the shortest, which is interesting considering its lower flammability. The less favourable performance of the phenolic laminate was attributed to heat-induced delamination and matrix cracking, which is more extensive than for the vinyl ester laminate. The phenolic matrix can contain a significant quantity of water, formed as the by-product of the cure reaction. This vaporises during heating, generating a sufficiently high internal pressure to produce delamination damage. Further water and other volatiles are produced as a result of ongoing postcure and ultimately resin decomposition.

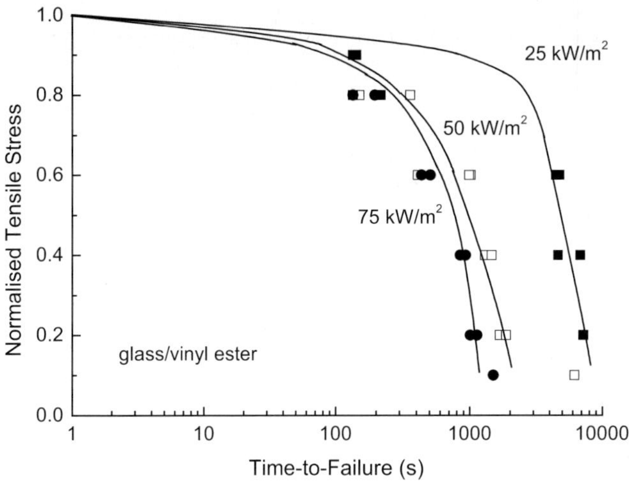



*(a)*

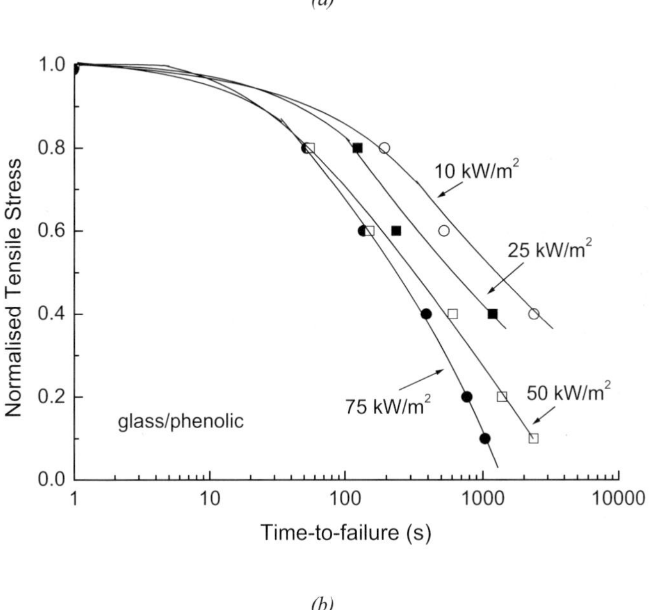



*(b)*

*Figure 6.18.   One-sided heat flux exposure results under tensile load at different heat fluxes on (a) glass/vinyl ester and (b) woven glass/phenolic laminates using a radiant heat source.*

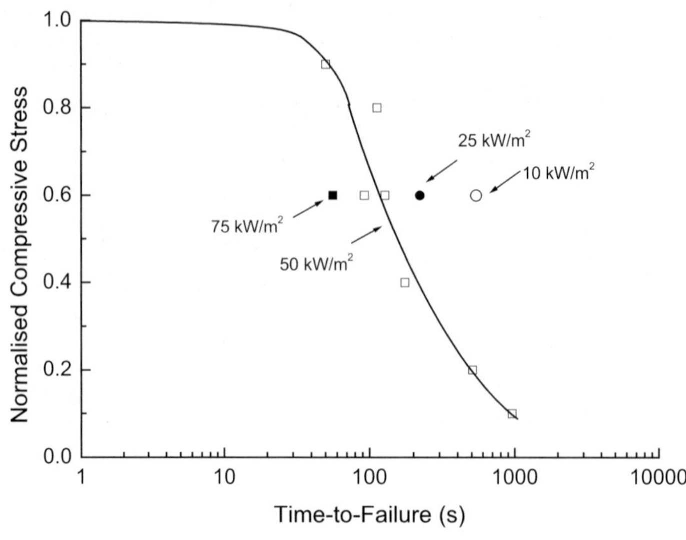



*(a)*

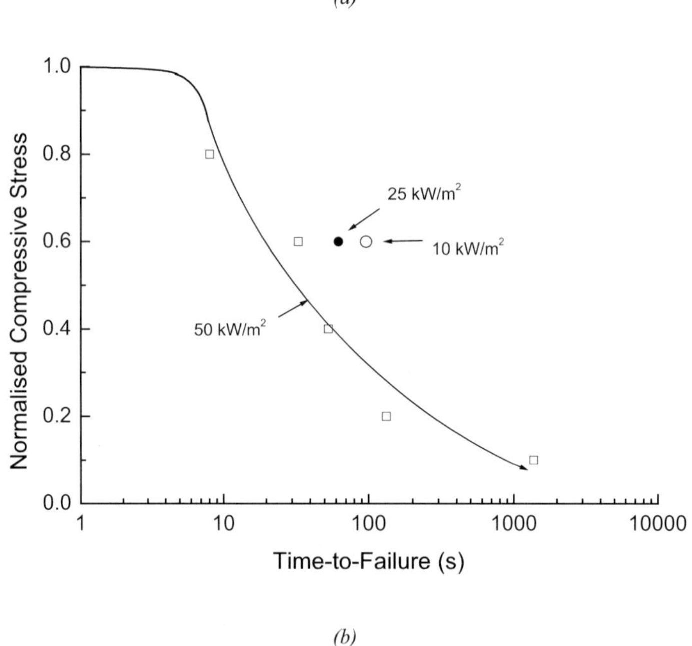



*(b)*

*Figure 6.19.   One-sided heat flux exposure results punder compressive  load at different heat fluxes on (a) glass/vinyl ester and (b) woven glass/phenolic laminates using a radiant heat source.*

Comparison of the tensile failure curves with those for compression loading underlines the significantly longer failure times for the tensile case. Figure 6.20 compares the failure times for a woven glass/vinyl ester laminate under tension and compression loading when exposed to the same heat flux (50 kW/m$^2$). The failure time for tension loading is about an order of magnitude longer than for compression loading. Failure of the laminate under tension loading is controlled by creep rupture of the fibers, whereas under compression the process is strongly influenced by thermal softening of the matrix, probably proceeding by a mechanism similar to that outlined for the isothermal case [42]. The loss in tensile strength of the glass fibers with increasing temperature is much more gradual than the loss in compressive strength of the polymer matrix, and this accounts for the laminate having longer failure times under tensile loading.

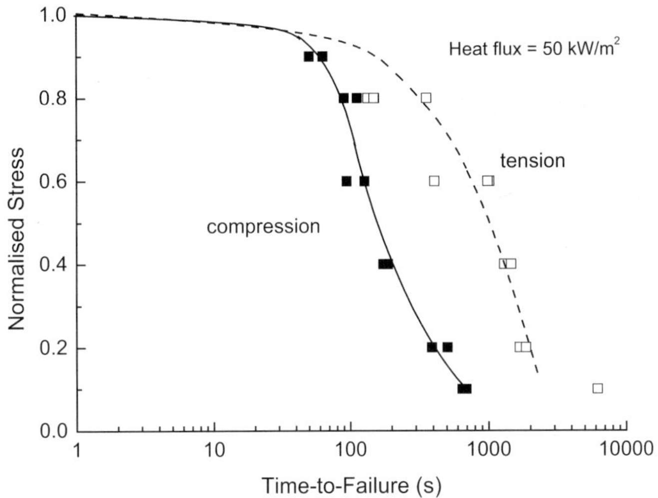


*Figure 6.20. Comparison of the failure times for a glass/vinyl ester laminate under tension or compression. Heat flux = 50 kW/m$^2$.*

Failure of continuous fibre laminates under tensile loading involves decomposition of the polymer matrix followed by tensile creep rupture of the fibres. Under compressive loading failure of laminates that are restrained against global buckling involves delamination and plastic kinking of the plies. Failure occurs by delamination cracking between the plies nearest the hot surface and plastic kinking of the tows away from the hot surface. Delamination cracks spread from the hot surface through the laminate with increasing time. The delaminated plies have little load-carrying capacity, so load is transferred to the cooler, delamination free area of the laminate, where failure occurs by plastic tow kinking. The low matrix stiffness, together with the out-of-plane displacement of the laminate away from the heat source, provides the conditions needed for a kink band to spread rapidly through the thickness. Figure 6.21 shows a photograph and schematic of a laminate at the time of failure, and it is apparent that failure occurred by rotation and plastic kinking of the tows. Plastic kinking occurs because of the low shear stiffness of the hot resin matrix, which allows the tows to rotate through a large angle from the load direction before ultimate failure and collapse.

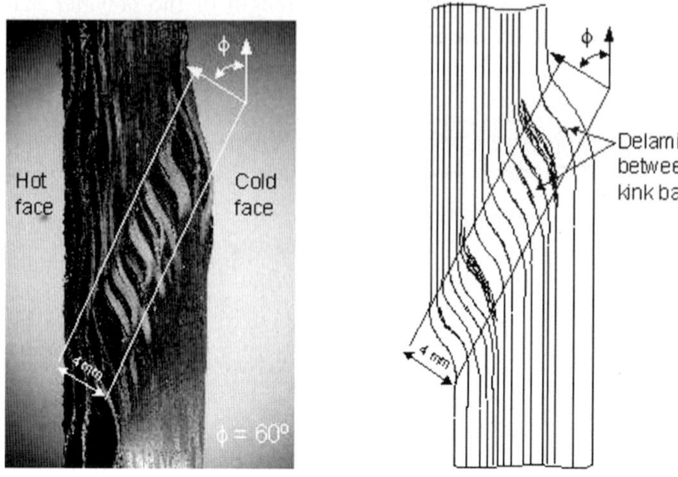


*Figure 6.21. Photograph and schematic of plastic tow kinking of a laminate under compressive loading.*

## 6.7    Modelling of Fire Resistance of Laminates Under Load

### 6.7.1   THERMAL EQUATIONS

In chapter 5 the modelling of heat flow through the thickness of a laminate was reviewed. The Henderson equation, a modified version of Laplace's equation [7,16-18,23,25,26,31,33,35], has been found to describe all the key features of this process, accounting, in particular, for the effects of resin decomposition. In the one-dimensional form, this relationship is:

$$\rho C_P \frac{\partial T}{\partial t} = \frac{\partial}{\partial x}\left(k\frac{\partial T}{\partial x}\right) - \rho\frac{\partial M}{\partial t}\left(Q_P + h_C - h_G\right) - \dot{M}_G\frac{\partial}{\partial x}h_G \qquad (6.35)$$

where $T$, $t$ and $x$ are temperature, time and through-thickness coordinates, respectively. $\rho$, $C_P$ and $k$ are the density, specific heat and thermal conductivity of the composite. $\dot{M}_G$ is the mass flux of volatiles. $h_C$ and $h_G$ are the enthalpies of the composite and evolved gas, respectively. $Q_P$ is the endothermic decomposition energy. The three terms on the right hand side relate respectively to heat conduction, the resin decomposition process (which is endothermic), and transport of heat towards the hot surface by volatile convection. It should be born in mind that the material property values in Equ. (6.35) are not generally constant, but evolve as functions of temperature and resin decomposition.

Thermal decomposition of the polymer matrix can often be approximated by a single reaction with Arrhenius temperature dependence:

$$\frac{\partial m}{\partial t} = -A\left[\frac{(m - m_f)}{m_o}\right]^n e^{(-E/RT)} \tag{6.36}$$

where $m$, $t$ and $T$ are the mass, time and temperature variables respectively; $A$, $E$ and $n$ are the rate constant, activation energy and order of the reaction, respectively; and $R$ is the gas constant. The majority of thermoset polymers and thermoplastic used as the matrix phase can be described in this way. Resins (such as phenolic) with a high aromatic content decompose in a more complex manner, leaving a higher proportion of carbonaceous char. In this case decomposition involves at least two stages: a primary condensation, sometimes with the evolution of water, followed at a higher temperature by char formation. Decomposition modelling then requires two successive rate processes, similar in form to Eqn. (6.36).

A finite difference version of the thermal model has been found to work well for different resin systems by comparing the calculated and measured thermal responses during furnace fire tests [21,25]. A finite element-based version of the model was also developed [23]. The one-dimensional nature of the thermal model is not a drawback for most applications involving laminated composites as these are generally shell-like in structure. However, if required, Eqn. (6.35) can be extended to include two- or even three-dimensional heat flow (see chapter 5 for further details). When this is carried out, a convenient approximation involves neglecting heat transfer due to the movement of volatiles, i.e. the last term in Eqn. (6.35). The effect of this term has been found to be small enough to allow it to be ignored in some cases [21].

## 6.7.2    LAMINATE CONSTITUTIVE RELATIONSHIP

Laminate theory describes the behaviour of a multi-ply composite [1-3]. Under isothermal conditions the applied forces and bending moments are related to the mid-plane strains and curvatures by:

$$\begin{bmatrix} \tilde{N} \\ \tilde{M} \end{bmatrix} = \begin{bmatrix} \tilde{A} & \tilde{B} \\ \tilde{B} & \tilde{D} \end{bmatrix} \begin{bmatrix} \tilde{\varepsilon}_0 \\ \tilde{k} \end{bmatrix} \tag{6.37}$$

where $\tilde{N}$ and $\tilde{M}$ are the matrices of normal loads and bending moments in the laminate coordinate system:

$$\tilde{N} = \begin{bmatrix} N_1 \\ N_2 \\ N_{12} \end{bmatrix} \quad and \quad \tilde{M} = \begin{bmatrix} M_1 \\ M_2 \\ M_{12} \end{bmatrix} \tag{6.38}$$

and $\tilde{\varepsilon}_0$ and $\tilde{k}$ are the mid-plane strains and curvatures:

$$\tilde{\varepsilon}_0 = \begin{bmatrix} \varepsilon_1 \\ \varepsilon_2 \\ \gamma_{12} \end{bmatrix} \quad and \quad \tilde{k} = \begin{bmatrix} k_1 \\ k_2 \\ k_{12} \end{bmatrix} \tag{6.39}$$

The $\tilde{A}$, $\tilde{B}$ and $\tilde{D}$ matrices can be defined as:

$$\tilde{A} = \sum_{k=1}^{n} \int_{h_{k-1}}^{h_k} \tilde{\bar{Q}}\, dz \qquad \tilde{B} = \sum_{k=1}^{n} \int_{h_{k-1}}^{h_k} \tilde{\bar{Q}} z\, dz \qquad \tilde{D} = \sum_{k=1}^{n} \int_{h_{k-1}}^{h_k} \tilde{\bar{Q}} z^2\, dz \tag{6.40}$$

where $\tilde{Q}$ is the matrix of ply stiffness constants transformed to the coordinate system of the laminate:

$$\tilde{Q} = \begin{bmatrix} \tilde{Q}_{11} & \tilde{Q}_{12} & \tilde{Q}_{16} \\ \tilde{Q}_{12} & \tilde{Q}_{22} & \tilde{Q}_{26} \\ \tilde{Q}_{16} & \tilde{Q}_{26} & \tilde{Q}_{66} \end{bmatrix} \tag{6.41}$$

These constants vary from ply to ply according to the orientation within each layer. In the present case they can also vary in the z-direction due to the effects of varying temperature and resin decomposition within the laminate: hence the integral in addition to the summation sign in equ.s (6.40).

For some problems it is convenient to employ a partially inverted form of Eqn. (6.37):

$$\begin{bmatrix} \tilde{\varepsilon}_0 \\ \tilde{M} \end{bmatrix} = \left[ \begin{array}{c|c} \tilde{A}* & \tilde{B}* \\ \hline \tilde{B}* & \tilde{D}* \end{array} \right] \begin{bmatrix} \tilde{N} \\ \tilde{k} \end{bmatrix} \tag{6.42}$$

This involves the following operations:

$$\tilde{A}* = \tilde{A}^{-1}$$
$$\tilde{B}* = -\tilde{A}^{-1}\tilde{B} = -\tilde{A}*\tilde{B}$$
$$\tilde{C}* = \tilde{B}\tilde{A}^{-1} = -\tilde{B}*^{T} \tag{6.43}$$

and
$$\tilde{D}* = \tilde{D} - \tilde{B}\tilde{A}^{-1}\tilde{B} = \tilde{D} - \tilde{C}*\tilde{B}$$

When the boundary conditions of the problem are expressed simply in terms of in-plane loads and moments it is often preferable to employ the fully inverted version:

$$\begin{bmatrix} \tilde{\varepsilon}_0 \\ \tilde{k} \end{bmatrix} = \left[ \begin{array}{c|c} \tilde{A}' & \tilde{B}' \\ \hline \tilde{B}' & \tilde{D}' \end{array} \right] \begin{bmatrix} \tilde{N} \\ \tilde{M} \end{bmatrix} \tag{6.44}$$

This involves the following further operations:

$$\tilde{D}' = \tilde{D}*^{-1}$$
$$\tilde{B}' = \tilde{B}*\tilde{D}*^{-1} \tag{6.45}$$
$$\tilde{C}' = -\tilde{D}*^{-1}\tilde{C}* = \tilde{B}'^{T} = \tilde{B}'$$

and
$$\tilde{A}' = \tilde{A}* - \tilde{B}*\tilde{D}*^{-1}\tilde{C}* = \tilde{A}* + \tilde{B}*\tilde{D}*^{-1}\tilde{B}*^{T}$$

## 6.7.3    THERMAL EXPANSION EFFECTS

The mechanical strain in a laminate under combined axial loading and one-sided heating is the sum of the stress-induced and thermally-induced deformations, so modifying Eqn. (6.37) to include these gives:

$$\begin{bmatrix} \tilde{N} \\ \tilde{M} \end{bmatrix} = \left[ \begin{array}{c|c} \tilde{A} & \tilde{B} \\ \hline \tilde{B} & \tilde{D} \end{array} \right] \begin{bmatrix} \tilde{\varepsilon}_0 - \tilde{\varepsilon}_0^{T} \\ \tilde{k} - \tilde{k}^{T} \end{bmatrix} \tag{6.46}$$

where $\tilde{\varepsilon}_0^{T}$ and $\tilde{k}^{T}$ are the thermally-induced distortions that occur in the absence of an externally applied load.

Alternatively, it is possible to analytically describe the matrices of thermally-induced loads and moments so that:

$$\tilde{F} = \sum_{k=1}^{n} \int_{h_{k-1}}^{h_k} \tilde{\overline{Q}} \tilde{\varepsilon}^T \, dz \quad \text{and} \quad \tilde{G} = \sum_{k=1}^{n} \int_{h_{k-1}}^{h_k} \tilde{\overline{Q}} \tilde{\varepsilon}^T z \, dz \tag{6.47}$$

where $\tilde{\varepsilon}^T$ is the matrix of thermal strains for each ply. Because the expansion coefficients may vary, both with temperature and $z$-coordinate, it is preferable to formulate these matrices in terms of thermal strain, rather than products of expansion coefficient and temperature change.

The thermal loads can be added to the existing loads and moments to give:

$$\begin{bmatrix} \tilde{N} \\ \tilde{M} \end{bmatrix} = \begin{bmatrix} \tilde{A} & \tilde{B} \\ \tilde{B} & \tilde{D} \end{bmatrix} \begin{bmatrix} \tilde{\varepsilon}_0 \\ \tilde{k} \end{bmatrix} + \begin{bmatrix} \tilde{F} \\ \tilde{G} \end{bmatrix} \tag{6.48}$$

Therefore:

$$\begin{bmatrix} \tilde{N} \\ \tilde{M} \end{bmatrix} - \begin{bmatrix} \tilde{F} \\ \tilde{G} \end{bmatrix} = \begin{bmatrix} \tilde{N} - \tilde{F} \\ \tilde{M} - \tilde{G} \end{bmatrix} = \begin{bmatrix} \tilde{A} & \tilde{B} \\ \tilde{B} & \tilde{D} \end{bmatrix} \begin{bmatrix} \tilde{\varepsilon}_0 \\ \tilde{k} \end{bmatrix} \tag{6.49}$$

This can be inverted, as discussed in relation to Eqn. (6.37), so the deformation with both thermal distortions and mechanical load present is given by:

$$\begin{bmatrix} \tilde{\varepsilon}_0 \\ \tilde{k} \end{bmatrix} = \begin{bmatrix} \tilde{A}' & \tilde{B}' \\ \tilde{B}' & \tilde{D}' \end{bmatrix} \begin{bmatrix} \tilde{N} - \tilde{F} \\ \tilde{M} - \tilde{G} \end{bmatrix} \tag{6.50}$$

The zero-load thermal distortions can be found by putting $\tilde{N}$ and $\tilde{M}$ equal to zero, so:

$$\begin{bmatrix} \tilde{\varepsilon}_0^T \\ \tilde{k}^T \end{bmatrix} = - \begin{bmatrix} \tilde{A}' & \tilde{B}' \\ \tilde{B}' & \tilde{D}' \end{bmatrix} \begin{bmatrix} \tilde{F} \\ \tilde{G} \end{bmatrix} \tag{6.51}$$

so, if required, Eqn. (6.46) may alternatively be expressed as:

$$\begin{bmatrix} \tilde{\varepsilon}_0 - \tilde{\varepsilon}^T \\ \tilde{k} - \tilde{k}_0^T \end{bmatrix} = \begin{bmatrix} \tilde{A'} & \tilde{B'} \\ \tilde{B'} & \tilde{D'} \end{bmatrix} \begin{bmatrix} \tilde{N} \\ \tilde{M} \end{bmatrix}$$

(6.52)

These relationships provide a basis for modelling thermal and fire-induced changes in laminates under stress. They can be used, for instance, to provide the input for a finite element analysis to model the behaviour of laminate shell structures under load.

The expansion coefficients of a unidirectional ply which are needed for the above calculations can be calculated from [2,9]:

$$\alpha_1 = \frac{1}{E_1}\left(\alpha_F E_F v_F + \alpha_M E_M v_M\right)$$

(6.53)

and

$$\alpha_2 = \left(1 + v_F\right)\alpha_F v_F + \left(1 + v_M\right)\alpha_M v_M - \alpha_1 v_{12}$$

(6.54)

In most practical cases, for instance those involving bending of flat laminates and composite products of relatively simple cross-section, it is possible to make significant simplifications to the above relationships. This is useful in modelling the fire response of plates and beams under load. Modelling of the effects of fire under load can be used to compare and validate experimental results and to perform sensitivity studies to determine which material parameters are most important in influencing behaviour.

In many cases, perhaps surprisingly, thermal expansion behaviour appears to be of relatively minor importance in its effect on mechanical response under load in fire. When a plate is loaded in uniaxial compression, for instance, and subjected to one-sided heat flux it is generally observed that failure involves some bending of the plate away from the heat source [36,44]. This appears to be true even when the overall failure mode is clearly compressive. This is counter-intuitive to thermal distortion considerations, which predict bowing of the sample towards the source of heat because of the thermal expansion of the hotter plies near to the heat source. Accurate measurements during fire exposure do indeed show that thermal bowing occurs towards the fire in the very early stages of the test. However, ply stiffness effects rapidly come to exceed those of thermal expansion, with the result that the bowing effect is soon reversed [36,44].

### 6.7.4   ORTHOTROPIC LAMINATES

Many laminates of practical interest possess orthotropic symmetry, allowing simplification because of the absence of shear/extensional coupling. For these the $\tilde{Q}$ matrix reduces to:

$$\tilde{Q} = \begin{bmatrix} \tilde{Q}_{11} & \tilde{Q}_{12} & 0 \\ \tilde{Q}_{12} & \tilde{Q}_{22} & 0 \\ 0 & 0 & \tilde{Q}_{66} \end{bmatrix} \tag{6.55}$$

Further simplification is possible for certain types of material, including some woven and multiaxial fabrics, which possess similar properties in the 1- and 2-directions, or which are designed as quasi-isotropic laminates. Laminates of this type show no in-plane shear distortions as a result of thermal strains, i.e. the thermal shear coefficient is zero. Provided there are no in-plane shear loads, that is often the case in simple bending situations, all in-plane shear effects can be neglected and the matrix rank reduced from 3 to 2.

### 6.7.5    MODELLING FIRE RESPONSE OF LAMINATES UNDER LOAD

Various mechanistic models have been developed to predict the fire response of composites under load [26,28-33,35,36,45,46]. The models include creep-based analysis [32], finite element analysis at the unit cell level [26], plasticity theory [45] and coupled fire-laminate analysis using computational fluid mechanics and structural analysis [46]. This section describes two modelling approaches that have been thoroughly evaluated using experimental fire data. The models are called the "two-layer model" for calculating the mechanical properties of laminates under combined tensile loading and one-sided heating and the "laminate model" for predicting the properties under tensile or compression loading and one-sided heating.

*Two-Layer Model*
A two-layer model has been successfully implemented to characterise the residual room temperature properties of laminates and sandwich composites after fire [28-30,33,35], as discussed in chapter 7. This model assumed a fire-damaged laminate to consist of two layers: an undamaged (virgin) layer having properties equal to those of the pristine material at room temperature, and another layer of severely damaged or 'char' material in which the mechanical properties are either negligible or a small fraction of those of the undamaged material. A schematic of the assumed damage state of a laminate using the two-layer model is given in Fig. 6.22. For simplicity, the tensile strength through the char layer is assumed to be constant. The tensile strength of the virgin layer is also assumed to be constant, and has the value of the tensile strength at room temperature[*]. In reality, however, the strength of the virgin layer is not constant, but is lowest at the char/virgin boundary and increases towards the unheated surface. In this respect, the two-layer model does not accurately represent the physical condition of a hot laminate exposed to fire. However, the use of this approach in conjunction with simple mechanics expressions for the overall properties of the two-layer structure is very

---

[*] This assumption is more accurate for the post-fire condition, where the properties of a virgin laminate are largely reversible and the polymer matrix can relax back into the room temperature state.

successful in providing an interpretation of residual properties after fire [28-30,33,35]. A full description of the model and the post-fire mechanical properties of composites is given in the next chapter.

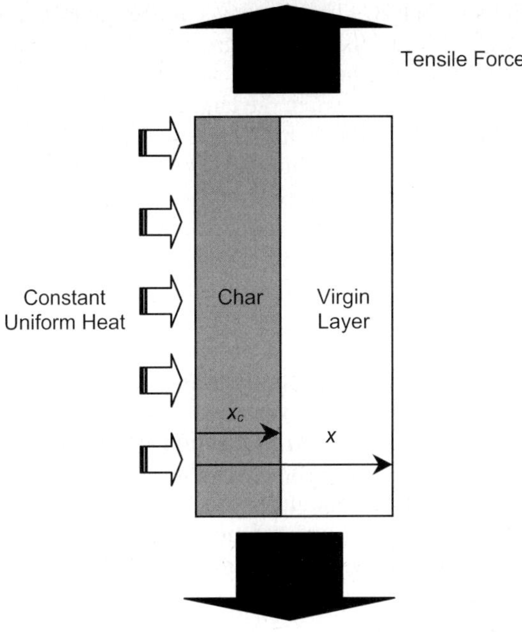


*Figure 6.22. Schematic of a laminate under axial tensile loading considered as a two-layer material consisting of char and virgin material.*

The relationship for calculating the residual tensile strength is simply:

$$\sigma_t = \left(\frac{x_o - x_c}{x_o}\right).\sigma_{t(o)} + \left(\frac{x_c}{x_o}\right).\sigma_{t(c)} \tag{6.56}$$

where $\sigma_{t(c)}$ is the tensile strength of the char layer, which in this analysis is assumed to be negligible based on experimental data [28]. $\sigma_{t(o)}$ is the original tensile strength of the laminate at room temperature. $x_o$ is the total thickness of the laminate. $x_c$ is the char thickness, which is calculated using Eqn. (6.35) and the temperature at which the polymer matrix decomposes to char. That is, $x_c$ is a location in the through-thickness direction at which the predicted temperature of the laminate reaches the temperature at which the matrix will decompose to char. The char temperature must be determined using TGA. When the residual tensile strength ($\sigma_t$) calculated using Eqn. (6.56) is reduced to the applied tensile stress, then the laminate is assumed to fail. The time taken for the residual strength to decrease to the applied stress is taken to be the time-to-failure.

Figure 6.23 compares the calculated and measured failure times of a woven glass/vinyl ester laminate at heat fluxes of 25, 50 and 75 kW/m$^2$. The failure times were calculated using the two-layer model assuming that the char region had no tensile strength when heated to the temperature that caused complete decomposition of the matrix, which for the vinyl ester matrix was ~440°C. The two-layer model gives a good estimate of the failure times at two highest heat fluxes. This agreement is encouraging considering the various assumptions and physical inaccuracies with the model. However, the model was unable to reliably predict failure when the laminate is exposed to the heat flux of 25 kW/m$^2$. At this heat flux the two-layer model predicted the laminate would have an infinite survival time because the heat flux was too low to completely decompose the polymer matrix to char and volatiles. In reality, the laminate failed after heating for a long time (above two hours) due to creep-induced rupture of the hot fibres. The model does not consider fibre creep, and therefore cannot be used to predict the failure times of laminates where long-term creep effects are important.

*Laminate Analysis Model*
This model provides a more accurate description than the two-layer model of the fire response of a composite structure. However, the laminate model requires a functional relationship for mechanical properties as a function of temperature, as described in section 6.4.1. The steps involved are shown in Fig. 6.24. The thermal model (Eqn. (6.34)), provides values of the temperature and residual resin content at all points through the depth of the laminate, as a function of time. This, along with the constitutive relationship for each ply, provides the input to the laminate analysis.

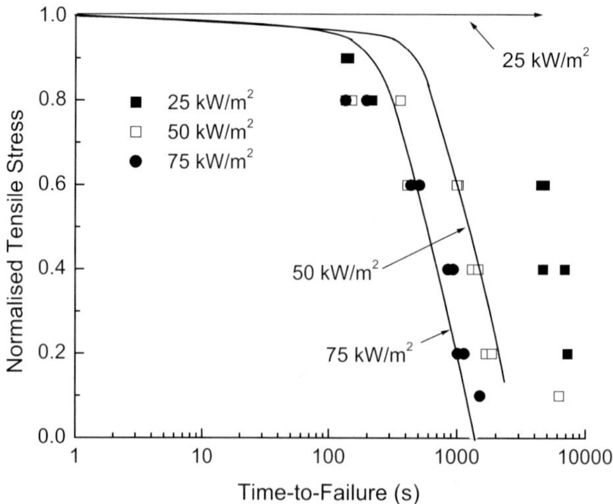


*Figure 6.23. Comparison of the calculated and measured times for a woven glass/vinyl ester laminate under tensile loading and one-sided heat flux. The curves were calculated using the two-layer model.*

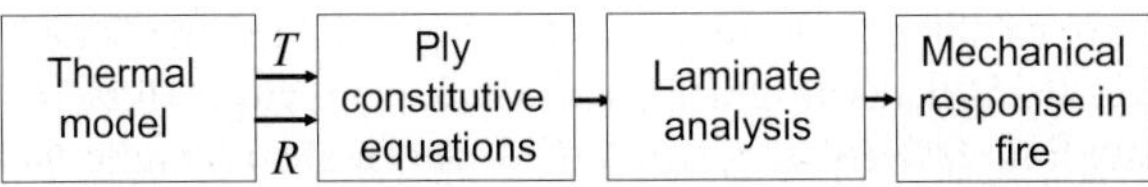


*Figure 6.24. Steps involved in laminate analysis model of a loaded composite structure in fire.*

Figure 6.25 shows an example of the evolution of the *A*, *B* and *D* matrix components of a laminate with time using the laminate model. This case is for a woven glass/vinyl ester laminate exposed to a one-sided heat flux of 75 kW/m$^2$. The *A* matrix components, which relate in-plane loads and deformations, decline rapidly with time over a period of about 200 seconds, reflecting the overall decline in the values of the elastic constants (Fig. 6.25a).  By contrast, the *B* matrix, which describes the interaction between in-plane loads and out-of-plane bending and twisting, is zero initially, rising to a maximum, before eventually declining over the same period (Fig. 6.25b).  The initial zero value of the *B* matrix reflects the symmetry of the properties about the central axis of a laminate with a symmetric lay-up.  The first effect of the heat flux is to produce an asymmetry of properties about the central axis: hence the increase in *B* matrix values.  The final fall reflects the ultimate decline in all the elastic constant values.  Finally the *D* matrix, which governs bending resistance, declines with time (Fig. 6.25c).  The influence of the developing asymmetry can be seen, however, from the shoulder on the *D* matrix curve, which coincides in time with the maximum in the *B* matrix.  The laminate Hooke's law components in Fig. 6.25 can be used to give a straightforward prediction of the laminate response in simple loading cases, such as in-plane stress or bending.  Alternatively, the *ABD* matrix data can form the input for a finite element analysis of structures with more complex geometry or loading.

Figures 6.26 and 6.27 show the evolution of the compressive and tensile strength of a laminate calculated using the laminate model compared with experimental values.  In the compressive case, the samples were held in a constrained compression jig, as shown in Fig. 6.16.  This allowed a significant compressive stress to be applied to the plate without Euler buckling. Predictions are shown for two modes of failure: compressive failure and local buckling. The local buckling load for a constrained plate of this type is given by an expression of the form [2, 47]:

$$\sigma_{Buckling} = \frac{c}{tb^2}\sqrt{D_{11}D_{22}} \tag{6.57}$$

where $\sigma_{Buckling}$ is the applied compressive stress at which local buckling occurs, $t$ is the plate thickness, and $b$ is the dimension of the square unsupported region. The constant, $c$, depends on the restraint conditions around the plate. Figure 6.26 shows the predicted loads and failure times are close to the measured values.  The predictions for material failure and local buckling are fairly similar; indeed they intersect at one point.

The shape of the predicted curve for buckling reflects the shape of the $D$ matrix evolution, in Fig. 6.25(c), as might be expected from Eqn. (6.57).  It is interesting to observe that buckling failure is predicted to dominate at higher stresses and shorter failure times, whereas material failure will dominate at lower stresses. Figure 6.27 shows the model predictions and experimental results for samples loaded in tension. Once again it can be seen that the predictions and experimental results agree well.

The results discussed in this chapter show that it is possible to achieve good agreement between model predictions and test data on composite under load in fire.  One effect that has not yet been taken into account is viscoelastic and viscoplastic time dependence of composites at elevated temperatures.  We believe the omission of these effects is probably justified in the current models because the path followed by the material in the thermal domain is much more critical than that followed in time.  The time periods covered here range from the order of 1 second to 1000 seconds.  Our justification in omitting time-dependence at this stage is one of simplification of computation requirements. Using isochronous creep data based on 1000 seconds, rather than on 1 second makes relatively little difference to the predictions reported here.  The approach presented here will be no doubt be extended to include full time-dependent analyses, of both the linear and non-linear kind, based on the recommendations of Schapery [9]. Indeed the foundations for doing this have already been established by Boyd et al. [45,46].

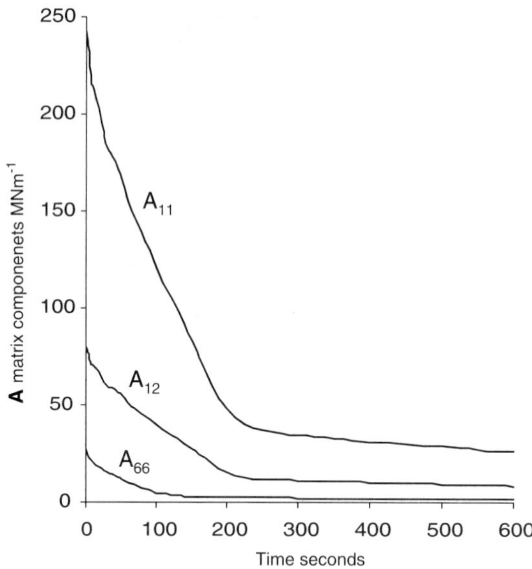


*(a)*

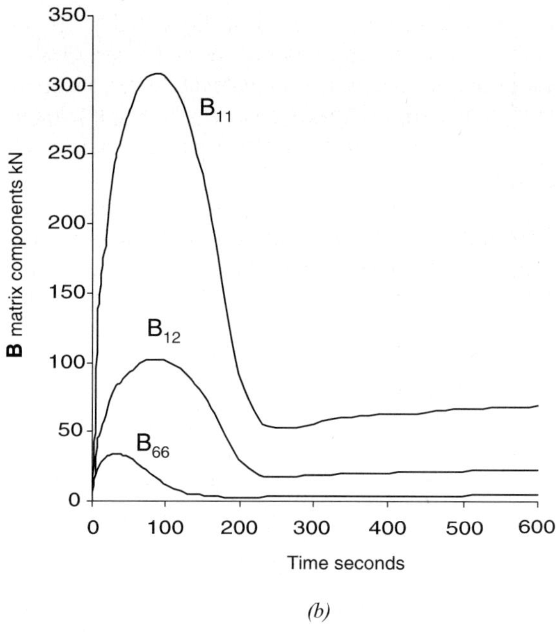


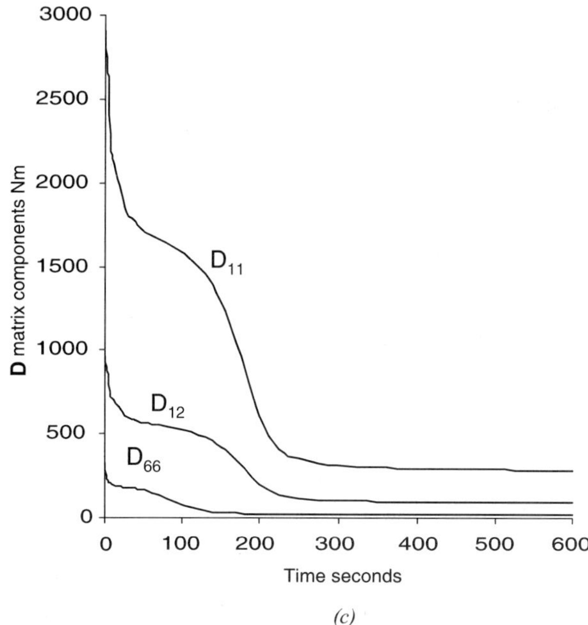


*Figure 6.25. Laminate model for a glass/vinyl ester laminate: predicted variation of the components of the laminate **A**, **B** and **D** matrices with time for a heat flux of 75 kW/m². (a) **A**, (b) **B** and (c) **D** matrix components.*

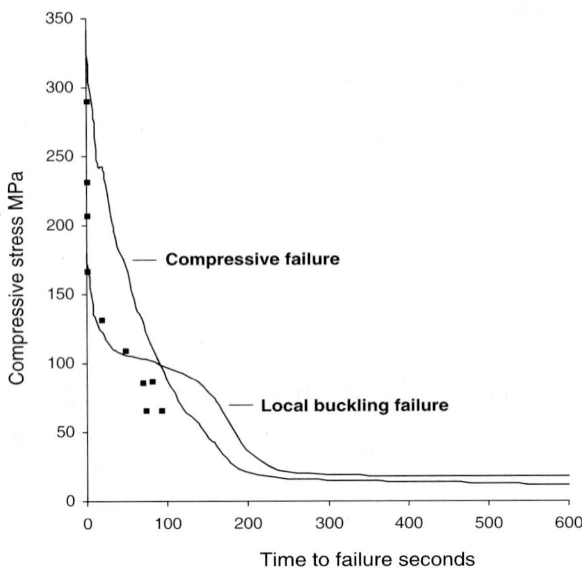


*Figure 6.26. Model predictions for the relationship between time-to-failure and applied compressive stress for a heat flux of 75 kW/m². (12 mm thick glass/vinyl ester laminate). Predictions are shown for both constrained buckling failure and compressive failure, along with experimental points.*

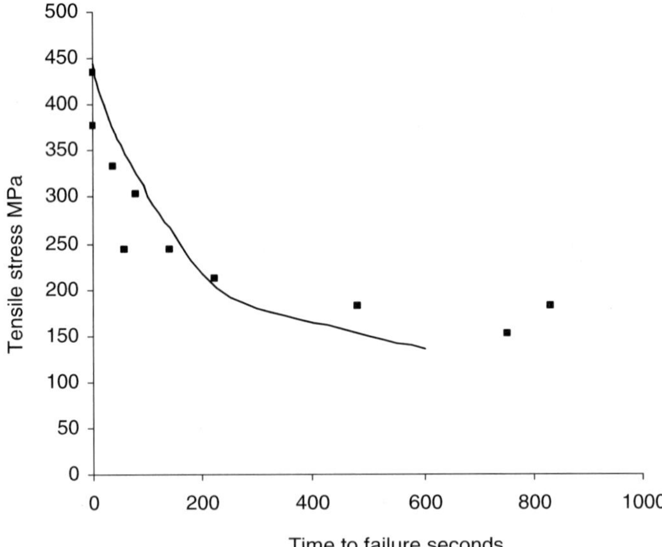


*Figure 6.27. Model prediction for the relationship between time-to-failure and applied tensile stress along with experimental points for a heat flux of 75 kW/m². (12 mm thick glass/vinyl ester laminate).*

## 6.8    Concluding Remarks

This chapter has reviewed the engineering basis for high temperature measurements of both elastic properties and the strength of composite laminates.  It has also discussed methods of carrying out small-scale tests under compressive or tensile load, subject to heat flux.  The basic requirements have been discussed for modelling behaviour under load in fire and it has been demonstrated that meaningful simulations can be achieved.

## References


1.    B.D. Agarwal and L.J. Broutman. *Analysis and Performance of Fiber Composites*, New York: John Wiley and Sons, 1990.

2.    G. Eckold. *Design and Manufacture of Composite Structures*, Cambridge: Woodhead Publishing Ltd., 1994.

3.    R.M. Jones. *Mechanics of Composite Materials*, Washington, D.C.: Scripta Book Company, 1994.

4.    A.P. Kulcarni and R.F. Gibson. Non-destructive characterisation of effects of temperature and moisture on elastic moduli of vinyl ester resin and E-glass/vinyl ester resin composite. In: *Proceedings of the American Society of Composites, 18th Annual Technical Conference*. Florida, 19th-22nd October 2003.

5.    C.A. Mahieux and K.L. Reifsnider. Property modeling across transition temperatures in polymer matrix composites: Part 1: Tensile properties. *Polymer*, 2001; 42:3281-3290.

6.    C.A. Mahieux and K.L. Reifsnider.  Property modeling across transition temperatures in polymers: Application to thermoplastic systems. *Journal of Materials Science*, 2002; 37:911-920.

7.    A.G. Gibson, Y.-S. Wu, J.T. Evans and A.P. Mouritz. Laminate theory analysis of composites under load in fire, *Journal of Composite Materials*, 2006, in press.

8.    N.G. McCrum, B.E. Read and G. Williams. *Anelastic and Dielectric Effects in Polymeric Solids*, London:Wiley, 1967.

9.    R.A. Schapery, Thermal expansion coefficients of composite materials based on energy principles. *Journal of Composite Materials* 1968; 2:380-404.

10.    J.J. Massot. Glass reinforced plastics heavy load flooring for offshore platforms, In: *Composite Materials in the Offshore Industry, Rueil-Malmaison, France, 3-4 November 1994*, Institut Français du Pétrole. 9.

11.    E. Greene, Fire performance of composite materials for naval applications, US Navy Contract N61533-91-C-0017, Structural Composites Inc., Melbourne FL USA, 1993.

12.    C.H. Bamford, J. Crank and D.H. Malan. The combustion of wood: Part I. *Proceedings of the Cambridge Philosophical Society,* 1946; 42:166.

13.    M.C. Adams, *American Rocket Society*, 1959; 29.

14.    H.C. Kung. A mathematical model of wood pyrolysis, *Combustion & Flame*, 1972; 18:185-195.

15.    E.J. Kansa, H.E. Perlee and R.F. Chaiken. Mathematical model of wood pyrolysis including internal forced convection, *Combustion & Flame*, 1977; 29: 311-324.

16.    J.B. Henderson, J.A. Wiebelt and M.R. Tant. A model for the thermal response of polymer composite materials with experimental verification. *Journal of Composite Materials*, 1985; 19:579-595.

17.    J.B. Henderson and T.E. Wiecek. A mathematical model to predict the thermal response of decomposing, expanding polymer composites. *Journal of Composite Materials*, 1987; 21:373-393.

18.    C.A. Griffis, J.A. Nemes, F.R. Stonesifer and C.I. Chang. Degradation in strength of laminated composites subjected to intense heating and mechanical loading. *Journal of Composite Materials*, 1986; 20: 216-235.

19.    J. Milke and A.J. Vizzini. Thermal response of fire-exposed composites. *Journal of Composites Technology & Research*, 1991; 13:145-151.

20.    U. Sorathia, C. Beck and T. Dapp. Residual strength of composites during and after fire. *Journal of Fire Science*, 1993; 11: 255-270.

21.     A.G. Gibson, Y-S. Wu, H.W. Chandler, J.A.D. Wilcox and P. Bettess. A model for the thermal performance of thick composite laminates in hydrocarbon fires. Revue de L'Institut Francais du Petrole, 1995; 50:69-74.

22.     U. Sorathia, R. Lyon, R. Gann and L. Gritzo, Materials and fire threat. *SAMPE Journal* 1996;32: 8-15.

23.     M.R.E. Looyeh, P. Bettess and A.G. Gibson. A one-dimensional finite element simulation for the fire-performance of GRP panels for offshore structures. International *Journal of Numerical Methods for Heat & Fluid Flow*, 1997; 7:609-625.

24.     M. Dao and R.J. Asaro. A study on failure prediction and design criteria for fiber composites under fire degradation, *Composites*, 1999; 30A:123-131.

25.     N. Dodds, A.G. Gibson, D. Dewhurst and J.M. Davies. Fire behaviour of composite laminates. *Composites*, 2000; 31A:689-702.

26.     J. Lua and J.O. O'Brien. Fire simulation for woven fabric composites with temperature and mass dependent thermal-mechanical properties. In: *Proceedings of Composites in Fire – 3*, University of Newcastle upon Tyne, 9-10 September 2003 (ed. A.G. Gibson), Paper 13, CompositeLink.

27.     G.A. Pering, P.V. Farrell and G.S. Springer. Degradation of tensile and shear properties of composites exposed to fire or high temperature. *Journal of Composite Materials*, 1989; 14:54-66.

28.     A.P. Mouritz and Z. Mathys. Post-fire mechanical properties of marine polymer composites. *Composite Structures*, 1999; 47:643-653.

29.     C.P. Gardiner, Z. Mathys and A.P. Mouritz. Tensile and compressive properties of GRP composites with localised heat damage. *Applied Composite Materials*, 2002; 9:353-367.

30.     A.P. Mouritz and Z. Mathys. Post-fire mechanical properties of glass-reinforced polyester composites. *Composite Science & Technology*, 2001; 61:475-490.

31.     D. Anilturk and C.S. Chan. *Journal of Composite Materials*, 2003; 37:687-700.

32.     J. Bausano, S. Boyd, J. Lesko and S. Case. Composite life under sustained compression and one sided simulated fire exposure: characterisation and prediction'. In: *Proceedings of the Third Conference on Composites in Fire*. Newcastle, UK, 9-10 September 2003. (ed. A.G. Gibson), Paper 19, CompositeLink.

33.     A.G. Gibson, P.N.H. Wright, Y-S. Wu, A.P.Mouritz, Z. Mathys and C.P. Gardiner. Modelling the residual mechanical properties of polymer composites after fire. *Plastics, Rubbers & Composites*, 2003; 32:81-90.

34.     P.G.B. Seggewiss. Properties of fire-damaged polymer matrix composites. In: *Proceedings of the Third Conference on Composites in Fire*. Newcastle, UK, 9-10 September 2003.

35.     A.G. Gibson, P.N.H. Wright, Y.-S. Wu, A.P. Mouritz, Z. Mathys and C.P. Gardiner. The integrity of polymer composites during and after fire, *Journal of Composite Materials*, 2004; 38:1283-1308.

36.     L. Lui, J.W. Holmes, G.A. Kardomateas and V. Birman. Compressive response of composites under combined fire and compressive loading. In: *Proceedings of Composites in Fire – 4, 15-16 September 2005*, (ed. A.G. Gibson) University of Newcastle upon Tyne, CompositeLink.

37.     S. Toll. *Micromechanics of Materials*, Gothenburg: Chalmers University, 1999.

38.     R.L. Coble and W.D. Kingery. Effect of porosity on the physical properties of sintered alumina. *Journal of the American Ceramic Society*, 1956; 39:381.

39.     A.M. Robinson, P.R. Novoa, A. Torres Marques and A.G. Gibson, *Dynamic mechanical behaviour of polymer composites*, to be published.

40.     R.C. Easby, *Fire Behaviour of Pultruded Composites*, Ph.D. thesis, University of Newcastle upon Tyne, 2006.

41.     T.N.A. Browne, *Fire Behaviour of Composite Laminates*, Ph.D. thesis, University of Newcastle upon Tyne, 2006.

42.     B. Budiansky and N.A. Fleck. Compressive failure of fiber composites. *Journal of the Mechanics & Physics of Solids*, 1993; 41:183-211.

43.     Boeing Specification Support Standard, Advanced Composite Compression Tests, BSS 7260, 1986.

44.     A.P. Mouritz, S. Feih, Z. Mathys and A.G. Gibson. Mechanical property degradation of naval composite materials. In: *Modeling of Naval Composite Structures in Fire*, ed. L. Couchman and A.P. Mouritz, Acclaim Printing, Melbourne, 2006.

45.     R.J. Asaro, P. Krysl and W.T. Ramroth. Structural fire integrity of FRP composites: analysis and design. In: *Modeling of Naval Composite Structures in Fire*, ed. L. Couchman and A.P. Mouritz, Acclaim Printing, Melbourne, 2006.

43.    C. Luo, W. Xie and P.E. DesJardin. Fluid-structure simulations of composite material response for fire environments.  In: *Modeling of Naval Composite Structures in Fire*, ed. L. Couchman and A.P. Mouritz, Acclaim Printing, Melbourne, 2006.

44.    W.C. Young, ed., Roark's Formulas for Stress and Strain, 6th Edition, New York: McGraw-Hill, 1989.

45.    S.E. Boyd, J.J. Lesko, S.W. Case and J.V. Bausano. The viscoelastic/viscoplastic characterisation of glass-vinyl ester composites under fire conditions, In: *Proceedings of Composites in Fire – 4, 15-16 September 2005*, (ed. A.G. Gibson) University of Newcastle upon Tyne, CompositeLink.

46.    S.E. Boyd, J.V. Bausano  J.J. Lesko and S.W. Case. Mechanistic approach to structural modelling of composites. In: *Modeling of Naval Composite Structures in Fire*, ed. L. Couchman and A.P. Mouritz, Acclaim Printing, Melbourne, 2006.

# Chapter 7

# Post-Fire Properties of Composites

## 7.1    Introduction

The fire reaction properties and thermal degradation mechanisms of polymer composites have been subjects of in-depth characterisation and analysis for many years because of the need for fire-safe materials.  The growing use of composites in aircraft, ships, civil construction and other applications has required knowledge of their reaction properties such as heat release rate, smoke production and toxic gas emission to ensure safety in the event of fire.  Furthermore, the thermal decomposition behaviour of organic resins, fibres and core materials that determine the fire reaction properties of composites are generally well understood.

Less well known are the fire resistant properties of composites, which are critically important when the materials are used in load-bearing structures. Deterioration of the mechanical properties of composites in a fire can seriously compromise structural integrity, and cause rapid creeping, buckling, collapse or some other failure.  Most of the effort to characterise the fire resistant properties of composites has centred on the deterioration of their load-bearing properties at elevated temperature or in fire (as described in chapters 4 & 6).  The residual mechanical properties of thermally degraded composites following fire are not as well understood.  After a fire is extinguished, it is important to know the residual properties of a burnt composite at room temperature in order to determine the mechanical integrity and safety of a fire-damaged structure. Progress towards characterising the post-fire properties has been slow, and there is much that is not known. With the use of composites in highly loaded structures on aircraft and ships and their increasing application in automobiles, bridges and offshore platforms, it is essential that a large database of post-fire mechanical properties is established and models for predicting the residual properties are available.

The residual mechanical properties of thermally degraded composite materials following fire are examined in this chapter. Empirical studies into the effects of heat flux and heating time of fire on the residual properties of laminates and sandwich composites at room temperature are described. Models are presented for predicting the mechanical properties of composites after fire. The chapter also examines the effectiveness of thermal barrier materials and flame retardant resins for minimising the impact of fire on the retained properties of composites. Gaps in our understanding of the post-fire mechanical properties of laminates and sandwich composites are identified, and topics requiring further research and modelling are proposed.


## 7.2      Post-Fire Properties of Laminates

### 7.2.1    POST-FIRE PROPERTIES OF THERMOSET LAMINATES

The post-fire mechanical properties of laminates with a thermoset polymer matrix have been determined for a variety of fire conditions ranging from low heat flux, short duration fires through to high temperature, long burning fires. The residual properties of the most popular laminates have been characterised, including carbon/epoxy and glass/phenolic composites used in aircraft and glass/polyester and glass/vinyl ester materials used in ships and civil infrastructure [1-22]. The post-fire properties of the less common thermoset laminates (eg. cyanate esters, bismaleimides, polyimides) are not as well known.

The post-fire mechanical properties of thermoset laminates have been investigated since the early 1980s, but as yet a standard approach for characterising the properties has not been devised and this is hindering progress in the field. The fire reaction properties of composite materials are measured using standardised test procedures to ensure reliable and reproducible data (as discussed later in chapter 11). For example, the mass loss, heat release rate, smoke density and gas emissions are measured using standard techniques such as the cone calorimeter under well-controlled heating conditions. Unfortunately, the same situation does not exist for measuring the post-fire mechanical properties. A standard fire test method for characterising the residual properties of laminates following fire has not been established. Most studies of the post-fire properties use the radiant heater of a cone calorimeter to burn laminate specimens [2-17,19,20]. The advantage of this test is that the heat flux and heating conditions are well-controlled and repeatable. However, the cone calorimeter can only burn small specimens and exposure to a cone heater represents an idealised fire condition. For example, the heating rate and final temperature are stable and there is no convective heat transfer from the heat source to the specimen. Other methods used to characterise the post-fire properties include gas burner, gas furnace and fuel fire tests [1,10-12,18]. This diversity makes it difficult to compare the post-fire properties of a composite tested using different fire methods. Despite this problem, the existing test methods have been used effectively to provide useful insights into the post-fire properties of thermoset laminates.

Sorathia and colleagues [2-4] have assessed the post-fire properties of a variety of thin thermoset laminates. The laminates were exposed to an incident heat flux of 25 kW/m$^2$ for twenty minutes in a cone calorimeter, and then their residual flexural strength was measured at room temperature. The retained flexural strengths are shown in Fig. 7.1, and all the laminates suffered a large loss in strength due to damage caused by the fire.

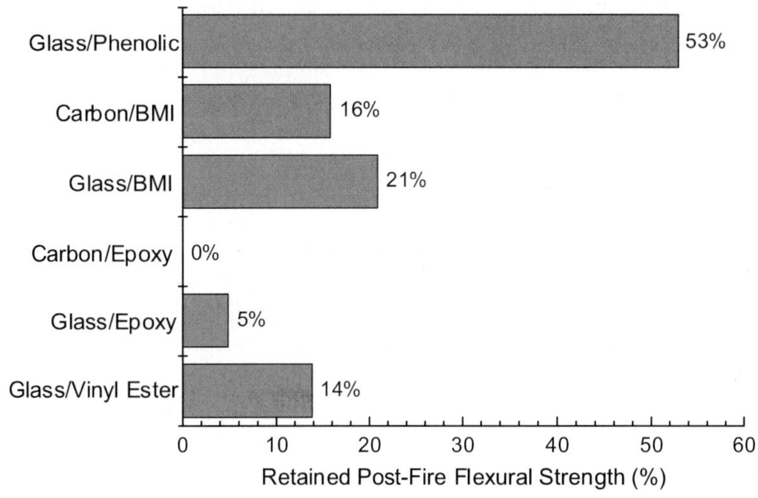


*Figure 7.1. Retained flexural strengths of various thermoset laminates following fire testing at the heat flux of 25 kW/m$^2$ for twenty minutes. Data from Sorathia [2].*

The lowest post-fire strengths given in Fig. 7.1 are for the epoxy matrix laminates, which is a concern because of their use in highly-loaded structures on aircraft, high performance marine craft and bridges. The effects of temperature and duration of a fire on the residual mechanical properties of epoxy composites have been studied by Pering and colleagues [1], Mouritz [13,19,20] and Seggewiss [21,22]. A disturbing feature is that the post-fire properties can diminish rapidly with increasing temperature or heating time of a fire due to the high flammability of the epoxy matrix. For example, Fig. 7.2 shows a large reduction to the post-fire tensile strength of a carbon/epoxy laminate following exposure to medium-to-high temperature gas fires (540 to 980°C) over a short time. Pering et al. [1] attribute this reduction to the rapid thermal decomposition of the epoxy matrix. Similarly, Mouritz [13,19,20] measured large reductions to the post-fire properties of carbon/epoxy, glass/epoxy and aramid/epoxy composites fire, and again rapid degradation of the epoxy matrix was the main cause for the sharp drop in the properties.

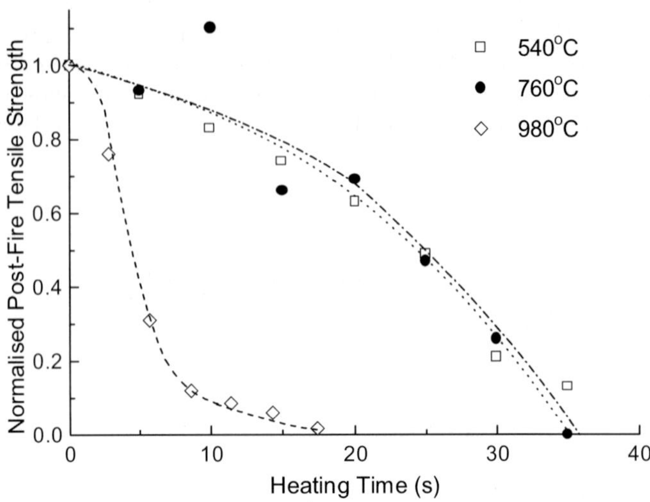


*Figure 7.2. Effects of temperature and time of a gas fire on the post-fire tensile strength of a 16-ply (1.9 mm thick) carbon/epoxy laminate. Data from Pering et al. [1].*

Large, rapid reductions to the post-fire tension, compression, flexure and interlaminar shear properties occur with increasing heat flux or duration of a fire. For example, Fig. 7.3 shows the large drop to the tensile strength of various thermoset laminates following increasing exposure time to a radiant heat source. The deterioration to the properties is due mostly to thermal decomposition of the polymer matrix. However, the reduced stiffness and strength of aramid and other organic fibre composites can also be caused by fibre decomposition in a fire.

Figure 7.1 shows that the phenolic matrix composite had the highest post-fire strength of the thermoset laminates examined by Sorathia et al. [3,4]. Despite the relatively high post-fire strength, the loss is surprising considering the excellent flame resistance and low fire reaction properties of phenolic laminates. The effect of fire on the residual mechanical properties of fire-resistant phenolic composites has recently been investigated for a range of fire conditions by Mouritz and colleagues [7,8,13,17,19,20]. It is found that the mechanical properties of phenolic composites can be severely degraded following a high temperature fire, and the magnitude of the loss is often similar to that suffered by more flammable thermoset laminates. For example, Fig. 7.4 shows the effects of increasing heat flux and heating time on the normalised residual flexural strength of a glass/phenolic composite. For comparison, also shown is the normalised residual strength of a highly flammable glass/vinyl ester laminate, which is representative of the post-fire behaviour of many thermoset composites. The glass/vinyl ester laminate had a similar resin content and was the same thickness as the phenolic laminate. The post-fire flexural strength of the phenolic laminate drops rapidly, and the reduction is similar to that experienced by the more flammable vinyl ester composite. Mouritz and Mathys [8] noted that the post-fire properties are severely degraded before the phenolic laminate had ignited. It is therefore apparent that phenolic laminates can suffer a substantial reduction to their post-fire mechanical properties despite the low flammability and excellent fire reaction properties of the resin matrix. It is important to note, however, that the loss in stiffness and strength following a fire is less than that experienced during fire.

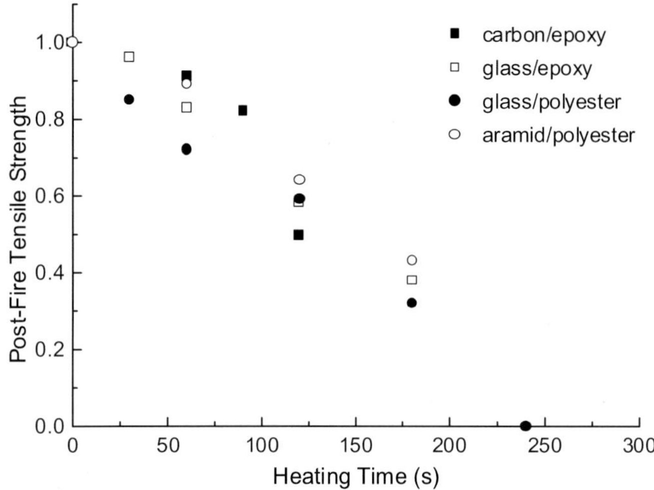


*Figure 7.3. Reductions to the tensile strength of various thermoset laminates after being exposed to a heat flux of 50 kW/m$^2$ for increasing time. The post-fire strengths have been normalised to the original strengths of the laminates. The fire tests were performed using the heating element to a cone calorimeter.*

The reduction to the post-fire mechanical properties of thermoset laminates is due to thermal degradation and damage caused by fire, as shown schematically in Fig. 7.5 and discussed in detail in chapter 2. Fire damage extends below the hot surface in zones termed the char region, decomposition region, and virgin region. The char produced by decomposition of the polymer matrix is the major cause for the reduced post-fire properties. Figure 7.6 shows the typical appearance of char produced by a epoxy matrix laminate that yields a high level of volatiles in fire. The char microstructure has a high void content and only a small amount of solid char. The amount of retained char is determined by the temperature, exposure time and type of resin, with less than ~10% of the original weight of polymers such as polyesters, vinyl esters and many epoxies reduced to char whereas aromatic-based resins such as phenolics and phthalonitriles yield substantially higher amounts of char.

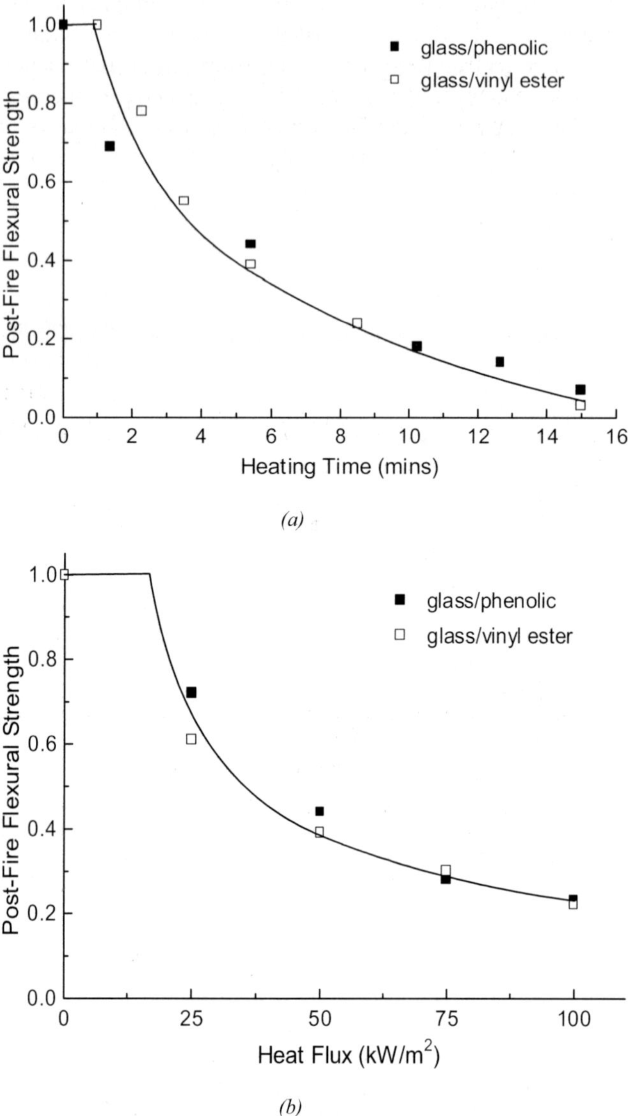



*Figure 7.4. Comparison of the post-fire flexural strengths of glass/phenolic and glass/polyester laminates after being exposed to a fire for increasing (a) time and (b) heat flux. The post-fire strengths have been normalised to the original strengths of the laminates. Data from Mouritz and Mathys [7].*

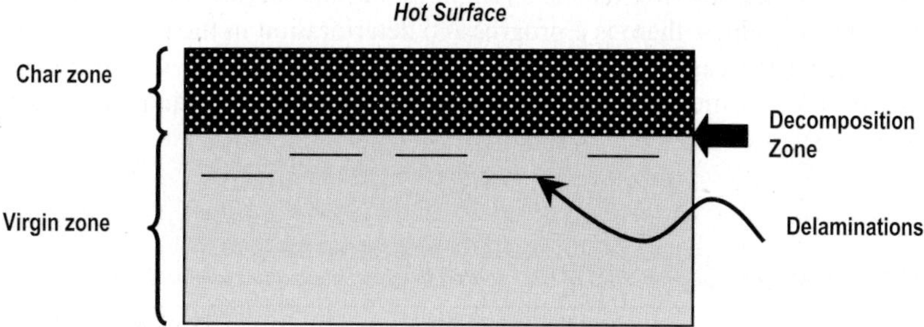


*Figure 7.5. Schematic showing the distribution of fire damage through a thermoset polymer laminate.*

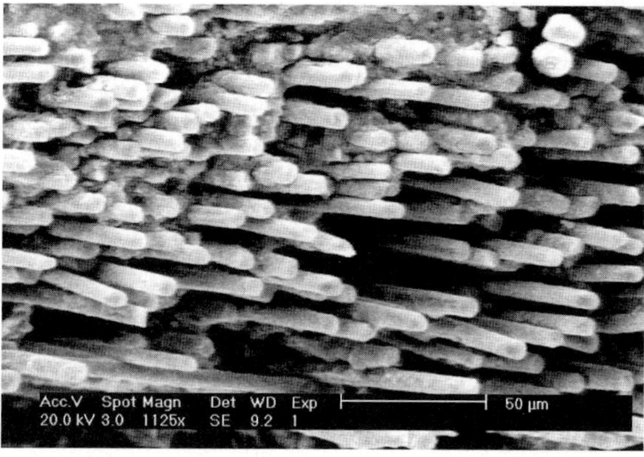


*Figure 7.6. Microstructure of the char in a thermoset laminate following a fire.*

The mechanical properties of char are very low, as shown in Table 7.1 that gives the modulus and strength values of two thermoset laminates after being completely reduced to char in a fire.  One of the laminates is glass/polyester, which yields little char during the thermal decomposition of the polymer matrix. The second material is glass/phenolic, which yields a substantially higher amount of char.  The char properties are low for both laminates for the different loading conditions; typically less than 10-20% of the original properties of a composite [7-9]. The strength in both flexure and compression is particularly low, which is expected as these are more dependant on load transfer in the matrix.

The size of the char region increases rapidly with the temperature and heating time of a fire, and therefore the post-fire properties of thermoset laminates are quickly degraded. Numerous studies have reported correlations between the thickness of the char layer and the residual mechanical properties [1,7,9,13]. Figure 7.7 shows the relationship between char thickness and the post-fire tensile strength of various thermoset laminates. Despite some scatter in the values, there is a progressive deterioration in the post-fire strength as the amount of char increases. An increase in the char thickness can also be correlated to reductions to the elastic modulus and compressive, flexural and shear strengths [1,7,9,13].

*Table 7.1. Mechanical properties of char (normalised to the original laminate properties).*

|  | Glass/Polyester Char | | Glass/Phenolic Char | |
|---|---|---|---|---|
|  | Modulus | Strength | Modulus | Strength |
| Tensile Properties | 0.09 | 0.17 | 0.21 | 0.098 |
| Flexural Properties | 0 | 0 | 0.27 | 0.073 |
| Compression Properties | 0 | 0 | - | - |

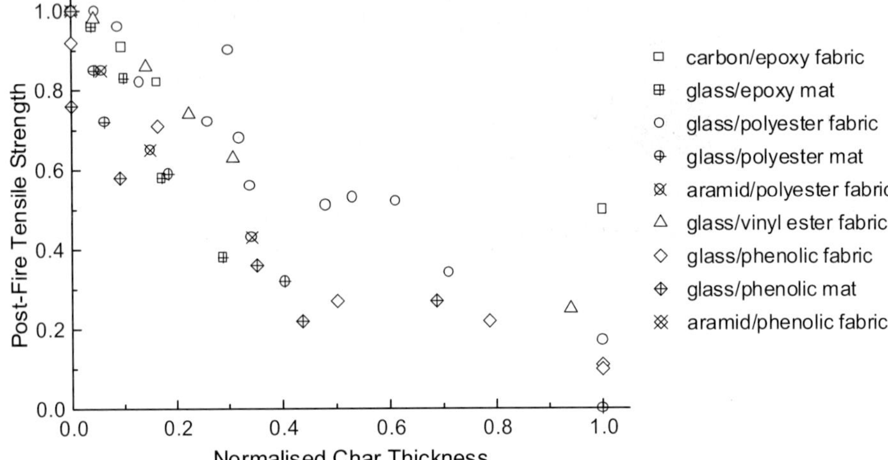



*Figure 7.7. Relationship between normalised post-fire tensile strengths and normalised char thickness for thermoset laminates. The normalised char thickness is the thickness of the char divided by the original thickness of the laminate. The normalised tensile strength is the post-fire strength divided by the original strength of the laminate.*

The decomposition region also contributes to the low post-fire properties of laminates. Thermoset resins inside the zone between the char region and virgin material experience some thermal decomposition in a fire that usually involves the loss of small volatile molecules and a reduction to the cross-linking density. In some cases more severe degradation can occur, including chain scission. The microstructure of degraded resin in the decomposition region to a thermoset laminate is shown in Fig. 7.8.   The resin matrix is heavily cracked, indicating that the post-fire properties of the interfacial zone are severely degraded.  Because the interfacial region is usually thin (<1-2 mm), it can be difficult to reliably measure the mechanical properties in this region. Kucner and McManus [24] assessed the post-fire properties of the decomposition zone in a carbon/epoxy laminate by microhardness testing following fire.  The change in post-fire hardness from the decomposition zone towards the cold face of the laminate is shown in Fig. 7.9.   The cold face location is at x = 0.  It is seen that the hardness near the decomposition region is only about 20% of the original hardness of the carbon/epoxy, and the hardness increases with distance away from this region into the virgin material.

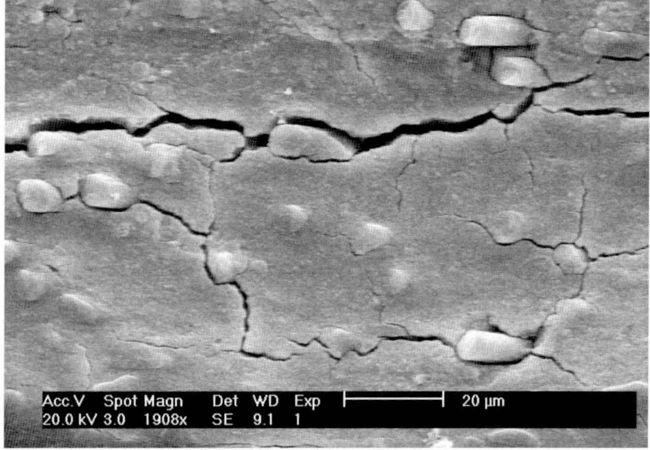



*Figure 7.8. Microstructure of the decomposition region in a thermoset laminate following a fire. The cracking is indicative of thermal degradation of the polymer matrix.*

The third type of fire damage often experienced by thermoset laminates is delamination cracking between the plies within the virgin region (Fig. 7.10). This can be attributed to the high internal pressure from the formation of volatile gases released by resin decomposition reactions and the vapourisation of moisture.  The steep thermal strain gradient due to the non-uniform temperature distribution through a composite can contribute to the delamination process.

Fire damage is consistently found to have a greater influence on the post-fire compressive properties compared to the tensile properties.  For example, Fig. 7.11 compares the post-fire tensile and compressive strengths of a glass/polyester laminate after being exposed to a constant heat flux for different times up to fifteen minutes.  The compressive strength is more dependent on the role of the matrix and hence is more sensitive to fire damage than the tensile properties. The presence of delaminations is also a contributing factor as buckling of delaminated plies can occur prematurely under compressive loading.

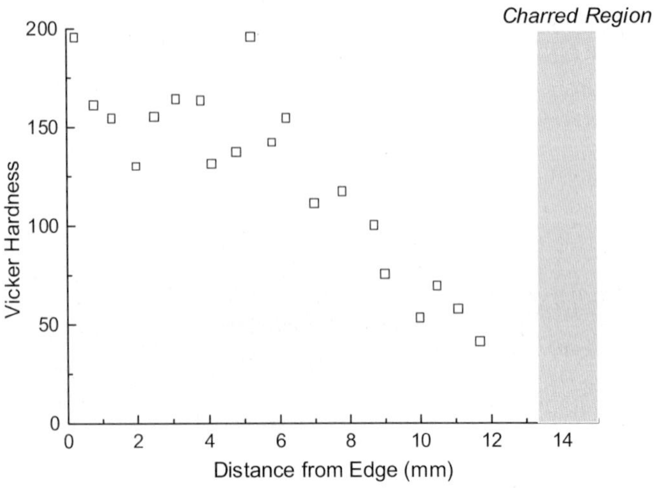


Figure 7.9. Profile of the Vickers hardness of a carbon/epoxy laminate with increasing distance from the char layer. Reproduced from Kucner and McManus [23].

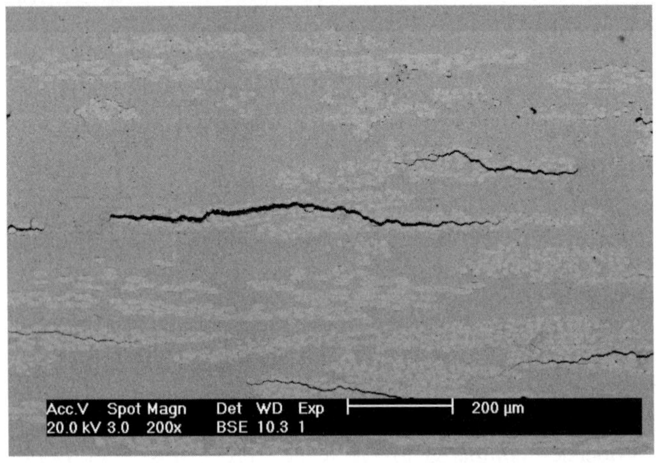


Figure 7.10. Delaminations in a fire-damaged thermoset laminate.

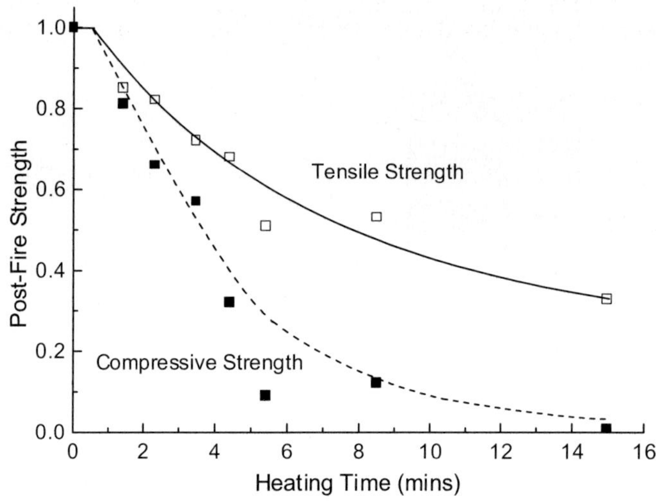


*Figure 7.11. Comparison of the post-fire tension and compression strengths of a glass/polyester laminate following exposure to a heat flux of 50 kW/m$^2$ for increasing times.*

## 7.2.2    POST-FIRE PROPERTIES OF THERMOPLASTIC LAMINATES

While the post-fire mechanical properties of thermoset laminates have been examined in detail, the residual properties of thermoplastic composites are not well understood. The post-fire flexural strengths of several thermoplastic composites after being exposed to a heat flux of 25 kW/m$^2$ for twenty minutes are shown in Fig. 7.12 [2-4]. Shown for comparison are the flexural strengths of two common thermoset laminates exposed to the same fire condition. The thermoplastic laminates are able to retain much higher strength than the thermoset composites for this fire condition. The only thermoset laminates with comparable post-fire strength to these thermoplastic composites are materials with aromatic-based thermoset resins such as phenolics or phthalonitrile.

The reason for the superior post-fire strength of thermoplastic composites has not been determined, although it is probably because these materials generally have higher decomposition temperatures, yield high amounts of char, and are less susceptible to delamination cracking than thermoset laminates. In addition, it is expected that melting and resolidification of the thermoplastic matrix will affect the post-fire properties by changing the degree of resin crystallinity.

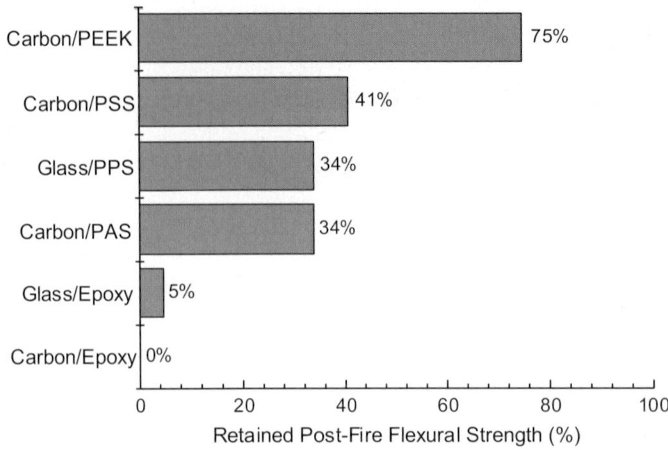


*Figure 7.12. Retained flexural strengths of various thermoplastic laminates following fire testing at the heat flux of 25 kW/m² for twenty minutes. Data from Sorathia [2].*

## 7.3    Modelling the Post-Fire Properties of Laminates

Empirical relationships between the post-fire mechanical properties of laminates and the amount of fire damage have been known for many years. Pering et al. [1] found in the early 1980s that the post-fire tensile properties of carbon/epoxy laminates correlated with their mass loss due to thermal decomposition while the post-fire shear properties could be empirically related to the thickness of the char zone. Although, it is only recently that mechanics-based models have been proposed for calculating the residual properties of composites following fire. Mouritz and Mathys [7,9] developed a 'two-layer' model to predict the post-fire tension, compression and flexure properties of laminates that had been exposed to uniform one-sided heating in a fire. The basis of the model is that a fire-damaged laminate is assumed to consist of two regions:
(i)      char region where the resin matrix is thermally degraded, and
(ii)     virgin (or unburnt) region.
These are illustrated in Fig. 7.13 with a char layer of constant thickness ($d_c$).

The model has several simplifying assumptions about the nature of the fire-induced damage and the effect of this damage on the post-fire properties. For example, delaminations that occur close to the boundary between the char and virgin regions are ignored. It is also assumed that the decomposition zone is much thinner than the char and virgin regions, and therefore its influence on the post-fire properties is ignored. The only fire damage that is considered to affect the post-fire properties is the char.

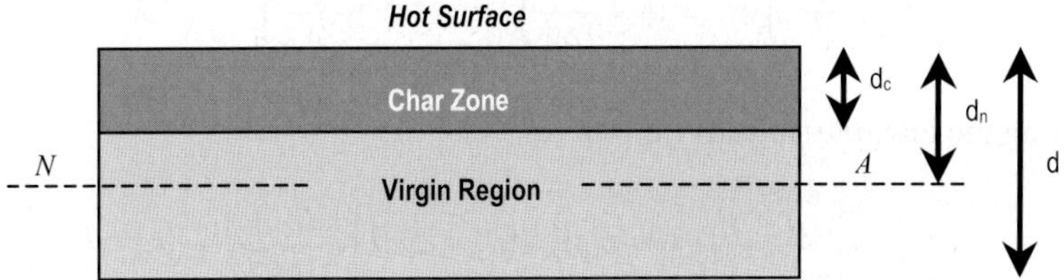



*Figure 7.13. Diagram showing the idealised damage profile through a laminate following a fire. NA refers to the neutral axis shown by the dashed line.*

A fire-damaged laminate is modelled as a two-layer material consisting of char and undamaged composite with each layer having different mechanical properties. The following equations were therefore proposed by Mouritz and Mathys [7,9] for the post-fire tensile modulus ($E_t$) and compressive modulus ($E_c$):

$$E_t = \left(\frac{d-d_c}{d}\right)E_{t(o)} + \left(\frac{d_c}{d}\right)E_{t(char)} \tag{7.1}$$

$$E_c = \left(\frac{d-d_c}{d}\right)E_{c(o)} + \left(\frac{d_c}{d}\right)E_{c(char)} \tag{7.2}$$

where $d$ is the original thickness of the laminate and $E_{t(o)}$ and $E_{c(o)}$ are the tensile and compressive moduli of the virgin (undamaged) region, which are assumed to be the same as the original moduli of the laminate. $E_{t(char)}$ and $E_{c(char)}$ are the tensile and compressive moduli of the char, which must be empirically determined. Typical char moduli values are given in Table 7.1. $d_c$ is the thickness of the char region, which can be measured experimentally or calculated using thermal models such as those proposed by Kucner and McManus [23] and Gibson et al.[24].

Mouritz and Mathys [7,9] also proposed an equation to estimate the post-fire flexural modulus based on Euler beam theory. When a slender isotropic beam is subjected to a bending stress then the bending moment, $M$, is determined by:

$$M = \int \frac{Ey^2}{\rho}bdy \tag{7.3}$$

where $E$ is the flexural modulus, $b$ is the breadth of the beam, $y$ is a depth coordinate and $\rho$ is the radius of curvature in bending. Expanding Eqn. 7.3 to represent a cross-section with fire damage as illustrated in Fig. 7.13 gives:

$$M = \frac{E_o b}{\rho} \int_0^{d-d_n} y^2 \, dy + \frac{E_o b}{\rho} \int_0^{d_n-d_c} y^2 \, dy + \frac{E_c b}{\rho} \int_{d_n-d_c}^{d_n} y^2 \, dy \qquad (7.4)$$

Completing the integration leads to:

$$\frac{\rho}{b} = \frac{E_o}{3M}\left[(d-d_n)^3 + (d_n - d_c)^3\right] + \frac{E_c}{3M}\left[d_n^3 - (d_n - d_c)^3\right] \qquad (7.5)$$

The post-fire flexural modulus of a fire-damaged laminate is then given by:

$$E_f = \frac{4E_{f(o)}}{d^3}\left[(d-d_n)^3 + (d_n - d_c)^3 + \frac{E_{f(char)}}{E_{f(o)}}\left[d_n^3 - (d_n - d_c)^3\right]\right] \qquad (7.6)$$

where $E_{f(o)}$ and $E_{f(char)}$ are the flexural moduli of the virgin material and char, respectively.

The distance from the exposed face to the neutral axis, $d_n$, is determined by equating the bending-induced forces either side of the neutral axis:

$$\frac{b}{\rho} \int_0^{d-d_n} E_{f(o)} y \, dy = \frac{b}{\rho} \int_0^{d_n-d_c} E_{f(o)} y \, dy + \frac{b}{\rho} \int_{d_n-d_c}^{d_n} E_{f(char)} y \, dy \qquad (7.7)$$

Integration of Eqn. 7.7 gives:

$$d_n = \frac{d_c^2 \left(E_{f(o)} - E_{f(char)}\right) - E_{f(o)} d^2}{2 d_c \left(E_{f(o)} - E_{f(char)}\right) - 2 E_{f(o)} d} \qquad (7.8)$$

Mouritz and colleagues [7,9,11-13,18-20] evaluated the accuracy of the two-layer model for determining the post-fire modulus of thermoset laminates. Figure 7.14 compares the experimental and theoretical post-fire flexural modulus of a glass/polyester laminate following increasing exposure time to a heat flux of 50 kW/m$^2$. The modulus values of the fire-damaged laminate is normalised to the original strength of the material. The theoretical post-fire modulus was calculated using the two-layer model using two approaches for determining the char layer thickness ($d_c$). The solid curve was calculated using char thickness values measured by visual inspection whereas the thickness values for the dashed curve were calculated with the thermal model proposed by Gibson et al. [24]. Both curves show good agreement with the measured post-fire modulus values, demonstrating the accuracy of the two-layer model. The model is able to accurately determine the post-fire modulus for most types of thermoset laminates following exposure to a wide range of heat fluxes [7,9,11-13,17-20].

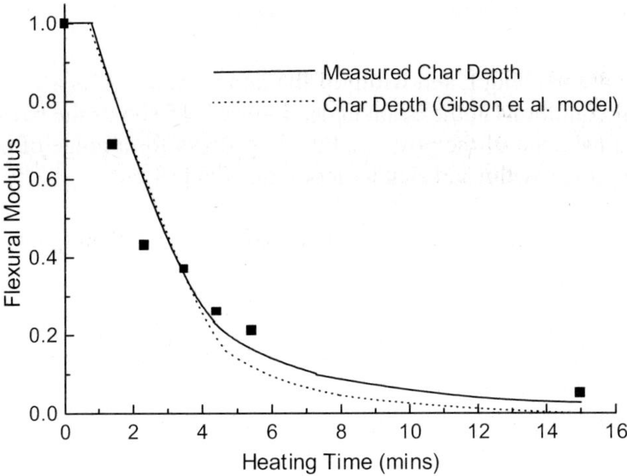


*Figure 7.14. Reduction to the post-fire flexural modulus of a slender glass/polyester beam following exposure to a heat flux of 50 kW/m² for increasing times. The curves show the theoretical reduction to the modulus.*

Mouritz and Mathys [7,9] have shown that the two-layer modelling approach can also be used to determine the post-fire strengths of laminates.  The post-fire tensile strength of a laminate can be determined by:

$$\sigma_t = \left(\frac{d-d_c}{d}\right)\sigma_{t(o)} + \left(\frac{d_c}{d}\right)\sigma_{t(char)} \qquad (7.9)$$

where $\sigma_{t(o)}$ and $\sigma_{t(char)}$ are the tensile strengths of the virgin laminate and char, respectively.

The relationship for post-fire compressive strength is governed by the failure mechanism of the burnt composite when loaded in axial compression.  When a laminate fails without buckling then this is simply:

$$\sigma_c = \left(\frac{d-d_c}{d}\right)\sigma_{c(o)} + \left(\frac{d_c}{d}\right)\sigma_{c(char)} \qquad (7.10)$$

where $\sigma_{c(o)}$ and $\sigma_{c(char)}$ are the compressive strengths of the original laminate and char, respectively.  However, thin laminate beams may fail by buckling, and in this case the Euler buckling load is calculated assuming ideal column behaviour of an isotropic material according to:

$$P_c = \frac{C\pi^2 Eb(d-d_c)^3}{12L^2} \tag{7.11}$$

where $L$ and $b$ are the length and width of the laminate, and $C$ is a constant determined by the restraint conditions at the beam ends.  Figure 7.15 shows the two-layer model can provide a good estimate of the post-fire buckling stress for a range of column and plate sizes defined by their width and slenderness ratio ($l/r$) [18]:

$$\left(\frac{l}{r}\right) = \frac{l\sqrt{12}}{(d-d_c)} \tag{7.12}$$

Mouritz and Mathys [7,9] also determined the post-fire flexural strength of laminates tested in quarter-point bending. Substituting in Eqn. 7.5 for the maximum bending moment that can be resisted by the unburnt layer ($M = PL/8$) under quarter-point bending, it can be shown that the flexural failure load is given by:

$$P_f = \frac{8\sigma_{f(o)}b}{3L^*}\left[(d-d_n)^2 + \frac{(d_n-d_c)^3}{(d-d_n)} + \frac{E_{f(c)}}{E_{f(o)}} \cdot \frac{\left[d_n^3 - (d_n-d_c)^3\right]}{(d-d_n)}\right] \tag{7.13}$$

where $L^*$ is the length of the support span, $b$ is the beam width, and $\sigma_{f(o)}$ is the flexural strength of the unburnt composite.

Figure 7.16 shows the reduction to the post-fire tensile, compressive and flexural strengths of a glass/polyester beam following exposure to a heat flux of 50 kW/m$^2$ for increasing time.  The curves show the theoretical reduction in the strengths when the char thickness is measured experimentally.  Good agreement is observed between the experimental and theoretical strengths, showing that the two-layer model can be used to accurately determine the post-fire strength.  The two-layer model has been adapted for determining the post-fire mechanical properties of laminates with a localised burn and a non-uniform char layer [14,15].

An appealing feature of the models proposed by Mourtiz and Mathys is that the post-fire modulus and strength of a laminate can be determined for any amount of char damage. The post-fire properties can be calculated by simply knowing the char thickness together with the mechanical properties of the original composite and char.  However, the properties of the char are usually significantly lower than the original properties of the laminate [7-9] (Table 7.1). It is therefore possible to obtain a reasonably accurate and conservative estimate of the residual mechanical properties by neglecting the properties of the char layer in the analysis.  Another virtue of the model is that it does not require any knowledge of the high temperature properties of the material, which is information that is frequently not available.  The model has been validated for thermoset laminates, although its accuracy in predicting the post-fire properties of thermoplastic composites has not been determined.

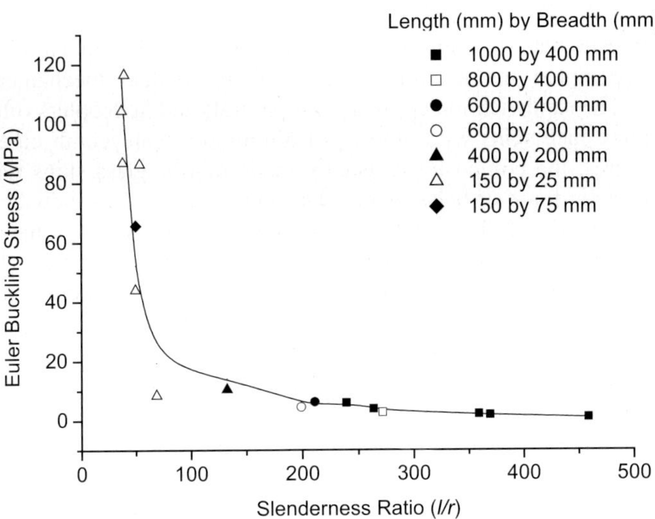



*Figure 7.15. Euler buckling strength as a function of slenderness ratio for various size glass/polyester plates and beams following exposure to a fuel fire. The solid line is the model calculation. Data from Gardiner et al. [18].*

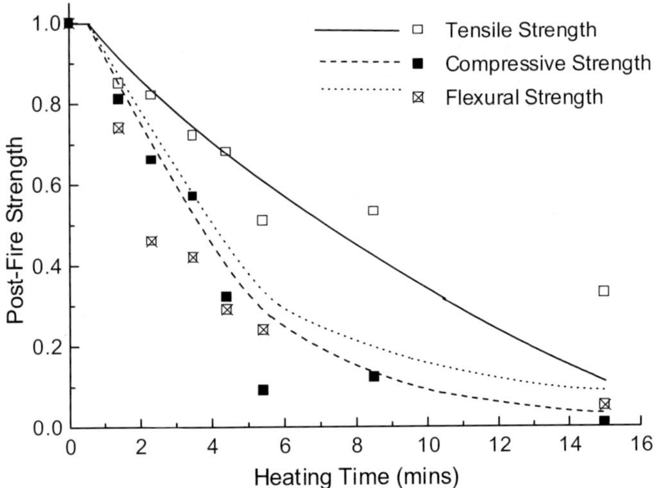



*Figure 7.16. Reductions to the tensile, compressive and flexural strength of a slender glass/polyester beam following exposure to a heat flux of 50 kW/m$^2$ for increasing times. The curve shows the theoretical reduction to the strength.*

## 7.4      Post-Fire Properties of Sandwich Composites

Sandwich composite materials are used extensively in aircraft, ship and civil structures, and this has prompted the need to evaluate their residual mechanical properties following fire.  The sandwich materials used in aircraft and helicopters often consist of thin carbon/epoxy face skins with Nomex or aluminium honeycomb core.  In marine and civil structures, the composite is usually made of fibreglass skins with a core of PVC foam, syntactic foam or balsa wood.  Despite such uses of sandwich composites, their residual properties following fire have not been thoroughly characterised. As reported in chapter 6, degradation to the mechanical properties of sandwich composites have been determined at elevated temperature or in fire, but there remains a poor understanding of their residual properties after the fire has been extinguished.

The first and only published study to date of the post-fire properties of aircraft sandwich composites was performed by McManus [25].  The residual tensile strength of a composite consisting of carbon/epoxy skins and Nomex honeycomb core was measured at room temperature after being exposed to a high temperature gas fire.  Figure 7.17 shows the residual strength of the sandwich composite, normalised to the original strength, plotted against exposure time for three different thicknesses for the skins.  The post-fire strength was reduced significantly, even after short times, due to the high flammability of the epoxy composite skins.  Not surprisingly, the thicker face skins retained higher strength at longer times, although even they were eventually severely weakened.  Despite this work, much remains unknown about the residual properties of aircraft sandwich composites, particularly when subjected to structural and aerodynamic loads experienced during flight. The effect of fire on the residual compressive properties and impact damage resistance of marine-grade sandwich composites used in ship and offshore platform structures has only been studied to a limited extent [26,27].

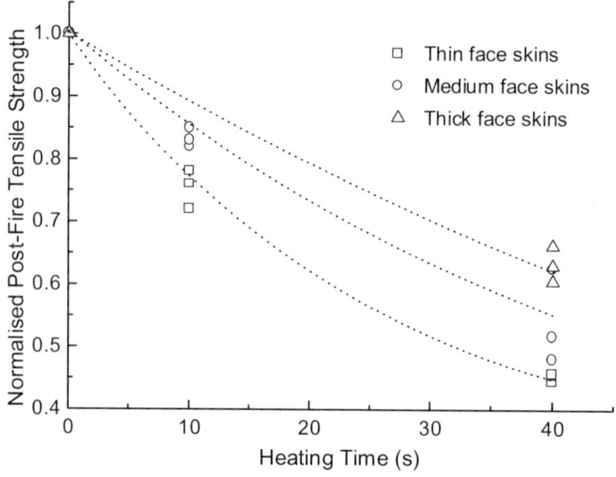


*Figure 7.17. Reduction to the tensile strength of an aircraft sandwich composite material following exposure to a gas fire.  Reproduced from McManus [25].*

## 7.5       Post-Fire Properties of Fire Protected Composites

Various methods are used to improve the fire resistance of composite materials. Thermal barrier materials such as ablative, ceramic and intumescent coatings can be highly effective in delaying ignition and thermal decomposition.  Composites with specially formulated flame resistant resins can also have improved fire reaction properties including long ignition times and reduced rates of heat release, smoke production and flame spread. Methods for improving the fire resistance of composites are described in the next chapter.

The impressive performance of thermal barrier materials and flame retardant resins has prompted an assessment of their effectiveness in retaining the mechanical properties of composites following fire [4-6,9,28].  Sorathia and colleagues [4-6,28] evaluated the use of fire barrier treatments for improving the post-fire properties of thermoset laminates. The fire barrier materials examined included ablative sheets, ceramic coatings, intumescent mats and films, phenolic skins and silicone foam. These materials were used to protect a thin glass/vinyl ester laminate exposed to an incident heat flux of 25 kW/m$^2$ for twenty minutes. Typical post-fire flexural strengths of the composite protected with the different thermal barrier systems are given in Fig. 7.18.  Almost all of the treatments reduced the impact of fire on the residual strength, with the most effective being ablative and phenolic systems that delay ignition and thermal degradation for the longest periods.

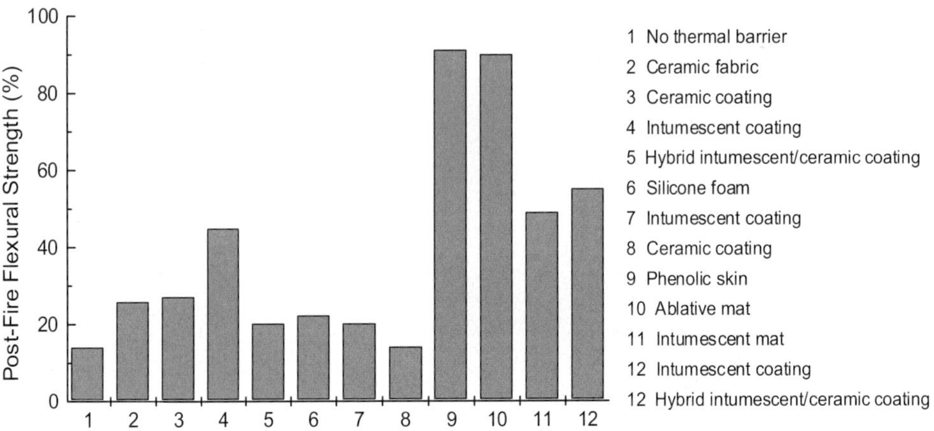


*Figure 7.18.  Effect of fire barrier materials on the post-fire flexural strength of a glass/vinyl ester laminate.  Data from Sorathia and Beck [5].*

Mouritz and Mathys [9] have also shown that thermal barriers can be highly effective in retaining the mechanical properties of laminates following fire, particularly intumescent and ceramic-based coatings.  Figure 7.19 shows the flexural strength of a glass/polyester laminate with thermal barrier coatings after being fire tested at different heat fluxes and times.  It is seen that the laminate with a fire barrier can survive higher heat fluxes and longer times before the strength is degraded.

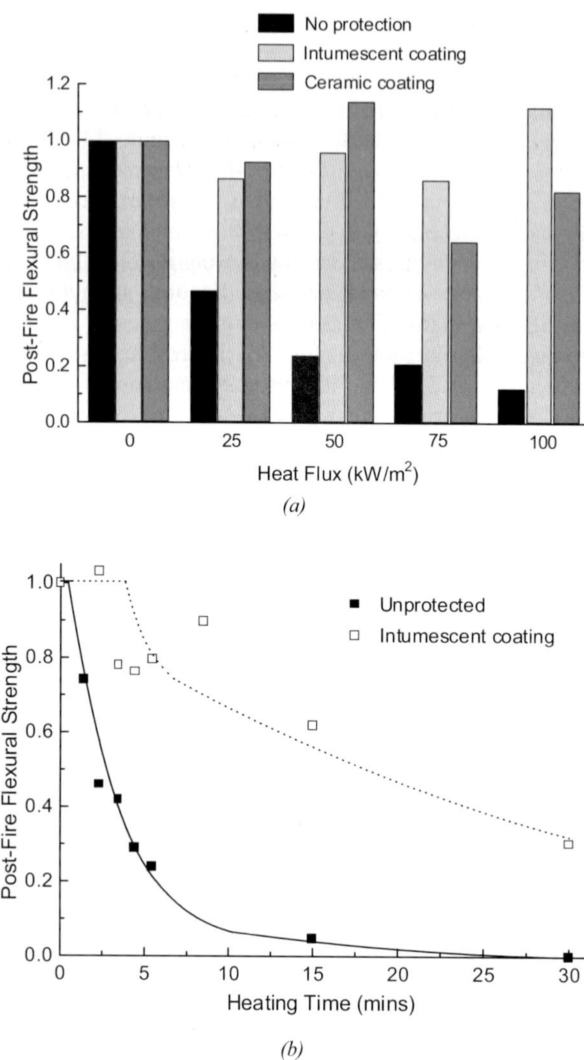


*Figure 7.19.  Effect of thermal barrier protection on the residual flexural properties of a glass/polyester laminate following increasing (a) heat flux and (b) heat time.*

In some cases the post-fire properties can actually be improved by using a thermal barrier. In Fig. 7.19a it is seen that the strength of the thermally protected composite is in some cases improved slightly after fire testing. This is due to the fire warming the composite to a moderate temperature that causes the resin matrix to post-cure, thereby raising the strength. While the effectiveness of thermal barrier materials have been determined, the ability of flame retardant resins to minimise the impact of fire on the properties of composites is not as well understood.


## 7.6    Concluding Remarks

The measurement and modelling of the post-fire mechanical properties of composites has advanced significantly in recent years, although gaps remain in our understanding of this important topic. A range of common thermoset laminates have been tested, and simple and effective models have been developed to predict the post-fire mechanical properties for uniform and localised fire damage. In contrast, there has been less attention given to thermoplastic laminates, and much remains unknown about their post-fire properties. Some useful studies of sandwich composites have been published, although like thermoplastic laminates, there is a need for further testing and analysis, particularly with respect to different types of loads and support conditions.


## References


1.    G.A. Pering, P.V. Farrell and G.S. Springer. Degradation of tensile and shear properties of composites exposed to fire or high temperature. *Journal of Composite Materials*, 1989; 14:54-66.
2.    U. Sorathia. Flammability and fire safety of composite materials. In: *Proceedings of the 1ˢᵗ International Workshop on Composite Materials for Offshore Operations*. Houston TX , 26-28 October 1993. pp. 309-317.
3.    U. Sorathia, T. Dapp and J. Kerr. Flammability characteristics of composites for shipboard and submarine internal applications. In: *Proceedings of the 36ᵗʰ International SAMPE Symposium*. San Diego, CA, 15-18 April 1991. pp. 1868-1878.
4.    U. Sorathia, C.M. Rollhauser, W.A. Hughes. Improved fire safety of composites for naval applications. *Fire & Materials*, 1992; 16:119-125.
5.    U. Sorathia, C. Beck and T. Dapp. Residual strength of composites during and after fire exposure. *Journal of Fire Sciences*, 1993; 11:255-270.
6.    R.E. Lyon, U. Sorathia, P.N. Balaguru, A. Foden, J. Davidovits and M. Davidovits. Fire response of geopolymer structural composites. In: *Proceedings of the 1ˢᵗ International Conference On Composites in Infrastructure: Fibre Composites in Infrastructure*. Tucson, AZ, 1996. pp. 972-981.
7.    A.P. Mouritz and Z. Mathys. Post-fire mechanical properties of marine polymer composites. *Composite Structures*, 1999; 47:643-653.
8.    A.P. Mouritz and Z. Mathys. Mechanical properties of fire-damaged glass-reinforced phenolic composites. *Fire & Materials*, 2000; 24:67-75.
9.    A.P. Mouritz and Z. Mathys. Post-fire mechanical properties of glass-reinforced polyester composites. *Composite Science & Technology*, 2001; 61:475-490.
10.   A.P. Mouritz, C.P. Gardiner, Z. Mathys, and C.R. Townsend. Comparison of post-fire properties of composites burnt by cone calorimetry and large scale fire testing. In: *Proceedings of the 13ᵗʰ International Conference on Composite Materials*. Beijing, 25-29 June 2001. Paper No. 1273.

11. C.P. Gardiner, A.P. Mouritz, Z. Mathys and C.R. Townsend. Post-fire flexural response of GRP composite ship panels. In: *Proceedings of the Eleventh International Offshore and Polar Engineering Conference (ISOPE-2001)*. Stavanger, Norway, 17-22 June 2001. pp. 160-167.

12. Z. Mathys, C.P. Gardiner, C.R. Townsend, S. Russo, and A.P. Mouritz. Modelling the mechanical properties of fire-damaged composite materials. In: *Proceedings of the Second International Conference on the Response of Composite Materials to Fire*. University of Newcastle, England, 12-13 September 2001.

13. A.P. Mouritz. Post-fire flexural properties of fibre-reinforced polyester, epoxy and phenolic composites. *Journal of Materials Science*, 2002; 37:1377-1386.

14. C.P. Gardiner, Z. Mathys and A.P. Mouritz. Tensile and compressive properties of GRP composites with localised heat damage. *Applied Composite Materials*, 2002; 9:353-367.

15. Z. Mathys, A.P. Mouritz, C.P. Gardiner and C.R. Townsend. Mechanical properties of GRP composites with localised heat damage. *International Journal of Materials Processing & Technology*, 2002; 17: 134-142.

16. A.G. Gibson, P.N.H. Wright, Y.S. Wu, A.P. Mouritz, Z. Mathys and C. Gardiner. Modelling the residual integrity of polymer composites after fire. In: *Proceedings of the 9th International Conference on Fibre Reinforced Composites*. Newcastle, UK, 26 – 28 March 2002.

17. A.G. Gibson, P.N.H. Wright, Y-S. Wu, A.P.Mouritz, Z. Mathys and C.P. Gardiner. Modelling the residual mechanical properties of polymer composites after fire. *Plastics, Rubbers & Composites*, 2003; 32:81-90.

18. C.P. Gardiner, Z. Mathys and A.P. Mouritz. Post-fire structural properties of burnt GRP plates. *Marine Structures*, 2004; 17:53-73.

19. A.P. Mouritz. Simple models for determining the mechanical properties of burnt FRP composites. *Materials Science & Engineering A*, 2003; 359:237-246.

20. A.P. Mouritz. Fire resistance of aircraft composite laminates', *Journal of Materials Science Letters*, 2003; 22:1507-1509.

21. P.G.B. Seggewiss. Properties of fire-damaged polymer matrix composites. In: *Proceedings of the Third Conference on Composites in Fire*. Newcastle, UK, 9-10 September 2003.

22. P.G.B. Seggewiss. Methods to evaluate the fire resistance of carbon fiber reinforced plastics. In: *Proceedings of the SAMPE Technical Conference*. Long Beach, CA, 16-20 May 2004.

23. L.K. Kucner and H.L. McManus. Experimental studies of composite laminates damaged by fire. In: *Proceedings of the 24th International SAMPE Technical Conference*. 17-20 October 1994. pp. 341-353.

24. A.G. Gibson, Y.S. Wu, H.W. Chandler, J.A.D. Wilcox and P. Bettess. A model for the thermal performance of thick composite laminates in hydrocarbon fires. Composite Materials in the Petroleum Industry, *Revue de l'Institut Français du Pétrole*, 1995; 50:69-74.

25. H.L. McManus. Prediction of fire damage to composite aircraft structures. In: *Proceedings of the 9th International Conference on Composite Materials*. Madrid, Spain, July 1993. p. 929-936.

26. A.P. Mouritz and C.P. Gardiner. Compression properties of fire-damaged polymer sandwich composites. *Composites*, 2002; 33A:609-620.

27. C.A. Ulven and U.K. Vaidya. Post-fire impact response of VARTM glass/vinyl ester balsa core sandwich composites. In: *Proceedings of the SAMPE Technical Conference*. Long Beach, CA, 16-20 May 2004.

28. U. Sorathia and C. Beck. Fire protection of glass/vinyl ester composites for structural applications. In: *Proceedings of the 41st International SAMPE Symposium*. 24-28 March 1996. pp. 687-697.

# Chapter 8

# Flame Retardant Composites

## 8.1    Introduction

This chapter gives an overview of methods to enhance the flame retardant properties of fibre reinforced polymer composite materials.  The methods used are extraordinary diverse, and vary in complexity from simple additive compounds blended into the polymer matrix or heat-resistant coatings, through to sophisticated methods that involve chemical modification of the matrix or heat-induced intumescence of the composite surface.  Also outlined are methods to improve the thermal stability and fire resistance of organic fibres used in composites.

A popular method to reduce the flammability of composites is the addition of inert fillers (eg. talc, silica) or thermally active fillers (eg. hydrated oxides) to the polymer matrix.  The types of fillers, their flame retardation mechanisms, and their efficacy when used in composite materials are outlined.  Following this the chemical and structural modification of organic polymers to improve their flammability resistance is described, with emphasis given to flame retardant mechanisms and fire reaction properties of brominated, chlorinated and phosphorated polymers.  The flammability properties of fire retardant fibres and inorganic polymers developed for fire resistant composites are also described.  The chapter also provides an overview of flame retardant polymer, thermal barrier and intumescent coatings that protect composite materials from fire.

Some of the methods described in this chapter have been used for many hundreds of years to reduce the flammability of combustible materials, having originally been used in clothing fabric and timber and more recently applied to polymers and polymer composites.  Other methods have been developed over the past ten to fifty years.  Several new methods to reduce flammability are currently under development, and offer the prospect of outstanding fire resistance in future composite materials.  Some of the more promising methods are described, including graft polymerisation of flame retardant compounds onto organic polymers and non-combustible inorganic structural polymers.


## 8.2      The Combustion Cycle

The basic combustion cycle of polymer composites is shown schematically in Fig. 8.1. The combustion process is described at length in chapters 1 and 2, and therefore is only briefly reviewed.  When a composite material is exposed to fire the polymer matrix and organic fibres (if present) will thermally decompose with the release of heat, smoke and combustion gases.   The gases consist of a mix of non-flammable vapours and flammable volatiles.  The temperature at which decomposition occurs depends on the chemical nature of the polymer and the fire atmosphere, although it is typically in the range of 300 to 500°C for most polymers and organic fibres used in composites.  The decomposition gases flow from the composite into the fire, where the flammable volatiles react with oxygen to produce highly reactive H· and OH· radicals.  These radicals have an important role in the chain reactions leading to the decomposition and sustained burning of polymers and other organic fuels.

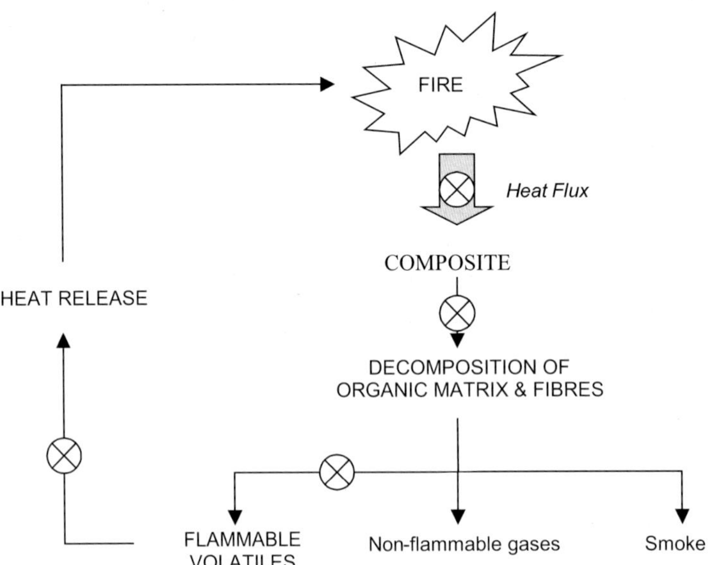


*Figure 8.1.  Combustion cycle for polymer composites in fire.  The symbol⊗ indicates the stages of the cycle where flame-retardant polymers disrupt the cycle.*

The pyrolysis reactions in a flame can be simply described by the $H_2$-$O_2$ scheme:

$$H^{\cdot} + O_2 \rightarrow OH^{\cdot} + O^{\cdot} \qquad\qquad (8.1)$$

$$O^{\cdot} + H_2 \rightarrow OH^{\cdot} + H^{\cdot} \qquad\qquad (8.2)$$

The main exothermic reaction that generates most of the thermal energy in a flame is:

$$OH^{\cdot} + CO \rightarrow CO_2 + H^{\cdot} \qquad\qquad (8.3)$$

The $H^{\cdot}$ radicals produced in reactions (8.2) and (8.3) feed-back into reaction (8.1), and thereby the combustion reaction sequence a self-propagating process or 'chain reaction' that continues as long as sufficient oxygen is available. The heat generated raises the temperature in the combustion zone of the flame, and this in turn accelerates the decomposition rate of the composite. Many of the polymer systems used, such as polyesters, vinyl esters and epoxies, release large amounts of flammable gas that can add greatly to the fuel load of a fire. With these materials the combustion process continues in the presence of air until the supply of flammable volatiles from the polymer matrix and organic fibres becomes insufficient to sustain burning, and this often occurs when the matrix and fibres are completely degraded.

The flammability of composite materials is reduced by breaking or slowing the chain-branching reactions (8.1) and (8.2) in the combustion cycle. Flame retardant polymers work by disrupting the cycle in one of three ways: modification of the thermal degradation process to reduce the amount and/or types of flammable gases; the generation of decomposition gases which 'quench' the flame by removing the $H^{\cdot}$ and $OH^{\cdot}$ radicals; and/or reduction in the temperature of the material by modifying its heat conduction and/or specific heat properties.

Flame retardant polymers are often classified as 'condensed phase' or 'gas phase' active, depending on whether they disrupt the decomposition of the polymer or combustion in the flame. The 'condensed phase' refers to the polymer, whether in the solid or molten state. Condensed phase activity encompasses several flame retardant mechanisms, which include:

- diluting the amount of combustible organic material by the addition of inert filler particles,
- reducing the temperature of the composite by the addition of filler that acts as a heat sink,
- reducing the temperature by the addition of fillers that decompose endothermically to yield water or other non-combustible products with a high specific heat capacity,
- reducing the heat release rate by using polymers that decompose via endothermic reactions,
- increasing the aromaticity of the polymer matrix in order that it decomposes into an insulating surface layer of carbonaceous char that slows heat conduction into the composite and reduces flammable gas emissions.

Polymer composites that are flame retardant in the gas phase operate by interfering with the combustion reaction, thus reducing both flame propagation and the amount of heat returned from the fire to the material. The most widely employed flame retardant mechanism in the gas phase is usually the release of bromine, chlorine or phosphorus based radicals that terminate the exothermic combustion reactions by removing H and OH radicals from the flame. Another common mechanism is the release of non-combustible vapours to dilute the concentration of H and OH gases in the flame, which also lowers the temperature. While there are many flame retardants that operate solely by a condensed or gas phase mechanism, the most effective retardants operate in both phases at the same time.


## 8.3    Flame Retardants for Composites

A diverse range of flame retardant materials have been developed for polymers and polymer composites, with between 150 and 200 different compounds being used [1-8][*]. Flame retardants are by far the largest group of additives used in polymers, and it is estimated by Georlette [9] that in 1997 about $US2 billion worth were used by the plastics industry.  Flame retardants account for about 27% of the plastic additive market, and this far exceeds other types of additives including heat stabilisers (15.6%), antioxidants (7.6%), lubricants (6%) and UV stabilisers (5%).

Flame retardants are classified as additive or reactive compounds.  Additive compounds are intimately blended into the polymer during processing, but do not chemically react with the polymer. The chemical composition of many of these compounds is based on the following elements: antimony, aluminium, boron, phosphorus, bromine or chlorine, which all confer a high level of flame retardation.  It is estimated that about 90% of all additive compounds are based on these elements, and are used in the form of antimony oxides, alumina trihydrate and boron oxides.  Less often, additive compounds are used that contain barium, zinc, tin, iron, molybdenum or sulphur.  Many additives exist as hydrated metal salts that decompose endothermically in a fire, and thus reduce the overall heat release rate of a polymer.  Some additive compounds also liberate water vapour during decomposition that dilutes the concentration of flammable gases released into the flame.

Reactive compounds are polymerised with a resin during processing to become integrated into the molecular network structure.  Reactive fire retardants are mainly based on halogen (bromine and chlorine), phosphorus, inorganic and melamine compounds.  Until recently, bromine and chlorine were the common retardants because of their potency in quenching flames. Halogen compounds confer flammability resistance by liberating reactive bromine or chlorine atoms into the flame where they

---

[*] Many references used in this chapter are excellent review articles, and the reader is encouraged to refer to these for further information and additional reference sources.

disrupt the combustion oxidation reactions of flammable volatiles.  At this time, however, there is strong pressure from governments and environmentalists to use non-halogenated flame retardants because of health and environmental concerns with bromine and chlorine vapour compounds. Phosphorus is a highly effective flame retardant that operates by reducing the amount of combustible gases released from a decomposing polymer by promoting char formation.

The selection of a flame retardant for a polymer composite is determined by several factors, including cost, chemical compatibility between the flame retardant compound and host polymer, decomposition temperature of the compound, and weight. Many flame retardant fillers reduce the mechanical properties of polymers, and this is another consideration when used in structural composites.  It is possible to ameliorate the adverse effects of the filler by surface treating the particles to promote chemical interaction with the polymer matrix. Some filler materials, while decreasing the flammability, increase the amount of smoke and toxic fumes given off by the decomposing material.  For these reasons, it is common practice to use a combination of flame retardants in polymer composites in order to maximise the flammability resistance while minimising the adverse effects on the mechanical properties, smoke and toxicity.


## 8.4        Flame Retardant Fillers for Composites

### 8.4.1    INTRODUCTION

Fillers are inorganic non-reactive compounds that are added to the polymer during the final stages of processing to reduce the flammability of the finished product.  The filler particles are under 10 μm in diameter, and often in the submicron range.  The particles are blended into the liquid resin and must be uniformly dispersed to ensure consistent flame retardant properties throughout the polymer.  Most polymers require a high loading of filler to show an appreciable improvement to their flammability resistance, and the minimum volume content is usually about 20% and the average content is typically 50 to 60%.  Fillers should only be used in polymers that are chemically compatible, otherwise the mechanical properties and environmental durability of the material can be severely degraded.  Fillers can have other deleterious effects on the properties, including an increase to the viscosity and a reduction to the gel time of the polymer melt which makes processing more difficult.  Many filler materials gradually break-down when exposed to moisture by hydrolysis, and this degrades their flame retardant action.  Despite these problems, fillers are often used because of their low cost, relatively easy addition into the polymer, and high fire resistance.  It is important to note that fillers are rarely used alone, but instead are used in combination with other flame retardants (such as organohalogen or organophosphorus compounds) to achieve a high level of flammability resistance.  There are two classes of fillers - 'inert' and 'active' flame retardants – which are distinguished by their mode of action.

## 8.4.2    INERT FLAME RETARDANT FILLERS

Inert fillers reduce the flammability and smoke yield of polymer composites by several mechanisms.  The dominant mechanism is reducing the fuel load by diluting the mass fraction of organic material in a composite by the addition of a non-combustible filler. Flame retardation and smoke suppression by this mechanism can only be achieved when the polymer content is reduced by a large amount, and for this reason filler loadings of 50-60% are often required.   Another important mechanism is the absorption of heat by the filler to reduce the burning rate of the polymer matrix.  To be an effective heat sink the filler must have a greater heat capacity than the polymer host.  Certain fillers also reduce the flammability by forming an insulating surface layer when the polymer is decomposed and vapourised from between the filler particles.  The layer reduces the rate of heat conduction to the underlying composite material, and thereby slows the decomposition rate of the polymer matrix. This surface layer can also obstruct the flow of combustible volatiles from the material into the flame, and thereby further reduce the rate of decomposition.

All fillers operate by reducing the mass content of polymer and most fillers also act as a heat sink.   Only a few types of fillers are capable of forming a surface layer that provides a high degree of thermal insulation and acts as an impervious gas barrier.  The inert fillers most commonly used in polymers and polymer composites are silica, calcium carbonate and carbon black for their ability to reduce the flammability and smoke yield by the fuel dilution/heat sink mechanisms.  In a few cases, simple hydrated clay silicates such as pumice, talc, gypsum and calcium sulphate dihydrate are used.

## 8.4.3    ACTIVE FLAME RETARDANT FILLERS

### 8.4.3.1  Introduction
Active fillers are more effective additives than inert fillers for reducing the flammability and smoke produced by polymer composites.  Active fillers operate as a heat sink and by diluting the mass fraction of organic matrix material in a composite.  The fillers also operate in the condensed phase by decomposing at elevated temperature via an endothermic reaction that absorbs a large amount of heat, and this has a cooling effect that slows the decomposition rate of the polymer matrix.  The decomposition reaction of the filler also yields a large amount of inert gases, such as water vapour and carbon dioxide, which diffuse into the flame where they dilute the concentration of flammable volatiles, H· and OH· radicals.  The dilution reduces the flame temperature, which in turn lowers the decomposition rate of the composite material.

The decomposition temperature of the filler is a critical factor in their efficacy as a flame retardant.  The decomposition temperature must be higher than the processing temperature of the polymer matrix, otherwise the filler will decompose during manufacture of the composite material. Composites containing a high temperature thermoplastic resin, such as polyphenylene sulphide or polyether ether ketone, must be processed in the range 300-400°C, and therefore require a filler material having a decomposition temperature above this range.  However, the decomposition temperature of the filler must be below the pyrolysis temperature of the polymer matrix, which for many resins used in composites is between 300 and 450°C.

A diverse range of metal oxides and metal hydroxides are used as active flame retardant fillers, although by far the most common is aluminium trihydroxide, $Al(OH)_3$.  Other types of aluminium oxide compounds are also used, as well as oxide compounds containing antimony ($Sb_2O_3$, $Sb_2O_5$), iron (eg. ferrocene, FeOOH, FeOCl), molybdenum ($MoO_3$), magnesium ($Mg(OH)_2$), zinc and tin.  The ability of these compounds to suppress combustion and smoke formation varies considerably, although their efficacy generally improves with their concentration in the polymer matrix.  As with inert fillers, high loading levels (of typically 20 to 60%) are needed to achieve a substantial reduction in flammability.  Nitrogen-based compounds are also highly effective flame retardants, and melamine and guanidine compounds have been used for many years to improve the flame resistance of woollen garments, cotton clothing and paper.  However, nitrogen-based additives are rarely used as flame retardants in polymer composites.

### 8.4.3.2  *Aluminium Trihydroxide Filler*

Aluminium trihydroxide (ATH), which is also known as alumina trihydrate, is the active flame retardant filler compound most often used in polymers and polymer composites [1-8,10].  The amount of ATH used is greater than the combined amount of all other flame retardant fillers.  ATH is popular for several reasons, most notably its low cost, good flame retardant properties, and non-toxic smoke.  ATH particles are usually 1 micron or finer, and must be uniformly dispersed through the matrix to ensure an even level of flame retardant through a composite.  It is often necessary to add 50 wt% or more of ATH to many resin systems to achieve good flammability resistance.  Unfortunately, this high loading degrades the mechanical and durability properties of most types of polymer composites.  To minimise these adverse effects, ATH is often used in lower concentration in combination with another flame retardant such as an organohalogen.

ATH is active in both the condensed and gas phases of the combustion process, and when used in a large amount is remarkably effective in suppressing flaming combustion and (in some cases) smoke.  The main condensed phase mechanism of ATH is the absorption of heat when the filler decomposes.  ATH decomposes between 220 and 400°C by the endothermic reaction:

$$2Al(OH)_3 \rightarrow Al_2O_3 + 3H_2O \tag{8.4}$$

This is a highly endothermic reaction that absorbs about 1 kJ of heat per gram of ATH.  The main endothermic peak of the reaction occurs at about 300°C, which means the reaction is absorbing the most heat at a temperature at which most polymers used in composites do not decompose.  Another important aspect of the reaction is the creation of water vapour formed from the hydroxyl groups bonded to the aluminium.  This water is released into the flame where it hinders combustion by diluting the concentration of flammable gases evolved from the polymer matrix and restricting the access of oxygen to the composite surface. An added benefit of the decomposition reaction is that no toxic or corrosive gases are produced, unlike some other flame retardant compounds (most notably organohalogens).

ATH also operates in the condensed phase as a heat sink that extends the time taken for the polymer matrix to reach its decomposition reaction temperature. ATH has a higher heat capacity than most organic resins, enabling it to absorb heat that promotes a 'cooling' effect in the host polymer. It is believed that another condensed phase mechanism of ATH is the promotion of char in some types of polymer. Improved flame retardation can also occur by the formation of $Al_2O_3$ in the decomposition reaction that produces a refractory layer in the polymer. Although, a very high loading of ATH is required to form a layer of aluminium oxide with the consistency and thickness needed to be an effective thermal barrier.

The efficacy of ATH as a flame retardant in polymer composites has been evaluated under a variety of fire scenarios [11,12]. Egglestone and Turley [11] measured the fire reaction properties of a glass/polyester composite with and without. The amount of ATH added to the polyester matrix was 50 wt%. Figure 8.2 shows the time-to-ignition values at different heat fluxes for the polyester laminate with and without ATH. For comparison, the ignition times for an inherently flame retardant glass/phenolic (without ATH) are given. The ATH extended the ignition time of the polyester composite, with the times being 20% to over 100% longer for the flame retardant laminate depending on the heat flux. Scudamore [12] measured similar improvements to the ignition time of a glass/epoxy composite when filled with ATH. However, it is shown in Fig. 8.2 that the phenolic composite has a higher resistance to ignition than the polyester laminate containing ATH. This indicates that the flame retardancy that can be achieved using ATH may not reach the level of an inherently fire resistant polymer such as phenolic resin.

ATH has been found to be particularly effective when used in combination with modified acrylic resins, which is available commercially as Modar™. These resins are formulated using a urethane oligomer dissolved in a methyl methacrylate solvent monomer. They are analogous in many respects to polyesters, and are cured in a similar manner using free radical initiators. Their distinguishing characteristics are low resin viscosity and rapid curing. The low viscosity permits the incorporation of exceptionally high levels of ATH, which enables fire reaction properties comparable to those of phenolic to be achieved. Indeed, highly ATH-filled resin systems are superior in terms of CO emission to those based on phenolic. Another benefit of Modar™ is it has better mechanical properties of phenolic resins, which makes it suitable for use in flame resistant structural composite materials.

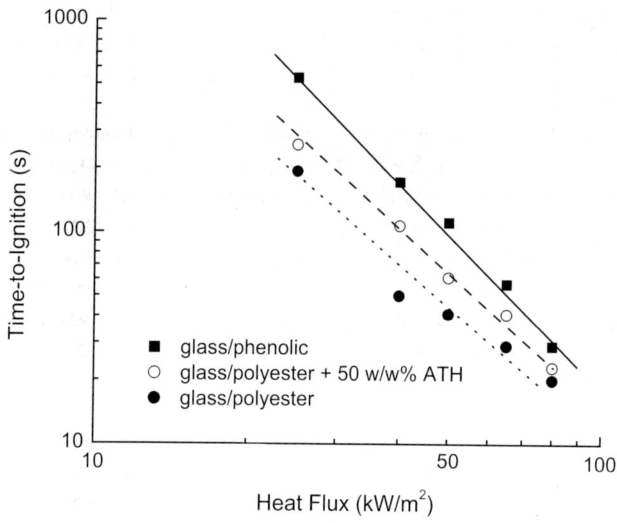



*Figure 8.2. Effect of heat flux on the ignition times of a glass/polyester composite with and without ATH filler and a glass/phenolic composite. Data from Egglestone and Turley [11].*

ATH is very effective in lowering the heat release rate of polymer composites, and thereby slowing the rate of pyrolysis. The effect of ATH on the peak and average heat release rates of a glass/polyester composite is shown in Fig. 8.3. The addition of ATH reduces the amount of heat released by 15 to 40%, depending on the heat flux. The lower heat release rate is attributed to three mechanisms: the dilution of flammable gases from the decomposing polyester by water vapour released by the decomposing ATH, the strong endothermic reaction of the ATH, and the 'heat sink' action of the ATH. Figure 8.3 also shows that the composite containing ATH is less sensitive to changes in heat flux, and therefore the filler becomes more effective compared to the unfilled material with increasing temperature.

ATH can also reduce the amount of smoke released by certain types of composite materials. Figure 8.4 compares the smoke parameter values for a glass/epoxy composite without and with ATH (67 w/w%) at different heat fluxes [12]. The smoke parameter is the combined units of smoke and heat release, and is defined as the product of specific extinction area (SEA) of smoke and the peak heat release rate. The parameter is used

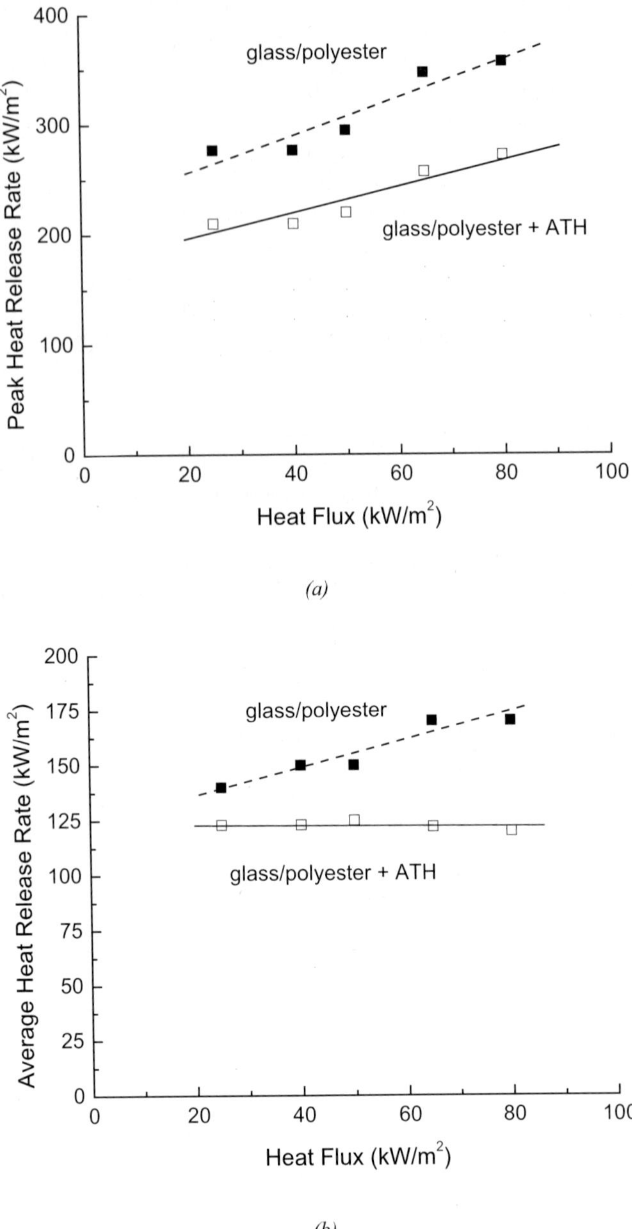


*Figure 8.3. (a) Peak and (b) average heat release rates of a glass/polyester composite with and without ATH filler. Data from Egglestone and Turley [11].*

because smoke production is largely dependent on the burning behaviour of the material, and it is indicative of the amount of smoke generated in a full-scale fire situation.  It is seen that the ATH was remarkably effective in reducing the amount of smoke.  It is also noteworthy that the amount of smoke released by the composite without the flame retardant increased with the heat flux, while the addition of ATH suppressed this trend with the smoke parameter showing little or no dependence on the heat flux.  However, ATH does not always reduce the smoke yield, and in some materials may increase the amount of smoke released from a burning composite [11]. Unfortunately, the amount of published information on the smoke properties of composites containing ATH is limited, and it is not possible to make general statements about the efficacy of this filler as a smoke suppressant.  Although, from the limited data that is available it appears that the ability of ATH to suppress smoke is dependent on the chemical nature of the polymer matrix, filler loading, and the temperature of the fire.

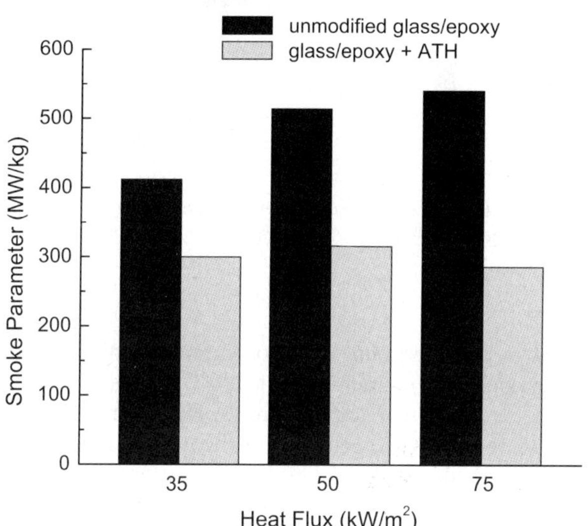


*Figure 8.4. Effect of ATH filler on the smoke parameter of a glass/epoxy composite. Data from Scudamore [12].*

ATH is an effective flame retardant in inherently fire resistant polymer composites, such as phenolic laminates, as well as in more flammable composites, such polyester, vinyl ester or epoxy laminates.  Scudamore [12] found that ATH is also an effective fire retardant for phenolic composites, as shown in Fig. 8.5.  This figure compares the fire reaction properties of a glass/phenolic composite without and with ATH (30 wt%)

measured at a heat flux of 50 kW/m$^2$. The ATH increased the ignition time and reduced the heat release rates and smoke yield.

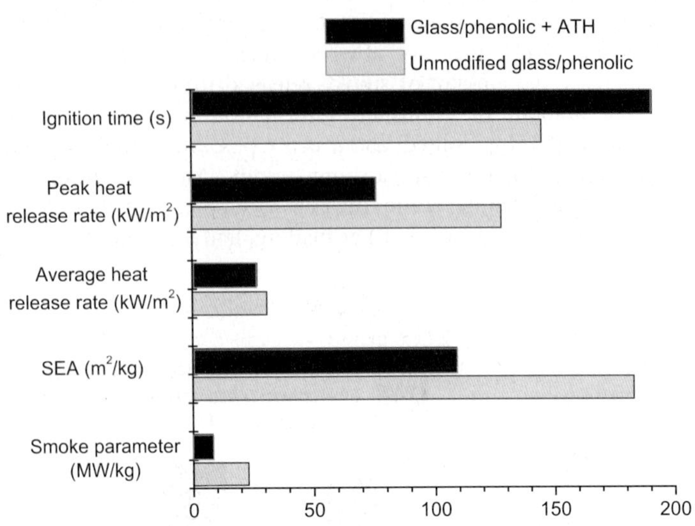


*Figure 8.5. Effect of ATH filler on the fire reaction properties of a glass/phenolic composite tested at an incident heat flux of 50 kW/m$^2$. Data from Scudamore [12].*

ATH cannot be used in polymers that need to be processed or cured at temperatures above the filler decomposition temperature of 220°C. This is not an issue for most thermoset resins that cure at lower temperatures. However, many high performance thermoplastics used in composites, such as polyphenylene sulphide or polyether ether ketone, need to be heated above the ATH decomposition temperature to reduce the viscosity for processing. In these materials, active filler compounds with higher thermal stability than ATH are required.

### 8.4.3.3   Magnesium Oxide Fillers

Magnesium hydroxide (Mg(OH)$_2$) and a hydromagnesite (Mg$_4$(CO$_3$)$_3$(OH)$_2$.3H$_2$O)/ hunite (Mg$_3$Ca(CO$_3$)$_4$) hybrid compound are active fillers suited to polymer composites that need to be processed above the decomposition temperature of ATH [4,6,13,14]. Magnesium hydroxide is thermally stable up to 330-340°C, and therefore can be used in most types of high temperature thermoplastics without decomposing during processing.

Both magnesium compounds act as flame retardants in a similar manner to ATH, with several flame retardant mechanisms occurring concurrently in a fire [14]. Magnesium compounds, as with ATH, need to be present in a large amount (30 to 60 wt%) to provide significant flame retardancy. The dilution of the polymer by using a high filler loading reduces the volume content of combustible organic material within a composite, thereby reducing its flammability. The magnesium compounds undergo a highly endothermic decomposition reaction that slows the heating rate of the host material in a fire. The reaction is simply:

$$Mg(OH)_2 \rightarrow MgO + H_2O \tag{8.5}$$

In addition, the hydroxyl groups bonded to the magnesium are converted in the reaction into water vapour, which dilutes the concentration of flammable organic volatiles and $H^{\cdot}/OH^{\cdot}$ radicals in the flame. The water in hydromagnesite is also released at elevated temperature that contributes to the dilution of the flame. The decomposition of magnesium compounds also yields magnesia (MgO) that has good insulating properties.

The flame retardant mechanisms of magnesium hydroxides are effective in prolonging the ignition time and reducing the amount of smoke produced by high temperature polymers. For example, Fig. 8.6 shows the effect of increasing $Mg(OH)_2$ loading on the oxygen index of several thermoplastics while Fig. 8.7 shows the reduction in smoke density and carbon monoxide gas from polyphenylene oxide (PPO) loaded with $Mg(OH)_2$. However, magnesium oxides are generally less effective than ATH for many engineering polymers used in composites, such as polyesters, vinyl esters and epoxies, because the decomposition temperatures of the filler and host matrix are similar. A further drawback is that magnesium oxides are considerably more expensive than ATH.

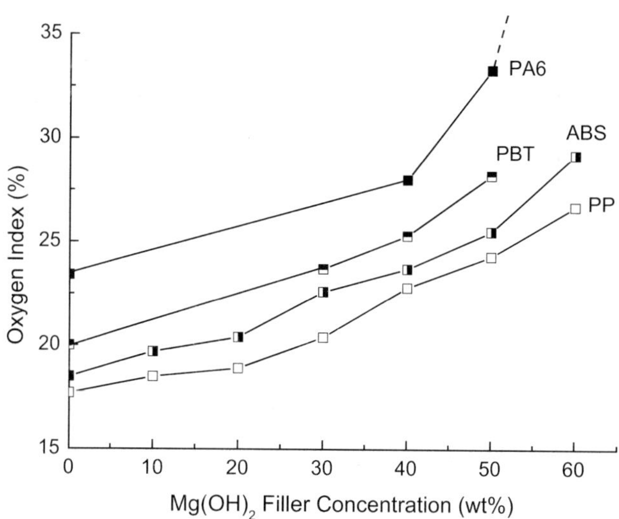


*Figure 8.6. Effect of magnesium hydroxide filler loading on the oxygen index of several thermoplastics. Data from Hornsby [14].*

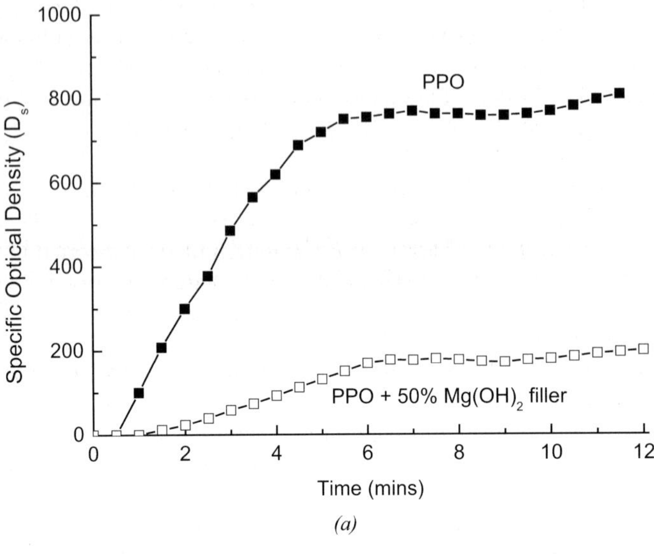

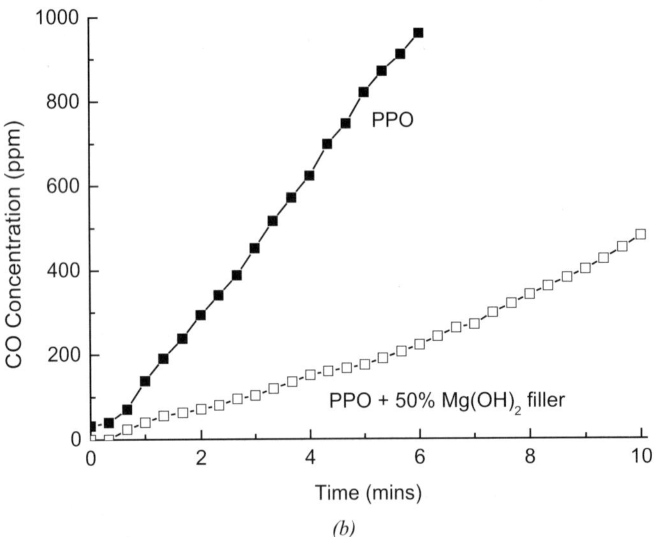

Figure 8.7. Effect of magnesium hydroxide filler on the (a) specific optical density and (b) carbon monoxide yield for polyphenylene oxide. Data from Hornsby and Watson [13,15].

### 8.4.3.4   Antimony Oxide Filler

Antimony oxide ($Sb_4O_6$) is another active filler used in polymers because of its good flame retardant properties [1,3,16]. When $Sb_4O_6$ decomposes at elevated temperature, the antimony is released into the gas phase where it reacts with atomic oxygen, water vapour and hydroxyl radicals. This produces reactive SbOH˙ and SbO˙ species that are highly efficient in scavenging the H˙ radicals driving the exothermic reactions in the flame. By eliminating H˙ radicals, the antimony compounds reduce the flame temperature that in turn slows the decomposition rate of the polymer. While antimony oxide can be used alone in polymers, more often it is used with flame retardant halogen compounds. The synergistic flame retardant mechanisms achieved by combining antimony oxide and halogens are described later in the chapter.

### 8.4.3.5   Iron-Based Fillers

A variety of organometallic and inorganic iron compounds, such as ferrous oxide, ferric oxides and ferrocene, are used as flame retardants and smoke suppressants in polymers [4]. However, the efficacy of these compounds in polymer composite materials has not been thoroughly evaluated. Ferrocene ($Fe(C_5H_5)_2$) and its derivatives are believed to decompose at elevated temperature which promotes the formation of char via a Lewis acid reaction process. The increased char yield reduces the amount of flammable organic volatiles released by a decomposing polymer, which lowers the fuel load in the flame.

Carty and White [17] studied the effect of FeOOH and other inorganic iron compounds on the char yield and fire reaction properties of unplastictised poly(vinyl chloride), acrylonitrile-butadiene-sytrene (ABS), and blends of PVC and ABS. Large increases in the char yield and limiting oxygen index together with substantial reductions in the heat release rate and smoke yield occurred when only a few parts per hundred of the iron compounds were added to the PVC and ABS/PVC blends. For example, Fig. 8.8 shows the effect of increasing FeOOH loading on the char yield of a 70%ABS/30%PVC blend heated to 500, 650 or 800°C in a vertical tube furnace. The char yield increases rapidly with relatively small additions of FeOOH, although the maximum char yield saturates at loading levels above about 10 pph. The activity of the filler becomes more effective as the temperature is reduced from 800 to 500°C, possibly due to a slower oxidation rate of the char. The addition of small amounts of FeOOH also causes a large reduction in the amount of smoke released from the ABS/PVC blend, as shown in Fig. 8.9, with the smoke density value being halved at the relatively low filler loading of 10 pph. Despite the high flame retardancy and smoke suppression of iron oxide compounds, they are used sparingly because of their high cost, high vapour pressure, and red/orange colour that is difficult to remove without the addition of a large amount of pigment filler.

    Fire Properties of Polymer Composite Materials

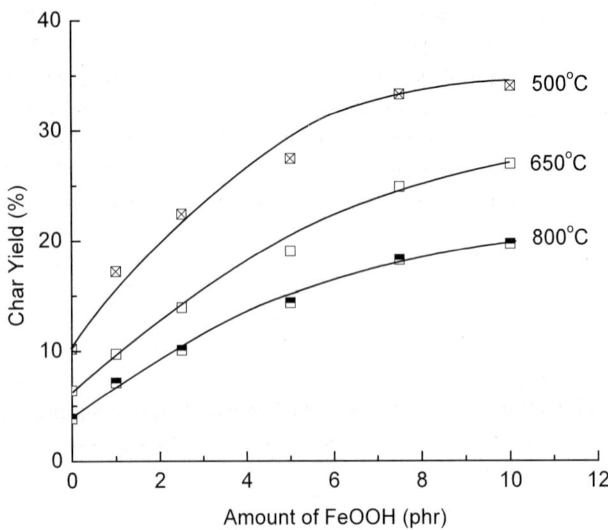


Figure 8.8. *Effect of FeOOH content  on the char yield of a 70ABS/30PVC polymer blend heated at different temperatures.  Data from Carty and White [17].*

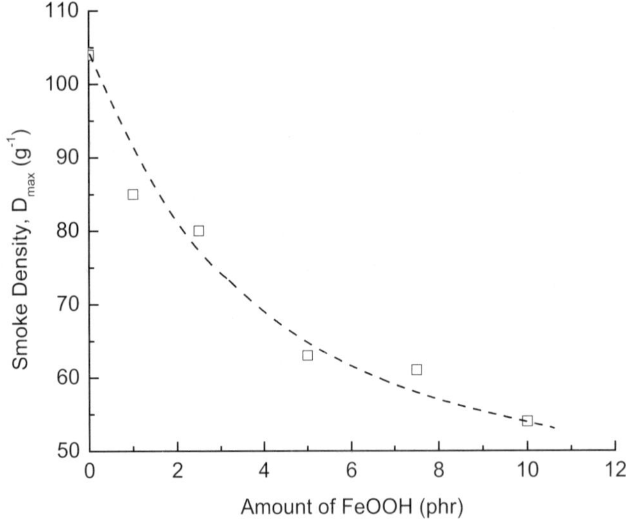


Figure 8.9.  *Effect of FeOOH content  on the smoke density of a 70ABS/30PVC polymer blend.  Data from Carty and White [17].*

### 8.4.3.6  Zinc Oxide and Borate Fillers

Zinc stannate ($ZnSnO_3$), zinc hydroxystannate ($ZnSnO_3.3H_2O$) and zinc borates are active flame retardants used occasionally in halogenated polymers [4,8]. Zinc borates are used more often than zinc stannate and zinc hydroxystannate because of their lower cost and good flame retardant properties in certain polymer systems. For example, Table 8.1 compares the fire reaction properties of a chlorinated polyester without and with zinc hydroxystannate (2 wt%) [18]. The filler improves many of the fire reaction properties, although there is an increase in the amount of carbon monoxide.

*Table 8.1.  Fire reaction properties of a chlorinated polyester with and without 2% zinc hydroxystannate. The properties were measured using a cone calorimeter operated at a heat flux of 50 kW/m². Data from Cusack [18].*

| Property | Virgin Polyester | Polyester + ZHS Filler | Percentage Improvement |
|---|---|---|---|
| Ignition time (s) | 24 | 22 | -8% |
| Peak heat release rate ($kW/m^2$) | 202 | 126 | +38% |
| Average heat release rate ($kW/m^2$) | 115 | 68 | +41% |
| CO yield (kg/kg) | 0.06 | 0.12 | -100% |
| $CO_2$ yield (kg/kg) | 0.70 | 0.69 | +1% |
| SEA ($m^2/kg$) | 825 | 415 | +50% |

Many different zinc borates, with the general formula $xZnO.yB_2O_3.zH_2O$, can be used as flame retardants, although the most common is $2ZnO.3B_2O_3.5H_2O$. This compound acts as a flame retardant in both the condensed and gas phases of the combustion process. Zinc borates decompose at elevated temperature and then react with chlorine-bearing volatiles released by the halogenated polymer to produce zinc chloride and zinc oxychloride free radicals. These compounds scavenge H· species in the flame that reduces the fire temperature. The water in zinc borates is also released at elevated temperature, which dilutes the concentration of H· radicals and organic volatiles in the flame and thereby contributes to a reduction in the flame temperature. After decomposition, the residual borate retained in the decomposed polymer may form a viscous layer on the surface that slows the mass transfer of flammable volatiles into the flame and protects the underlying carbonaceous char from oxidation.

Other types of borates can also be used as flame retardants, most notably boric acid ($H_3BO_3$) and borax ($Na_2B_4O_7.10H_2O$) [3]. Dehydration of borates occurs at elevated temperature, and the liberated water vapour causes the polymer to swell in an intumenscent-type process. The formation of an intumescent type layer at a polymer surface slows the rate of heat conduction into the material, thereby reducing the decomposition reaction rate. Cullis and Hirschler [3] report that borax is an excellent flame retardant whereas boric acid is effective in suppressing glowing combustion but provides only moderate protection against flaming combustion.

### 8.4.3.7 Intumescent Flame Retardant Fillers

A novel approach to improve the flame resistance of composite materials is the addition of intumescent fillers to the polymer matrix [19-23]. Kovlar and Bullock [19] established this method by adding small intumescent particles to a glass reinforced phenolic composite. The filler was an ammonium polyphosphate/pentaerythritol compound which intumescences at elevated temperature. The flame retardant mechanism of the composite is shown schematically in Fig. 8.10. When exposed to heat the intumescent particles decompose in a reaction that yields a large amount of non-flammable, non-toxic gases that remain trapped in the phenolic. As these gases accumulate they cause the soften polymer to foam and swell. (The decomposition process of intumescent materials is described later in the chapter). As the composite continues to heat-up the foamed phenolic itself decomposes, and this produces a highly porous char layer that insulates and protects the underlying virgin composite material. The fibres strengthen the intumescent char and prevent it from spalling or flaking.

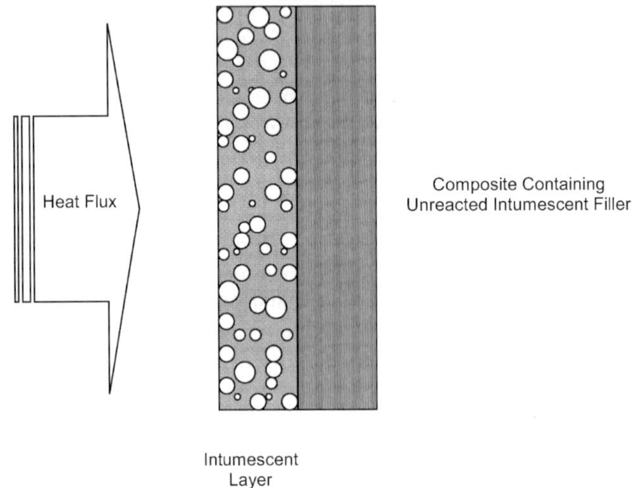


*Figure 8.10. Mechanism of flame resistance in a composite material containing intumescent filler particles. Adapted from Kovlar and Bullock [19].*

Kovlar and Bullock [19] discovered that the foaming process must occur while the polymer is in a soft viscous state. If the filler particles decompose at a temperature below the glass transition temperature of the polymer, then the matrix is too rigid to foam and swell. Instead, the high pressure generated by the rapid accumulation of the gases may cause delamination cracking in the rigid composite, thereby degrading structural performance. Kovlar and Bullock [19] also found that if decomposition of the filler particles occurs at too high a temperature, then the gases escape from the composite without forming an intumescent layer. To ensure a high degree of fire protection, the reaction temperature of the intumescent particles must be above the glass transition temperature but below the decomposition temperature of the polymer matrix.

Kovlar and Bullock [19] measured a large improvement to the fire resistance of phenolic composites when filled with intumescent particles. Figure 8.11 shows the rise in back-face temperature with heating time for a glass/phenolic composite with and without intumescent filler. Both materials were exposed to a propane flame (with a temperature of about 1100°C) for five minutes. The composite containing the intumescent filler shows a much slower rise in back-face temperature. This demonstrates the ability of intumescent filler particles to produce an insulating layer that dramatically slows the rate of heat conduction. However, a limitation of this method is that it is only effective for polymer composites that yield a large amount of char capable of forming a stable, continuous intumescent layer. Polymers that produce a small amount of char, such as polyesters and epoxies, are unable to form a flame retardant intumescent phase in a high enough quantity to form a protective surface layer.

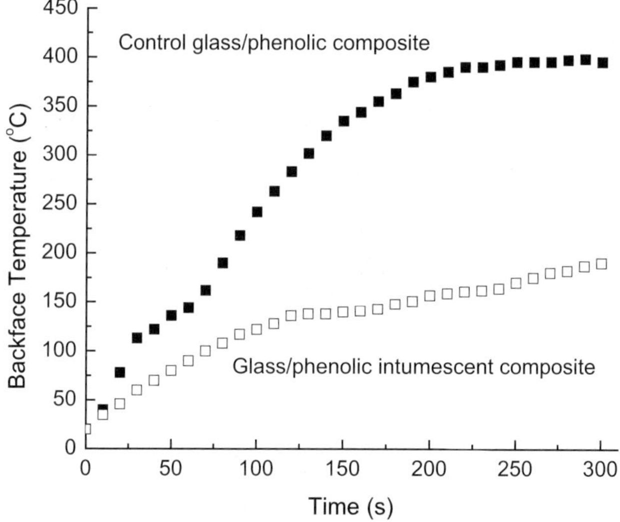


*Figure 8.11. Comparison of the back-face temperature-time profiles for a glass/phenolic composite with and without intumescent filler particles. Data from Kovlar and Bullock [19].*

Kandola et al. [20-23] have also developed fire resistant composite materials using intumescent particles. An important development is that the technique can produce a protective intumescent layer in polymer composites that yield only a small amount of char. The polymer matrix contains intumescent filler particles together with milled polysilicate fibres (Visil) to maximise the fire resistance. Kandola and colleagues found that when a phosphate-based intumescent decomposes at elevated temperature in the presence of polysilicate fibres then the two components interact to produce a rigid char-bonded structure with excellent thermal stability and fire resistance. It is believed that a reaction between melamine phosphate from the intumescent compound and silicic acid from the fibres promotes the formation of char as the matrix decomposes. Kandola and colleagues have shown that polyester, vinyl ester, epoxy and phenolic composites containing phosphate intumescent and polysilicate fibre fillers have much lower peak heat release rates and reduced smoke yields compared with the unmodified composite material.


## 8.5      Flame Retardant Organic Polymers for Composites

### 8.5.1    BACKGROUND

A variety of flame retardant polymers have been developed over the past twenty years, and many of these are suitable for use in fibre composites. The incorporation of bromine, chlorine or phosphorus into the molecular structure of a polymer is the most common method used to improve the flammability resistance of thermoset resins and thermoplastics. The flame retardant mechanisms and flammability properties of these polymers are described in sections 8.5.3 and 8.5.4. The incorporation of nano-sized particles into a resin is another approach for improving fire resistance. Polymer nanocomposites are rapidly emerging as an important class of flame retardant materials, and the next chapter is devoted entirely to a description of their fire reaction properties. A number of other methods can be used to produce flame retardant polymers, including chemical modification of the molecular network structure by graft copolymerisation. This section outlines some of the methods used to produce flame resistant polymers, and describes several important flame retardant organic and inorganic polymer systems being developed for use in composite materials.

### 8.5.2    POLYMERISATION FOR FLAME RESISTANCE

Structural modification of the polymer chains is an effective technique for improving flammability resistance [7]. As outlined in Chapter 2, the thermal stability of a polymer is determined by the bond energy (resonance stability) between atoms on the main chain. Polymers containing large amounts of hydrogen, nitrogen or oxygen display high flammability because of their low bond enthalpy with carbon. The thermal stability of a polymer can be improved by increasing the strength of the chain bonds. Thermal stability is increased by incorporating aromatic and hetrocyclic ring structures with high resonance stabilisation energies into the main chain, and minimising the presence of H, N and O. Not only is the decomposition temperature of a polymer increased by this modification, but the mass fraction of flammable volatiles is reduced which lowers the heat release rate. Figure 8.12 shows the relationship between the density of aromatic groups in the back-bone of polymers against their percentage yields of volatile gas and char [24].  There is a linear correlation between the density of aromatic groups and char yield, with a corresponding linear reduction in volatiles. Increasing the aromaticity of a polymer generally results in better flame retardant properties.  For example, Fig. 8.13 shows that there is a linear relationship between the char yield of polymers and their limiting oxygen index value due to a reduction in the amount of flammable volatiles that provide fuel to sustain flaming combustion [25].

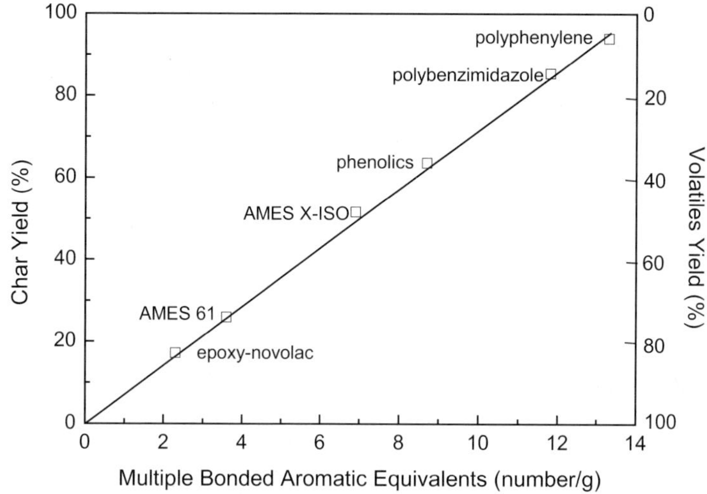


*Figure 8.12.  Relationship between aromatic content and yields of char and volatiles. Adapted from Parker and Kourtides [24].*

The bond strength between the chains is another important factor controlling the thermal stability of thermoset polymers.  Polymers that are able to form a highly cross-linked three-dimensional network structure usually exhibit high thermal stability because rupture and reformation of the cross-links promotes char formation at the expense of flammable volatiles. Examples of highly aromatic, highly cross-linked organic polymers with excellent flame retardant properties are polyphenylenes, poly(p-phenylene oxides) and polybenzimidazoles.  A problem with these resins, however, is that they are not readily processed into fibre composites because of their high processing (softening) temperatures.  Furthermore, these polymers are expensive and therefore are not viable in high commodity composite products.

Many thermoset resins used in composites can be made more flame retardant by increasing the aromatic content.  This is achieved by simply co-polymerising the resin with an aromatic monomer.  Examples of flame retardant resins made by copolymerisation with aromatic monomers are epoxy-nolovacs, phthalonitriles, and aromatic polyesters, vinyl esters, polycarbonates and cyanate esters.

A more common approach to increase the flammability resistance of polymers is by the incorporation of halogen or phosphorus monomers into the chain by copolymerisation, and these are described in the next two sections.

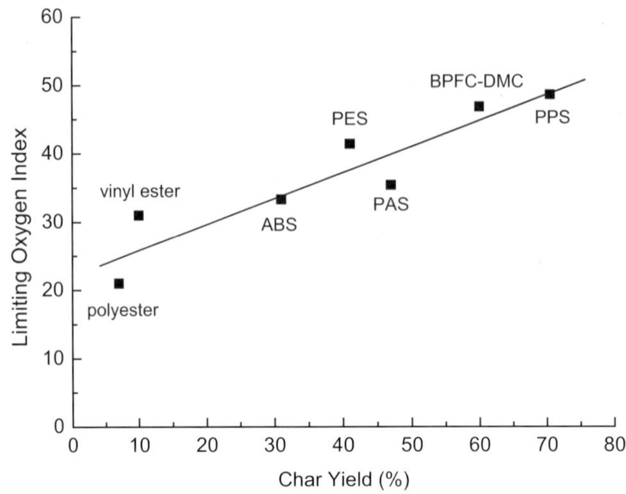


*Figure 8.13. Relationship between char yield and oxygen index for polymers. The char yield is defined as the residual mass measured using TGA at 800°C in an inert atmosphere. Data for thermoplastics from van Krevelan [25].*

### 8.5.3   HALOGENATED POLYMER COMPOSITES

The chemical modification of polymers using organohalogen compounds is one of the more common and effective methods for reducing the flammability of composite materials [1,2,4,6,7,9,26,27]. Halogen-based compounds contain bromine or chlorine which are extremely active flame retardant elements in the gas phase of the combustion process. Halogenated polymers are made by incorporating halogen into the molecular network structure of the resin via co-polymerisation. Bromine is often incorporated into polymers using monomers such as decabromobiphenyl ether, tribromophenol and tetrabromobisphenol. For example, flame retardant polyesters are produced by a polymerisation reaction between polyester and a brominated monomer such as tere-bromophtalic anhydride, dibromoneopentyl glycol or terebromobisphenol A. As another example, flame retardant epoxy can be produced using a brominated monomer derivative of bisphenol A (tetrabromobiphenol A). An example of a brominated monomer and the molecular structure of a polymerised epoxy resin containing this monomer are shown in Fig. 8.14.  Chlorinated monomers can also be incorporated into polymers by polymerisation.  For example, monomers such as chlorendic anhydride are used to produce flame retardant polyester and vinyl ester resins.  Chlorinated polymers are generally less flame retardant than brominated polymers for an equivalent halogen content, but are used more often because of their lower cost.

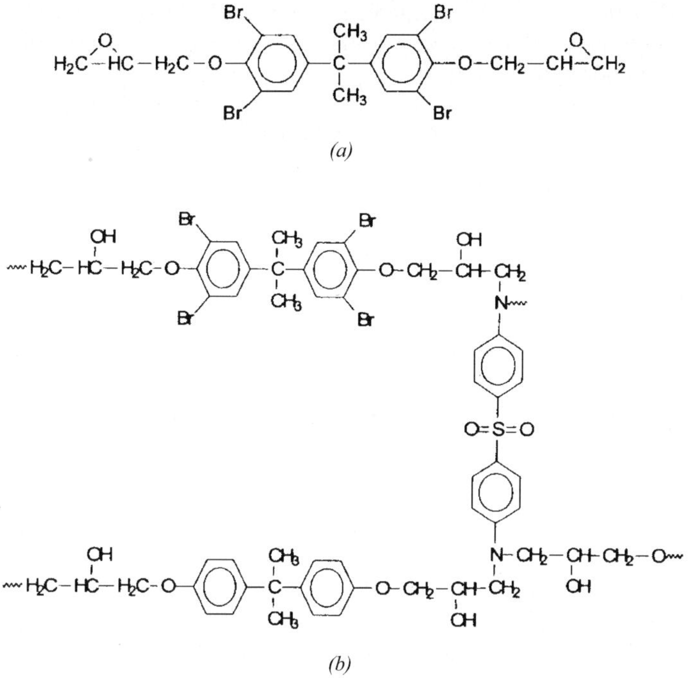

Figure 8.14.  Chemical structure of (a) brominated monomer and (b) brominated epoxy.  (M.P. Luda; A.I. Balabanovic; G. Camino; J. Anal. App. Pyro., 65, **2002**, 25-40; reference 28. Reproduced with permission of Elsevier.)

The main flame retardant action of halogenated polymers is disruption of the gas phase reactions that control the flame temperature of a fire.  Reactive halogen species are released from a decomposing brominated or chlorinated polymer into the flame where they terminate the exothermic decomposition reactions of organic volatiles, and thereby lower the temperature. As mentioned earlier, the main exothermic reaction that provides most of the thermal energy of a flame is:

$$OH^{\cdot} + CO \rightarrow CO_2 + H^{\cdot}$$

The flame retardant mechanism begins with the release of halogen atoms (Br⋅ or Cl⋅), from polymers that do not contain hydrogen. The formation of halogen atoms can be described by the general reaction:

$$MX \xrightarrow{\text{heat}} M\cdot + X\cdot \tag{8.6}$$

where MX is the organohalogen molecule in the polymer, X⋅ is the halogen radical (Br⋅ or Cl⋅) and M⋅ is the residue of the organic molecule. The halogen radicals diffuse from the polymer into the combustion zone of the flame where they react with hydrocarbon fuel (RH) to produce hydrogen halide:

$$RH + X\cdot \rightarrow R\cdot + HX \tag{8.7}$$

where HX is hydrogen halide.

For halogenated polymers that contain hydrogen, the decomposition reaction is different to that described by reactions 8.6 & 8.7. Instead, the reaction of halogenated polymers containing hydrogen promotes the formation of hydrogen halides (HBr or HCl) rather than halogen atoms. Most halogenated polymers used in composite materials contain hydrogen, including polyesters, vinyl esters, phenolics and most types of structural thermoplastics. The production of hydrogen halides in a polymer is described by the reaction:

$$MX \xrightarrow{\text{heat}} M\cdot + HX \tag{8.8}$$

The hydrogen halides diffuse into the flame where they disrupt the combustion process of fuel by inhibiting the exothermic chain branching reactions. The rate of these reactions is controlled by the concentration of H⋅ and OH⋅ radicals in the combustion zone of the flame. Hydrogen halides are believed to react with H⋅ and OH⋅ to produce atomic halide radicals:

$$HX + H\cdot \rightarrow H_2 + X\cdot \tag{8.9}$$

and

$$HX + OH\cdot \rightarrow H_2O + X\cdot \tag{8.10}$$

The atomic halide radicals are much less active than H⋅ and OH⋅ in a flame, and this slows the combustion process. Lewin and Weil [6] report that the rate of reaction 8.9 occurs at about twice the rate of reaction 8.10, and therefore the removal of H⋅ is the main flame retardant mechanism. Both chlorine and bromine halides are excellent scavengers of H⋅ and OH⋅ radicals, although bromine compounds are generally more effective. This is because the strength of the H-Br bond (366 kJ/mol) is lower than the

H-Cl bond (432 kJ/mol), and so dissociation of bromine halides in the flame occurs more readily.

An important aspect of the flame retardant mechanism is that the release of halogen radicals must occur at a lower temperature than the decomposition of the non-brominated part of the polymer.  This is necessary to slow the release rate of flammable organic volatiles from the non-brominated sections of the polymer chain, thereby starving the flame of this fuel.  For example, Luda et al. [28] found that thermal degradation of a brominated epoxy resin in an inert atmosphere occurs first by decomposition of the brominated part of the polymer at about $300^{\circ}C$ with the release of halogen volatiles.  The decomposition of the non-brominated part of the epoxy, with the subsequent release of flammable volatiles into the flame, does not commence until about $330^{\circ}C$.  The flame inhibiting effect of the halogen lowers the flame temperature, and thereby slows the decomposition of the non-brominated parts of the polymer chain.

Volatile organic fragments released from a decomposing halogenated polymer also react with atomic halogen radicals in the flame to produce organohalide gases.  These gases are believed to further assist in the deactivation of H and OH radicals [9].  It is speculated that another flame retardant mechanism is a 'blanketing effect' at the flame-solid interface due to the high flow rate of heavy halogen gases from the polymer [6,9].  This limits the access of oxygen to the surface and slows the oxidation rate of the organic residue in the decomposed polymer, and thereby promotes the formation of an insulating char layer.  Grassie and Hirschler [2] report that the ignition temperatures of certain brominated polymers are increased due to the vapour lowering the oxygen content near the polymer surface. However, conclusive evidence of this blanketing effect for halogenated polymers does not exist, and there is doubt about its importance as a flame retardant process.

A large improvement to the fire reaction properties of composite materials can be achieved by using a brominated polymer.  For example, the reaction properties of glass reinforced laminates containing a general-purpose polyester or a brominated polyester are given in Table 8.2 [12].  These properties were measured using the cone calorimeter technique.  The ignition times of the brominated composite were more than double the times of the non-brominated material.  Furthermore, the peak heat release rate, average heat release rate and smoke parameter values for the brominated composite were less than half the values for the non-brominated laminate.  Improvements to the fire reaction properties of other types of brominated and chlorinated polymers and polymer composites have been reported [eg. 1,2,8,6,9,28-30], and the magnitude of the improvement is generally better than that achieved using flame retardant filler compounds.

The efficacy of brominated polymers as flame retardants is strongly dependent on their bromine content.  The bromine content must be above 20 wt% to impart significant flammability resistance. Unlike brominated resins, no simple correlation exists between the halogen content and flammability resistance of chlorinated polymers. Most polymers need a chlorine content of more than 25 wt%, however increasing the chlorine content above this level does not necessarily result in a corresponding improvement to the flammability resistance.

Table 8.2. *Comparison of the fire reaction properties of a non-brominated and brominated glass/polyester composite. Data from Scudamore [12].*

| Property | Heat Flux | Non-brominated Composite | Brominated Composite | Percentage Improvement |
|---|---|---|---|---|
| Ignition time | 35 kW/m$^2$ | 41 s | 93 s | 127% |
|  | 50 kW/m$^2$ | 25 s | 62 s | 148% |
|  | 75 kW/m$^2$ | 13 s | 31 s | 139% |
| Peak HRR | 35 kW/m$^2$ | 327 kW/m$^2$ | 112 kW/m$^2$ | 292% |
|  | 50 kW/m$^2$ | 374 kW/m$^2$ | 159 kW/m$^2$ | 235% |
|  | 75 kW/m$^2$ | 471 kW/m$^2$ | 174 kW/m$^2$ | 271% |
| Average HRR | 35 kW/m$^2$ | 78 kW/m$^2$ | 38 kW/m$^2$ | 205% |
|  | 50 kW/m$^2$ | 115 kW/m$^2$ | 49 kW/m$^2$ | 235% |
|  | 75 kW/m$^2$ | 109 kW/m$^2$ | 83 kW/m$^2$ | 131% |
| Smoke Parameter | 35 kW/m$^2$ | 338 MW/kg | 94 MW/kg | 360% |
|  | 50 kW/m$^2$ | 374 MW/kg | 155 MW/kg | 241% |
|  | 75 kW/m$^2$ | 457 MW/kg | 175 MW/kg | 261% |

The flammability resistance of brominated and chlorinated polymers can be modified considerably when used with a flame retardant filler. The combination of fillers with halogens can have 'additive', 'antagonistic' or 'synergistic' effects on the flame retardant properties of a polymer system. An additive effect is when the flame retardant efficiency of the total polymer system is equal to the combined efficiencies of the halogen and filler. That is, the halogen and filler operate independently to increase the flammability resistance, and there is no interaction between the two that promotes or decreases the flame retardant action. An example of the additive effect is a halogenated polymer containing certain inert fillers. The halogen improves the flammability resistance in the gas phase while the filler operates in the condensed phase by reducing the fuel load of the polymer and acting as a heat sink. Both compounds act independently to increase the fire performance of the polymer system. An antagonistic effect is when the efficiency of the polymer system is less than the additive efficiencies of the individual components. The halogen and filler interfere with the flame retardant action of each other, thereby decreasing the overall flammability resistance of the polymer.

The best situation is when the halogen and filler have a synergistic flame retardant action. This occurs when the flame retardant efficiency of the polymer system is greater than the additive efficiencies of the halogen and filler. A wide variety of reactive compounds can be used as a synergistic filler in halogenated polymers. The compounds include bismuth oxide, molybdenum oxide, tin oxide and boron compounds, although the most common is antimony oxide ($Sb_2O_3$). This compound is a weak flame retardant when used alone in a non-halogenated polymer, but when used in brominated resin its flame retardant efficiency is improved dramatically.  The improvement is due to synergistic interactions between the flame retardant mechanisms of the halogen and antimony oxide.  During thermal decomposition of a halogenated polymer containing antimony oxide the halogen-bearing volatiles react with antimony volatiles in the gas phase to produce antimony halide or oxyhalide compounds.  In the case of chlorinated polymer systems, the antimony oxide reacts in the solid phase with hydrogen chlorine gas liberated from the polymer to produce antimony oxychloride:

$$Sb_2O_3 + 2HCl \rightarrow 2SbOCl + H_2O \qquad (8.11)$$

Cullis and Hirschler [3] report that the antimony oxychloride breaks down via a series of temperature-dependent reactions with the formation of antimony trichloride:

$$5SbOCl\ (s) \rightarrow Sb_4O_5Cl_2\ (s) + SbCl_3(g) \quad T = 520\text{-}555K \qquad (8.12a)$$

$$4Sb_4O_5Cl_2(s) \rightarrow 5Sb_3O_4Cl(s) + SbCl_3(g) \quad T = 685\text{-}750K \qquad (8.12b)$$

$$3Sb_3O_4Cl(s) \rightarrow 4Sb_2O_3(s) + SbCl_3(g) \quad T = 750\text{-}840K \qquad (8.12c)$$

The antimony trichloride vapour diffuses from the decomposing polymer into the flame, where it inhibits the combustion process by scavenging $H^{\cdot}$ and $OH^{\cdot}$ radicals.  For example, the elimination of $H^{\cdot}$ is believed to occur by the reactions:

$$SbCl_3 + H^{\cdot} \rightarrow SbCl_2 + HCl \qquad (8.13a)$$

$$SbCl_2 + H^{\cdot} \rightarrow SbCl + HCl \qquad (8.13b)$$

$$SbCl + H^{\cdot} \rightarrow Sb + HCl \qquad (8.13c)$$

In addition to this flame retardant process, antimony oxide improves the flammability resistance by diluting the organic content of a polymer system.  Under some circumstances it may be possible for the solid residue of the filler compound to form a protective inorganic coating on the polymer surface.  This combination of gas-phase and solid-phase processes is responsible for the synergistic flame retardant efficiency of halogenated polymers filled with antimony oxide.  Another feature of these polymers that is their physical properties and durability often improve with the addition of antimony oxide.  At the same time, however, the tensile and impact properties can be degraded and there can be a large rise in the amount of smoke and toxic decomposition products (eg. HCl) evolved during polymer combustion.

The improved fire performance derived from the addition of antimony oxide and other fillers have been evaluated for a variety of halogenated thermoset polymers and thermoplastics. The flammability resistance improves with an increasing amount of additive up to critical loading level, above which adding more will result in no further improvement or degrade the flammability resistance. Casu et al. [31] examined the effect of antimony trioxide loading on the flammability properties of a glass/poly(butylene terephthalate) composite. Figure 8.15 shows that the limiting oxygen index value for the composite increased steadily with filler content up to about 20%, at which point the flammability resistance was nearly doubled. The improvement is attributed to the removal of H and OH radicals in the flame by HBr and SbBr$_3$ volatiles as well as the increased charring of the polymer matrix due to the bromination [31,32]. However, increasing the filler content caused a rise in the amount of smoke and carbon monoxide released from the material.

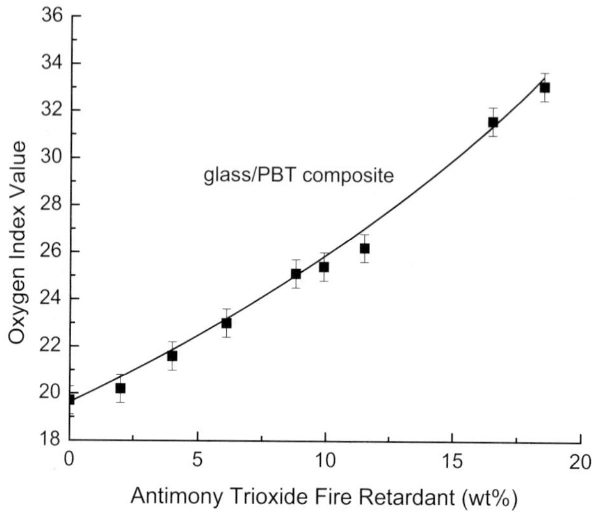


*Figure 8.15. Effect of antimony trioxide content on the limiting oxygen index for a glass/poly(butylene terephthalate) composite. Data from Casu et al. [31].*

The effect of adding metal oxides on several fire reaction properties of a brominated vinyl ester that is used in fibreglass composites is shown in Fig. 8.16 [33]. The fillers reduced the flame spread index and, with the exception of ATH, increased the limiting oxygen index. Of these additives, antimony trioxide showed the strongest synergist interaction with the brominated resin; with a reduction in the flame spread index of over 400% and an increase in the limiting oxygen index of about 25%. The influence of the fillers on the specific optical density of smoke is mixed. Alumina trihydrate, antimony pentoxide and zinc borate lowered the smoke density while the antimony trioxide caused a slight increase, possibly due to the high content of water vapour in the smoke.

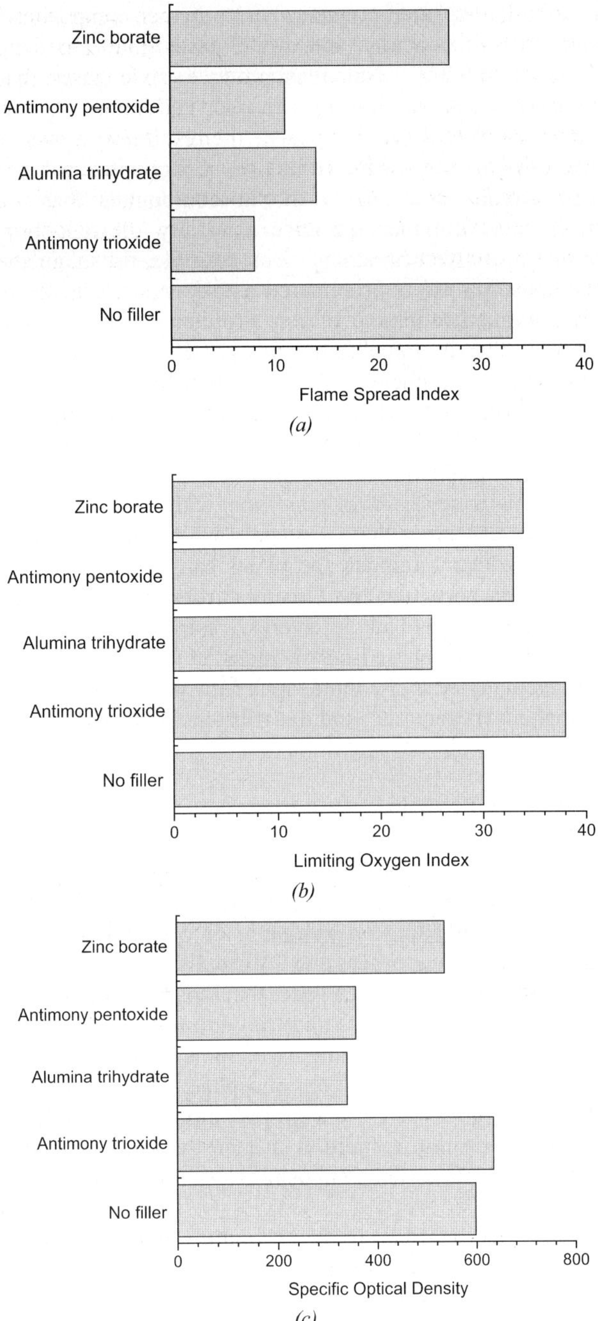



*Figure 8.16. Effect of flame retardant fillers on the (a) flame spread index, (b) limiting oxygen index and (c) specific optical density of a brominated vinyl ester. Data from Morchat and Hiltz [33].*

A major concern with halogenated polymers and polymer composites is the release of smoke containing corrosive, acidic and toxic gases that are serious health and environmental hazards [4,9,33]. Halogens produce toxic gases that are extremely hazardous, particularly in confined spaces with poor ventilation. Chlorinated polymers release copious amounts of HCl gas that attacks the respiratory system and eyes, which impairs the ability of people to escape from fire. Chlorinated polymers also produce chlorinated dibenzodioxins and related dioxin compounds that are highly toxic. Decomposing brominated polymers produce a variety of toxic bromine-containing organic volatiles that also affect breathing. For example, the major gas products from the thermal decomposition of a brominated epoxy resin are brominated phenols; bromine-containing aromatic/aliphatic ethers; propilenyl bromine; methyl, ethyl and propyl bromides; and other bromine complexes [28,29]. Of particular concern is the formation of polybrominated diphenyl ethers and other brominated aromatics that break down in the flame to form toxic and environmentally stable brominated dibenzodioxine and dibenzofuran compounds [34]. Exposure to a high concentration of dioxins can cause a variety of health problems, including cancers, skin discolouration, skin rashes and chloracne, which is a severe skin condition with acne-type lesions. A further concern is that dioxins can enter the eco-system where they remain stable for many years during which time many animals and plants can be affected. For these reasons, environmental groups (in particular the Greens parties in Germany and Scandinavia) have pressured Governments to ban or severely restrict the use of halogens. These polymers are gradually being phased-out in many countries, and  are being replaced with more environmentally-friendly flame retardant polymers containing brominated indan, tris(tribromophenyl) cyanurate, and tris(tribromoneopentyl).

## 8.5.4  FLAME RETARDANT PHOSPHORUS POLYMER COMPOSITES

The flammability resistance of polymers and polymer composites can be greatly improved by the addition of phosphorus [2,5,6,27,35-37]. The most common method for adding phosphorus is blending an inorganic or organic-based phosphorous filler compound into a polymer during processing. Virtually any phosphorus compound can provide some degree of fire resistance, and the most common types are elemental phosphorus, ammonium polyphosphates and trialylphosphates. Phosphorus can also be incorporated into the molecular structure by copolymerisation of the resin with a reactive organophosphorous monomer (eg. phosphate esters, polyols, phosphates) or halogenated phosphate (eg. tris(1-cloro-2-propyl) phosphate, tris(2,3-dibromopropyl) phosphate). The polymerisation method is used to produce many varieties of flame retardant polymers suitable for use in composite materials.

Phosphorus acts as a flame retardant in the gas and/or condensed phase, depending on the chemical nature and thermal stability of the host polymer. The gas phase mechanism dominates in most thermoplastics and non-oxygenated thermoset polymers. This mechanism involves the release of phosphorus radicals from the polymer at elevated temperature, although to be effective the volatilisation process must occur below 350-400°C or otherwise the polymer itself will decompose.  A variety of phosphorus radicals can be released into the flame, depending on the temperature and composition of the phosphorus-containing flame retardant.  For example, the decomposition of triphenylphosphate $[(C_6H_5)_3PO]$ results in the release of PO radicals and smaller amounts of P, $HPO_2$ and $P_2$ volatiles.  PO has a strong affinity to H and OH radicals, and terminates combustion processes in the flame by the reactions:

$$PO + H \rightarrow HPO \tag{8.14a}$$

$$PO + OH \rightarrow HPO + O \tag{8.14b}$$

$$HPO + H \rightarrow H_2 + PO \tag{8.14c}$$

The HPO produced in this reaction sequence is inherently less reactive than the H and OH radicals it replaces, and thereby flaming combustion is suppressed.  A secondary flame retardant mechanism operating is the gas phase is a blanketing effect at the hot surface of the polymer.  Many of the phosphorus-containing volatiles released from a decomposing polymer are relatively heavy, and these form a vapour-rich phase at the polymer surface that restricts the access of oxygen.

When phosphorus compounds are used in oxygenated and hydroxylated organic polymers they act mainly as a flame retardant in the condensed phase.  Phosphorus in these polymer systems promotes the formation of char that reduces the amount of flammable volatiles released into the flame. Phosphorus can also accelerate heat loss in some thermoplastics by promoting melting and dripping.  Further information on the modes of action of phosphorus flame retardants can be found in a comprehensive review by Granzow [26].

As mentioned, the efficacy of phosphorus as a flame retardant is strongly dependent on the chemical nature of the polymer.  Some polymer systems, such as cyanate esters, show no improvement to their flammability resistance when phosphorus is added, even in large amounts.  Phosphorus is also largely ineffective in polymers that do not contain hydroxyl groups because a large amount (50-90%) is lost by evaporation as $P_2O_5$ and other phosphrous oxides [6].  As a rule, the flame retardant efficiency of phosphorus increases with the oxygen content of the polymer.

The effect of phosphorus content on the flame retardant behaviour of polymers has been extensively studied, and it is often found that flammability resistance usually improves with increasing phosphorus content.  For example, Bannister et al. [39] found that the fire reaction properties of an epoxy-based composite material improved rapidly with phosphorus content.  Figure 8.17 shows the effect of phosphorus content on the vertical burn time of a carbon/epoxy composite.  The burn time is defined by the time taken for the specimen standing in a vertical orientation to self-extinguish after removal of an ignition flame.  The burn time drops rapidly with increasing phosphorus content up to about 3%.  Adding more phosphorus above this level only causes a further slight reduction in burn time, revealing that a large improvement to the flammability resistance can be achieved with a small loading of the fire retardant. Bannister et al. [38] also examined the effect of phosphorus on the heat release rate properties of a carbon/epoxy composite when exposed to an external heat flux of 50 kW/m$^2$. Both the peak and average heat release rates decreased with an increase in the amount of phosphorus, as shown in Fig. 8.18. Again, a large improvement is achieved with a modest amount of phosphorus, with only 1.5 wt% reducing the peak and average heat release rates by 25% to 30%.

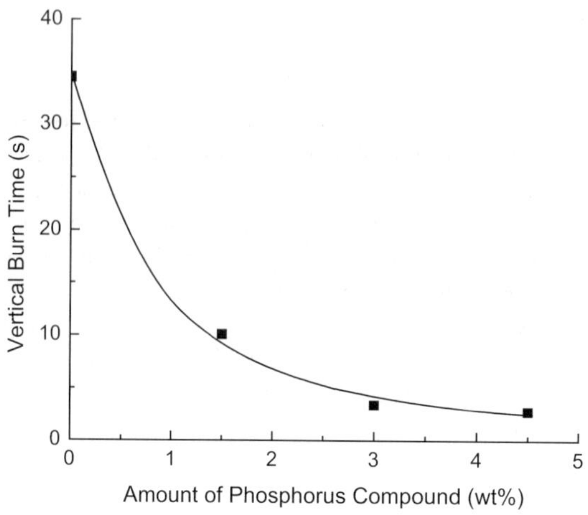

*Figure 8.17. Effect of phosphorus content on the vertical burn time of a carbon/epoxy composite. Data from Bannister et al. [38].*

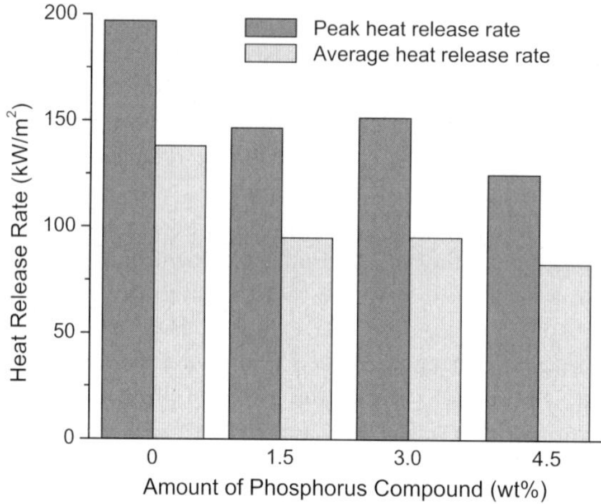


*Figure 8.18. Effect of phosphorus content on the peak and average heat release rates of a carbon/epoxy composite. Data from Bannister et al. [38].*

The flame resistance of polymers can be greatly improved when phosphorus and halogen compounds are used together.  A synergistic flame retardant response occurs in many polymers containing both phosphorus and bromine or chlorine, and the synergistic effect is more pronounced when the two elements exist in the same organic molecule [2].  When the molecule decomposes at elevated temperature a variety of highly flame retardant volatiles, such as phosphorus halides and oxyhalides, are produced.  These compounds lower the flame temperature by terminating the radical chain reactions.  Phosphorus halides and oxyhalides are much heavier volatiles than the hydrogen halides produced by conventional halogenated polymers, and therefore they remain in the combustion zone of a flame for a longer time which makes them more durable scavengers of H and OH radicals.

Table 8.4 gives two examples of the synergism derived from combinations of phosphorus and halogens.  The amount of phosphorus by itself, halogen by itself, and combinations of phosphorus and halogen to achieve the same degree of flame resistance in a polyester and epoxy are given in the table.  It is apparent that phosphorus alone is a more efficient flame retardant than bromine or chlorine in both polymers.  Furthermore, the amount of phosphorus and halogen needed to achieve a given degree of flammability resistance is reduced considerably when used in combination. Unfortunately, however, when phosphorus and halogens are used together the smoke toxicity is greater than when phosphorus is the only flame retardant because of the formation of chlorinated or brominated toxins.

*Table 8.4.  Amounts of phosphous, halogen, and phosphorus with halogen combinations required to achieve an arbitrary degree of flame retardant in an polyester and epoxy. Data from Grassie and Hirschler [2].*

| Polymer | Phosphorus alone (%) | Halogen alone (%) | Phosphorus + Halogen (%) |
|---|---|---|---|
| Polyesters | 5% | 25% Cl<br>12-15% Br | 1.0%P + 15-20% Cl<br>2.0%P + 6% Br |
| Epoxy | 5-6% | 26-30% Cl<br>13-15% Br | 2.0% P + 6% Cl<br>2.0%P + 5% Br |

## 8.5.5    GRAFT COPOLYMERISATION FOR FLAME RESISTANCE

Graft copolymerisation is an emerging technique for the production of flame retardant polymers [39].  This technique basically involves attaching a monomer, which is a strong char former, onto the polymer chain.  The copolymerisation process can occur via two routes known as 'grafting from' and 'grafting onto'.  The former process involves the polymer reacting with an initiator compound to create radical sites along the polymer chain.  The monomers are then chemical bound to the chain by the radicals. The grafting onto process occurs when the monomer reacts with the initiator to form a radical that is then bonded to the polymer chain. Regardless of the process, it is essential that the monomer thermally decomposes at a lower temperature than the host polymer, and yields a large amount of char that provides protection of the polymer. Wilkie and colleagues [42] have found that several types of inorganic salts are suitable monomers that can be grafted to thermoplastics. These salts include alkaline earth salts, alkali metal salts, and salts of methacrylic and acrylic acids. When grafted to a thermoplastic, these monomers decompose at elevated temperature to produce an insulating and adherent char on the polymer surface that provides fire protection. In addition, the decomposition reaction of the salts produces a significant amount of carbon dioxide that serves to quench the flame.

Table 8.5 shows an example of the efficacy of the graft copolymerisation technique in improving the fire resistance of polymers [40]. The fire reaction properties of virgin ABS and grafted ABS measured using the cone calorimeter operated at a heat flux of 25 kW/m$^2$ are compared. The ABS is grafted with 21% sodium methacrylate, and this resulted in large improvements to all the properties. Graft copolymerisation is a promising technique for producing flame retardant polymers, however it is an emerging technology that requires further development. While a variety of flame retardant thermoplastics can be produced using the technique, the graft copolymerisation of engineering thermoset polymers commonly used in structural composites requires further research and development.

*Table 8.5. Comparison of the fire reaction properties of virgin and grafted ABS. Data from Susuki and Wilkie [40].*

|  | Virgin ABS | Grafted ABS | Improvement (%) |
|---|---|---|---|
| Time-to-ignition (s) | 285 | 460 | 161% |
| Peak heat release rate (kW/m$^2$) | 901 | 259 | 348% |
| Time to burnout (s) | 670 | 1400+ | 209%+ |
| Mass loss rate (mg/s) | 170 | 40 | 425% |
| Mass loss (% @ 20 mins) | 92 | 37 | 249% |

## 8.6    Flame Retardant Inorganic Polymers for Composites

The use of specialty inorganic polymers in composites is another emerging flame retardant technology. Several inorganic polymers, most notably geopolymers, POSS® (polyhedral oligometric silsesquioxanes) and FyreRoc® (inorganic metallosilicates), have considerable promise because they can be processed and cured into fibrous composites under conditions similar to organic resins, are reasonably inexpensive, and have outstanding flammability resistance [41-44].

Geopolymers are potassium aluminosilicate compounds prepared from a two-part system consisting of an alumina liquid suspension and silica powder.  When the components are mixed into a paste the geopolymer has a sufficiently low viscosity to enable it to be applied as a binder onto reinforcing fabrics.  Geopolymers cure into a rigid glassy matrix at reasonably low temperatures ($<150^{\circ}$C) that makes them suitable for processing. When cured, the general chemical composition of geopolymers is $K_n\{-(SiO_2)_z-AlO_2\}_n\cdot wH_2O$ where $z \gg n$ [41,42].  Geopolymers are extremely resistant to fire, and will not ignite, spread flame, release heat or cause flashover. For example, Lyon et al. [45] evaluated the fire reaction properties of carbon/geopolymer and glass/geopolymer composites, where the empirical composition of the geopolymer was $Si_{32}O_{99}H_{24}K_7Al$, and found their fire reaction properties were superior to those of conventional organic matrix composites. Table 8.6 compares the fire reaction properties of the carbon/geopolymer with several carbon/thermoset and carbon/thermoplastic composites tested under the same fire conditions, and the geopolymer is superior in every property.  Geopolymers remain largely unaffected by fire at temperatures below $\sim1000^{\circ}$C.  Chemically bound water is released as steam at temperatures between 250-625$^{\circ}$C, which may have an additional quenching effect on the flame.

*Table 8.6.  Fire properties of carbon fibre composites, including geopolymer. Data from Lyon et al. [45].*

| Polymer Matrix | Time-to-ignition (s) | Weight Loss (%) | Peak HRR (kW/m$^2$) | Average HRR (kW/m$^2$) | Smoke (m$^2$/kg) |
|---|---|---|---|---|---|
| Geopolymer | 0 | | 0 | 0 | 0 |
| Epoxy | 94 | 24 | 171 | 93 | - |
| Phenolic | 104 | 28 | 177 | 112 | 253 |
| PPS | 173 | 16 | 94 | 70 | 604 |
| PEEK | 307 | 2 | 14 | 8 | 69 |

POSS[®], which is short for polyhedral oligometric silsesquioxanes, are a new hybrid inorganic-organic polymer nanocomposite produced by a sol-gel process. The inorganic component is an intermediate silica ($SiO_{1.5}$) between that of silica ($SiO_2$) and silicone ($R_2SiO$). The chemical diversity of POSS is very broad and a large number of POSS monomers and polymers are available or undergoing development.  The molecular structure of POSS[®], shown in Fig. 8.19, has a cage-like configuration.  The X represents one or more reactive organo-functional groups that are bonded onto organic polymers by polymerisation or grafting.  The inorganic component of POSS[®] (the $SiO_{1.5}$ cage) provides excellent thermal and oxidation stability, and provides high flammability resistance when used in composite materials.

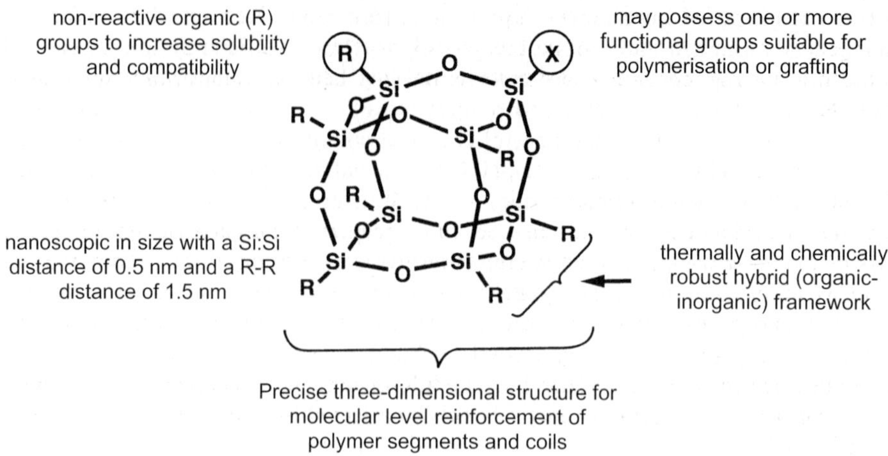


*Figure 8.19. The chemical structure of POSS®. From Sorathia [43].*

## 8.7 Flame Retardant Fibres for Composites

While a great deal of attention is devoted to the development of flame retardant polymers for use in composites, comparatively little is done to improve the flammability resistance of the fibre reinforcement. Glass fibre, which is the most common reinforcement, is not combustible while carbon fibres remain stable (in the absence of oxygen) over the temperature range of most fires. However, the organo-sizing and binding agents used on these fibres do contribute to the production of smoke and volatiles released by a decomposing composite, as described in Chapter 3. The contribution of the size to the flammability of composites is insignificant, and therefore the need to develop flame retardant sizing agents is not necessary.

Recently, continuous basalt-based fibres have become available [46]. These fibres, produced by a spinning process developed in the former USSR, are roughly similar in composition to S-2 glass but less expensive. Their mechanical properties are somewhat better than those of E-glass, and substantial improvements in fire properties are claimed with fire resistance up to 1500°C. Basalt fibres will undoubtedly find use in a range of fire protection applications, with and without organic matrices.

There is a more urgent requirement to improve the fire resistance of organic fibres. Aramid fibre, which is the most commonly used organic fibre in engineering composites, generally has better flammability resistance than many polymer systems used as the matrix material. The reduction in the volume content of the polymer with the addition of aramid fibres often results in reduced heat release and smoke formation due to the higher char yield of the reinforcement (see Chapters 2 and 3).  However, the development of aramid fibres with enhanced flame retardance has been slow, and is not commercially available for use in composite materials.  The heat resistance and flammability of high performance fibres has recently been the subject of a substantial review [47].

Various techniques have been developed to produce flame retardant polyethylene fibres [7].  These include chemical modification to the polyethylene chain with the addition of flame retardant halogen by chlorination or oxidation chlorophonylation or flame retardant phosphorus-based monomers by radiation grafting.  The flammability resistance can also be improved by increasing the cross-linking density between the polyethylene chains by using low molecular-weight silane oligomers or irradiation with $\gamma$-rays and electrons.  However, the improvements to the fire properties of composites containing flame retardant polyethylene fibres have not been characterised.  The next chapter describes the development of flame retardant fibres made of polymer nanocomposites.

## 8.8      Fire Protective Surface Coatings

### 8.8.1    INTRODUCTION

A common method to protect composites from fire is to use an insulating coating.  The ideal coatings should possess the following properties: non-flammability, low thermal conductivity, strong adhesion (with similar expansion coefficient) to the composite substrate, environmentally durable, wear resistant, light-weight, thin and inexpensive. There are hundreds of coating materials that are commercially available for use on composites, although none possess all of the properties required for an ideal coating.

There are three major classes of insulating coatings: flame retardant polymers, thermal barriers, and intumescent coatings.  Flame retardant polymers are inherently fire resistant organic resins (eg. brominated polymers) or inorganic materials (eg. geopolymers) that are applied as a thin film (usually less than 5 mm) over the composite substrate.  These polymers delay ignition and flaming combustion of the substrate due to their high thermally stability and, in the case of inorganic polymer coatings, low thermal conductivity.  Thermal barrier coatings are usually ceramic-based materials that are non-flammable and have low heat conducting properties.  Examples of these coatings include ceramic (eg. silica, rockwool) fibrous mats and ceramic (eg. zirconia) plasma-sprayed films.  Intumescent materials provide fire protection by undergoing a chemical reaction at elevated temperature that causes the coating to foam and swell. This reaction process produces a highly porous, thick char coating that has very low thermal conductivity.  The ability of these three classes of coatings to protect composite materials against fire is described in this section.

Another class of coatings are ablative materials that provide thermal protection by removing heat from the hot surface by spalling or melting. Ablative materials are rarely used as fire protective coatings on composites [48], and are more commonly used for the protection to polymer composites in high temperature applications, such as rocket nozzles and heat shields to re-entry spacecraft. Due to the limited use of ablative materials as a fire barrier, they are not described in this chapter.

## 8.8.2    FLAME RETARDANT POLYMER COATINGS

Flame retardant organic and inorganic polymer coatings are often used to delay pyrolysis of composites [eg.49,50]. A variety of organic polymers can be used, with the most common being phenolics, brominated resins and alkyd resins, often with a high loading of flame retardant filler. Organic polymer coatings are usually applied by brushing or spraying the liquid resin directly onto the composite substrate or coating the tool surface with the flame retardant resin and then over-laminating with the composite in a process similar to the application of a gel coat. Recent developments in liquid moulding processing allow the possibility of manufacturing a composite with a flame retardant polymer coating in a one-stage operation. This involves injecting a flame retardant resin into the surface region of the laminate to produce the coating while at the same time a conventional polymer is injected into the bulk laminate. The benefits of organic polymer coatings include moderate cost, light-weight and (usually) good chemical compatibility with the composite substrate that ensures good adhesion. However, a limitation of the application techniques is that the maximum coating thickness is usually very thin (under 2-3 mm), which can only provide short-term protection against high temperature fires.

Phenolic coatings are an effective and low cost method to improve the flammability resistance of composites with a highly combustible polymer matrix. Phenolic resins improve the fire performance due to their low yield of flammable volatiles that delays ignition and forms an insulating char layer. However, for a phenolic to create an effective insulating layer it is essential that a thick coating be used. Sorathia and colleagues [48,49] evaluated the fire reaction properties of composites protected with phenolic coatings, and good improvements in the flammability resistance were observed. For example, Fig. 8.20 shows the effect of a phenolic coating on the peak heat release rate and smoke extinction area properties of a glass/vinyl ester tested at different heat fluxes. The phenolic not only reduced these properties, but also reduced the average heat release rate, total heat release, total mass loss, and increased the ignition time.

Some commercial organic coatings do not provide much protection when the film is too thin. Figure 8.21 compares the time-to-ignition values at different heat fluxes for a glass/polyester composite with and without a thin coating (0.5 mm) of halogenated polyester. The ignition times are not improved by the coating, and at the lower heat fluxes the times are actually reduced slightly. The problem with very thin coatings is that the amount of brominated gas released by the flame retardant polymer is insufficient to scavenge a significant amount of the H⋅ and OH⋅ radicals in the flame and thereby terminate the chain branching reactions. To ensure a high level of fire protection, it is necessary to coat the composite with a thick layer of flame retardant resin to provide a high concentration of brominated volatiles to quench the flame.

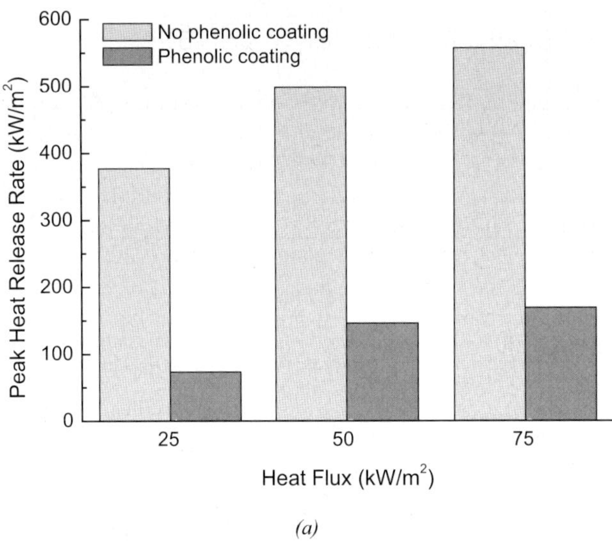

*(a)*

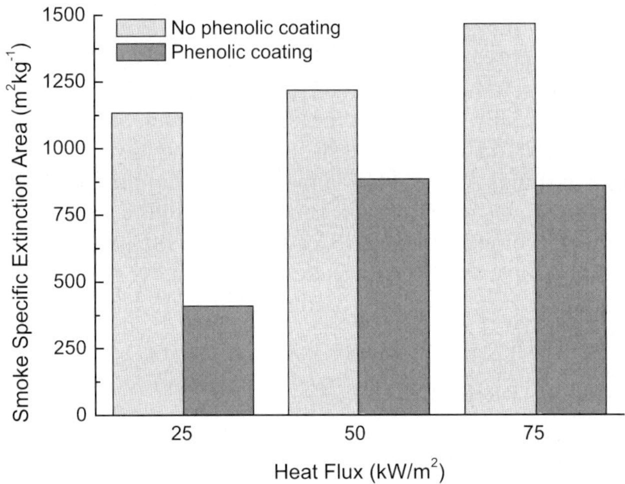

*(b)*

*Figure 8.20. Effect of a phenolic skin on the (a) peak heat release rate and (b) smoke extinction area for a glass/vinyl ester. These properties were measured at a heat flux of 75 kW/m². Data from Sorathia and Beck [48].*

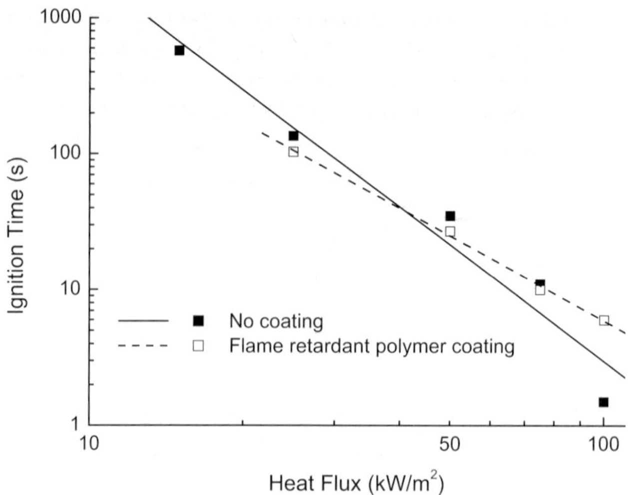



*Figure 8.21. Effect of heat flux on the time-to-ignition for a glass/polyester composite with and without a thin flame retardant (brominated) polyester coating.*

The use of inorganic polymers as flame retardant coatings is a recent development, and the number of materials that are used is currently limited to a few. There is considerable interest in using geopolymers, POSS® and Tecnofire® (which is a char-forming graphite mat) as coating materials, and several other inorganic polymer systems are being developed as coatings. Inorganic polymer coatings provide better fire protection than organic flame retardant polymers because of their higher resistance to pyrolysis and heat conduction. The most common method for applying inorganic coatings is by brushing the uncured polymer directly onto the composite. However, some inorganic resins are viscous which makes it difficult to brush, although the high viscosity makes it is possible to apply coatings up to 8-10 mm thick. The fire protection offered by inorganic coatings improves with their thickness, and therefore the ability to apply thick coatings is important. For example, Fig. 8.22 shows that the peak heat release rate of a sandwich composite drops rapidly with increasing thickness of a geopolymer coating [49]. It is essential, however, that the coating is strongly bonded to the substrate to avoid spalling and flaking.

## 8.8.3    THERMAL BARRIER COATINGS

Thermal barrier coatings provide fire protection by having excellent insulating properties and (in some coating systems) heat reflective properties that direct heat back towards the fire [49,52,53]. The most commonly used coatings consist of mineral fibre or ceramic wool mats, and these are bonded using a high-temperature adhesive to the composite substrate. It is also possible to bond the mat directly onto a composite while the polymer matrix is still curing. Two examples of commercial thermal barrier materials are Rockwool® and Structogard®, and these are widely used for fire insulation in ships and buildings.  Figure 8.23 shows the effectiveness of a ceramic fibre barrier material in providing thermal protection to a glass/polyester composite.  The thermal barrier increases greatly the ignition resistance of the composite, even at very high heat flux.  Furthermore, the minimum incident heat flux needed to ignite the composite is increased from about 15 to 35 kW/m$^2$ due to the excellent insulation of the coating.

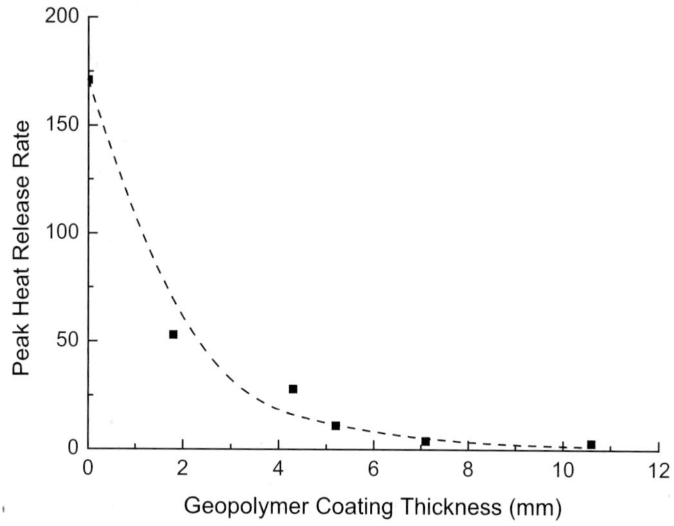


*Figure 8.22.  Effect of geopolymer coating thickness on the peak heat release rate of a balsa core sandwich composite.  Data from Giancaspro et al. [50].*

The protection provided by thermal barriers also improves the fire resistive properties of composites, including increased burn-through time and higher structural properties in fire [49,53]. For example, Fig. 8.24 compares the char growth rates in a glass/polyester composite with and without a thermal barrier coating. The onset of charring occurs after about 30 seconds in the unprotected composite, and at longer heating times the char grows through the material until it is completely burnt-through after 17 minutes.  A long delay in charring and burn-through occurs when the composite is protected with the thermal coating.  Charring does not commence until after 11 minutes and the materials did not completely burn-through after 30 minutes.

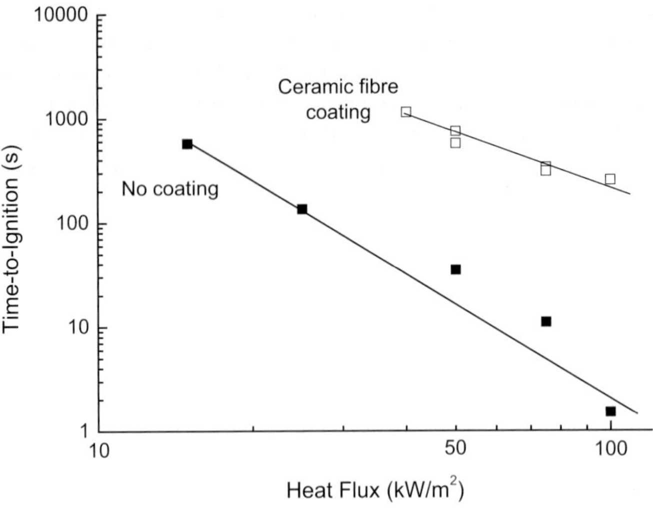



*Figure 8.23.  Effect of incident heat flux on the ignition time of a glass/polyester composite with and without a ceramic fibre thermal barrier.*

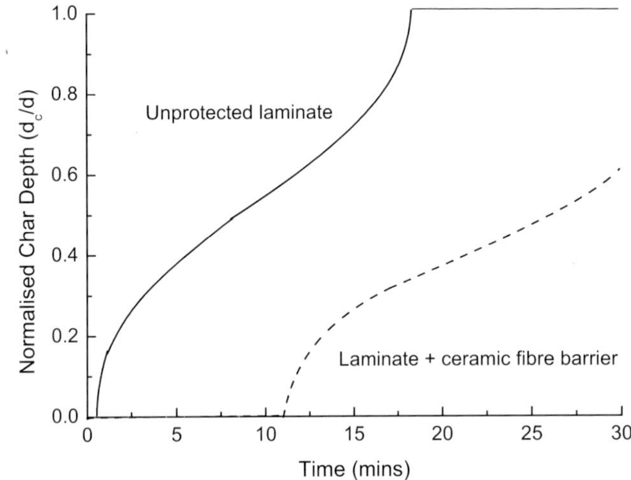



*Figure 8.24.  Effect of exposure time on the amount of charring to a glass/polyester composite with and without a ceramic fibre thermal barrier.  The normalised char depth is the char thickness (d$_c$) divided by the original thickness of the composite (d).  The composite was fire tested in a cone calorimeter at an incident heat flux of 50 kW/m$^2$.*

While ceramic fibre coatings provide a high level of fire protection, they have a number of drawbacks when used on composites.  One major disadvantage is that most coatings must be very thick (at least 10-20 mm) to provide long-term protection against a high temperature fire, and this adds to the weight and bulk of a composite structure. A further problem is that many coatings, particularly those that are extremely good insulators, are expensive.  The cost of using the coatings is increased further because they need to be bonded to the composite structure, which can be labour-intensive.  A concern with using these coatings on ships is they absorb spilled fuel or flammable liquid cargo, which can prolong the combustion process in a fire.

Non-fibrous ceramic coatings are another class of thermal barriers that do not have many of the problems associated with fibre mats.  These coatings are applied as a thin film directly onto the composite substrate using techniques such as liquid or plasma spraying.  Virtually any type of ceramic can be applied provided it can adhere strongly to the composite substrate, with the most common being zirconium oxide and alumina. There is limited information about the efficacy of ceramic coatings, but what data is available shows good fire protection.  For example, Fig. 8.25 shows the improvement to the fire reaction properties of a carbon/epoxy composite when coated with a thin layer (1.25 mm) of zirconium oxide [49].  Despite the improved fire resistance, ceramic coatings are used sparingly on composites because they can be expensive and prone to cracking and spalling due to the mismatch in their thermal expansion properties with the substrate.

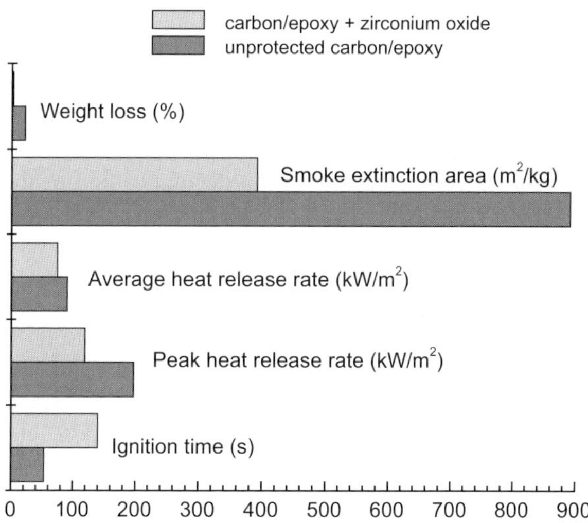


*Figure 8.25.  Fire reaction properties of a carbon/epoxy composite with and without a zirconium oxide coating.  The properties were measured at a heat flux of 75 kW/m$^2$.  Data from Sorathia et al. [49].*

### 8.8.4　INTUMESCENT COATINGS

Intumescent coatings provide fire protection by undergoing an endothermic decomposition reaction process at elevated temperature that causes the material to swell and foam into a highly porous, thick and thermally stable char layer [3,54-58]. The high void content and thickness of the coating allows it to act as an insulation barrier to the underlying composite against flame and heat.

An intumescent coating can be applied by painting or spraying a liquid compound onto a composite. The compounds cure in air over several hours into a solid intumescent film. The maximum coating thickness that can be achieved with this method is under about 5 mm. Thicker coatings are applied by bonding of fibrous intumescent mat directly onto the substrate using high-temperature adhesive paste.

The fire protection provided by intumescent coatings occurs by three reaction processes:
- the coating material decomposes,
- inert gases evolved from the decomposition reaction are produced at a high enough rate to drive back hot convective air currents, and most importantly,
- the coating expands into a highly porous char layer with a high resistance to heat conduction from the flame into the underlying composite substrate.

Intumescent coatings consist of a mix of compounds that each has a role in the intumescent process. The four main types of compounds are a carbon-rich (carbonific) compound, inorganic acid or acid salt, organic amine or amide, and a blowing agent (spumific). For intumescence to occur these compounds must undergo a series of decomposition reactions and physical processes almost simultaneously, but within a proper sequence. The order of these processes is given in Fig. 8.26. If the time between the processes is too long or they do not occur in the correct order, then the coating will fail to intumescence.

The intumescence process commences with the decomposition of the inorganic salt or acid salt within the coating. The decomposition temperature of the acid must be sufficiently high that normal external heating (eg. warming from direct sun-light) does not cause the coating to intumesce in the absence of fire, but the temperature must be below the pyrolysis temperature of the composite substrate. Furthermore, the acid must decompose before any other compound in the coating to ensure dehydration of the carbonific compound. The acid compounds used include zinc borate, linear high-molecular-weight ammonium phosphate, melamine phosphate, organic esters, and salts of ammonium, amide or amine. These compounds decompose at between 100 and $250°C$, which is below the pyrolysis temperature of most organic resins used in composites. The decomposition reaction of the acid is catalysed by organic amides or amines, such as urea, melamine, dicyandiamide and their derivatives.

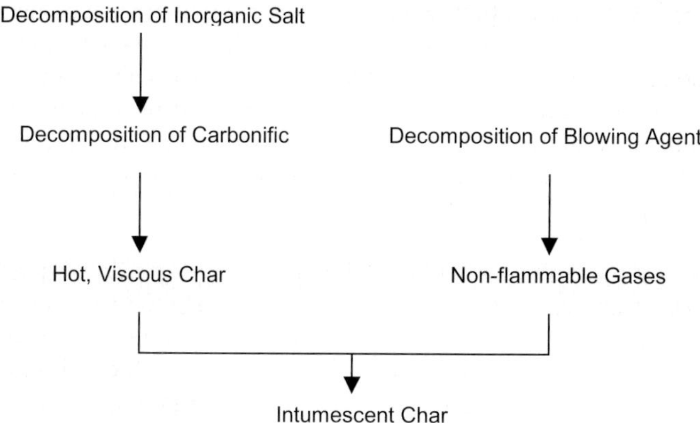



*Figure 8.26. Sequence of intumescent reaction processes.*

The next stage is decomposition of the carbonific by a dehydration reaction with the decomposed inorganic salts. This reaction converts the carbonific into a carbonaceous char. The carbonific is a carbon-rich polyhydric compound that yields a large amount of char, and is usually a polycarbonate (such as starch or polyhydric alcohol) or phenol (such as phenol-formaldehyde). The hot and viscous char is then expanded with the decomposition of the blowing agent. Expansion of the char is dependent on decomposition of the carbonific and blowing agent occurring at the same temperature, otherwise the coating will fail the fully intumesce. The blowing agent decomposes via an endothermic reaction that produces a copious amount of non-flammable gases that cause the char melt to swell. Blowing agents are usually nitrogen compounds such as urea, dicyandiamide, guanidine, melamine and glycine that yield ammonia, carbon dioxide and water vapours. Chlorinated paraffin is also an effective agent that yields hydrogen chloride, carbon dioxide and water vapours. The gases collect into small bubbles that cause the char to foam and swell. The coating eventually solidifies into a thick, multicellular material that slows the rate of heat conduction from the fire into the composite substrate.

A typical example of an intumescent coating is shown in Fig. 8.27, and it has expanded to many times the original thickness. A good intumescent coating expands 50 to 200 times, and forms a fine-scale multicellular network with a cell size of 20 to 50 μm and wall thickness of 6-8 μm [3,59]. Quershi and Krassowski [59] found that the addition of graphite flakes to an intumescent coating improve the fire resistance. The flakes expand up to 100 times on heating that produces a more effective insulating layer. The cell size can be controlled by the addition of inert fillers that assist the cell nucleation process. The addition of fillers such as titanium oxide and silica reduce the average diameter of the cells. The exact chemical composition of the compounds used in commercial intumescent coatings are closely guarded by manufacturers, and while there is a wide choice of compounds from which to formulate intumescent compositions, only a few are used in practice [54]. In addition to the compounds that control the intumescent process, the coating may contain other additives for purposes other than intumescence. For example, coatings may contain antioxidants, thickeners, coalescing agents, pigments for colouring, and milled fibres for structural reinforcement.

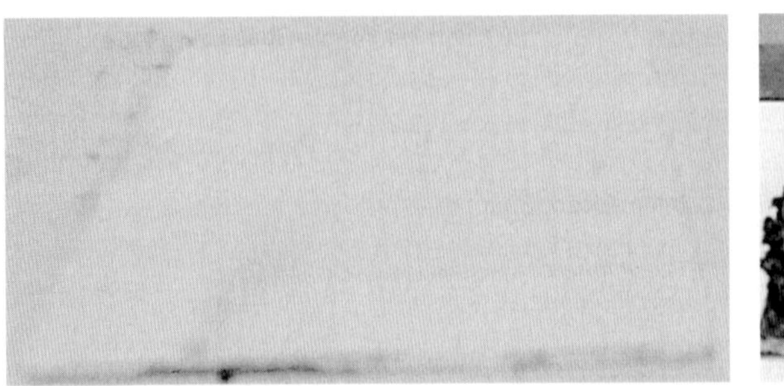

(a)                                                                    (b)

*Figure 8.27. Intumescent coating (a) before fire testing (thickness = 0.8 mm) and (b) after fire testing (thickness = 52 mm). Photographs supplied courtesy of Z. Mathys (DSTO).*

Intumescent coatings are excellent heat insulators that slow the rate of heat transfer into composite laminates in fire. Figure 8.28 compares the temperature rise in a composite panel with and without such a coating. The temperature was measured at the back-face of the panel. Thermo-chemical models, including that described in reference [60] have been developed to predict the thermal protection provided by intumescent coatings.

Intumescent coatings can be extremely effective in delaying combustion, suppressing flame spread, reducing the heat release rate, and lowering the smoke density of composite materials [48,52,53]. For example, Fig. 8.29 shows the effect of incident heat flux on the ignition time of a glass/polyester composite with and without an intumescent coating. One coating is a thick intumescent mat while the other coating was a thin intumescent film. The onset of ignition was delayed considerably by the intumescent coating, with the thick mat providing greater protection because it was able to swell more than the film.

While intumescent coatings are effective at protecting composite materials from heat and flame, they have several disadvantages. A major problem with many commercial coating products is that they do not bond strongly with the substrate, and often fall off during swelling, exposing the underlying composite directly to the flame [49]. This is a common occurrence when a coating is applied to vertical (eg. walls) or overhead (eg. ceiling) structures. It is essential that the coating is bonded strongly to the substrate and has mechanical strength to ensure adequate fire protection. Further problems with intumescent coatings can include their incompatibility with certain manufacturing processes, poor aesthetic features, poor durability, rapid ageing by weathering (eg. UV radiation, moisture absorption), and low resistance to wear and erosion.

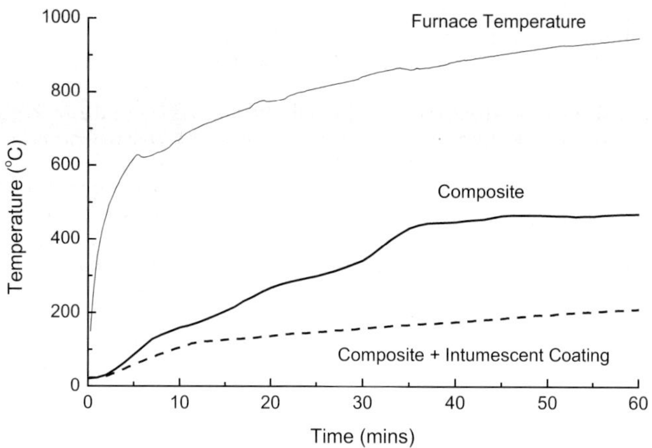


*Figure 8.28.  Comparison of the back-face temperature-time profiles of a composite panel with and without an intumescent coating when exposed to fire.  Data supplied courtesy of Z. Mathys (DSTO).*

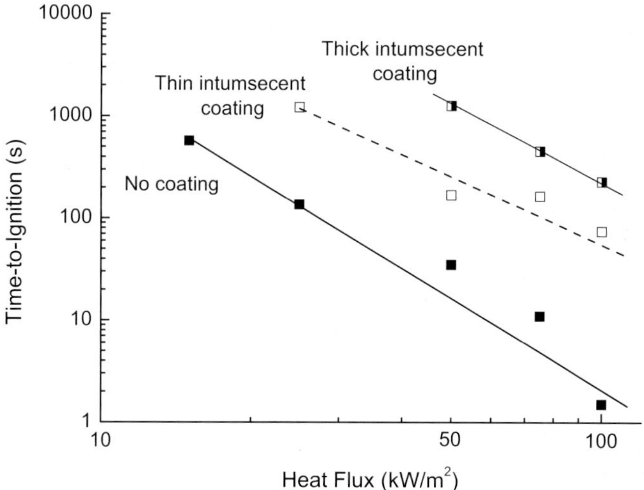


*Figure 8.29.  Plots of heat flux against time-to-ignition for a glass/polyester composite with and without an intumescent coating.*

## References


1. R.G. Gann, R.D. Dipert and M.J. Drews. Flammability. In: *Encyclopedia of Polymer Science*, Vol. 7, Wiley-Interscience, 1987, pp. 154-205.

2. S.L. Madorsky. *Thermal Degradation of Polymers*. New York: Robert E. Kreiger, 1985.

3. C.F. Cullis and M.M. Hirschler. *The Combustion of Organic Polymers*. Oxford: Clarendon Press, 1981.

4. J.R. Ebdon and M.S. Jones. Flame retardants (overview). In: *Polymeric Materials Encyclopaedia*, ed. J.C. Salamore, Boca Raton: CRC Press, 1995, pp. 2397-2411.

5. A.R. Horrock and D. Price (eds). *Fire Retardant Materials*, Cambridge: Woodhead Publishing Limited, 2001.

6. M. Lewin and E.D. Weil. Mechanisms and modes of action in flame retardancy of polymers. In: *Fire Retardant Materials*, ed. A.R. Horrocks and D. Price, Cambridge: Woodhead Publishing Limited, 2001, pp. 31-68.

7. P. Joseph and J.R. Ebdon. Recent developments in flame-retarding thermoplastics and thermosets. In *Fire Retardant Materials*, ed. A. R. Horrocks and D. Price, Cambridge: Woodhead Publishing Limited, 2001, pp. 220-263.

8. G.L. Nelson and C.A. Wilkie (eds.). *Fire and Polymers: Materials and Solutions for Hazard Prevention*, ACS Symposium Series 797, Washington DC: American Chemical Society, 2001.

9. P. Georlette. Applications of halogen flame retardants. In: *Fire Retardant Materials*, ed. A.R. Horrocks and D. Price, Cambrdige: Woodhead Publishing Limited, 2001, pp. 264-292.

10. S.C. Brown and M.J. Herbert. New developments in ATH technology and applications. In: *Proceedings of Flame Retardants 92*, London: Elsevier Applied Science, 1992, pp. 100-109.

11. G.T. Egglestone and D.M. Turley. Flammability of GRP for use in ship superstructures. *Fire & Materials*, 1994; 18:255-260.

12. M.J. Scudamore. Fire performance studies on glass-reinforced plastic laminates. *Fire & Materials*, 1994; 18:313-325.

13. P.R. Hornsby and C.L. Watson. A study of the mechanism of flame retardance and smoke suppression in polymers filled with magnesium hydroxide. *Polymer Degradation & Stability*, 1990; 30:73-87.

14. P.R. Hornsby. The application of magnesium hydroxide as a fire retardant and smoke-suppressing additive for polymers. *Fire & Materials*, 1994; 18:269-276.

15. P.R. Hornsby and C.L. Watson. Mechanism of smoke suppression and fire retardancy in polymers containing magnesium hydroxide filler. *Plastics & Rubber Processing & Applications*, 1989; 11:45-51.

16. D. Price, G. Anthony and P. Carty. Introduction: polymer combustion, condensed phase pyrolysis and smoke formation. In: *Fire Retardant Materials*, ed. A.R. Horrocks and D. Price, Cambridge: Woodhead Publishing Limited, 2001, pp. 1-30.

17. P. Carty and S. White. Char formation in polymer blends. *Polymer*, 1994; 35:343-347.

18. P.A. Cusack. Evaluation of the fire-retardant properties of zinc hydroxystannate and antimony trioxide in halogenated polyester resins using the cone calorimeter method. *Fire & Materials*, 1993; 17:1-6.

19. R.F. Kovar and D.E. Bullock. Multifunctional intumescent composite firebarriers. In: *Proceedings of the 4th Annual Conference on Recent Advances in Flame Retardancy of Polymeric Materials*, May 1993, pp. 87-98.

20. A.R. Horrocks, B.K. Kandola, P. Myler and D. Blair. UK Patent Application 99929178.3. 9 December 1999.

21. B.K. Kandola, A.R. Horrocks, P. Myler and D. Blair. Flame retardant properties of novel fibre-reinforced composite materials. In: *Flame Retardants 2000*, London: Interscience Communications Ltd, 2000, pp. 217-225.

22. B.K. Kandola, A.R. Horrocks, P. Myler and D. Blair. The effect of intumescents on the burning behaviour of polyester-resin-containing composites. *Composites*, 2002; 33A:805-817.

23. B.K. Kandola, A.R. Horrock, P. Myer and D. Blair. Mechanical performance of heat/fire damaged novel flame retardant glass-reinforced epoxy composites. *Composite*, 2004; 35A:863-873.

24. J.A. Parker and D.A. Kourtides. New fireworthy composites for use in transportation vehicles. *Journal of Fire Science*, 1983; 1:432-458.

25. D.W. van Krevelan. Entzundlichkeit und flammhemmung bei organischen hocjpolymeren und ihre beziehungen zur chemischen struktur. *Chemie-Ing. Techn*, 1975; 47:793-803.

26. A. Granzow. Flame retardation by phosphorus compounds. *Accounts of Chemical Research*, 1978; 11:177-183.

27.  M. Lewin. Physical and chemical mechanisms of flame retarding of polymers. In *Fire Retardancy of Polymers: The Use of Intumescence*, ed. M Le Bras, G. Camino, S. Bourbigot and R. Delobel, The Royal Society of Chemistry, London, 1997, pp. 1-32.

28.  M.P. Luda, A.I. Balabanovich and G. Camino. Thermal decomposition of fire retardant brominated epoxy resins. *Journal of Analytical & Applied Pyrolysis*, 2002; 65:25-40.

29.  M. Nakao, T. Nishioka, M. Shimizu, H. Tabata, K. Ito. In: *ACS Symposium Series 407*, ed. J.H. Lupinsky and R.S. Moore, 1989, p. 421.

30.  E. Kicko-Walczak. New polyester resins with reduced flammability and smoke evolution capacity. *Fire & Materials*, 1999; 22:253-255.

31.  A. Casu, G. Camino, M.P. Luda and M. de Giorgi. Mechanisms of fire retardance in glass fibre polymer composites. *Makromolecular Chemistry Macromolecular Symposium*, 1993; 74:307-310.

32.  A. Casu, G. Camino, M. De Giorgi, D. Flath, A. Laudi and V. Morone. Effect of glass fibres and fire retardant on combustion behaviour of composites, glass fibres-poly(butylene terephthalate). *Fire & Materials*, 1998; 22:7-14.

33.  R.M. Morchat and J.A. Hiltz. Fire-safe composites for marine applications. In: *Proceedings of the 24th SAMPE Technical Conference*, 1992, pp. T153-T164.

34.  A. Rappe. in *Proceedings from the Flame Retardants '92 Conference*, London: Elsevier Applied Science, 1992, p. 133.

35.  S.K. Brauman and N. Fisman. Phosphorus flame retardance in polymers. III. Some aspects of combustion in polymers. *Journal of Fire Retardation and Chemistry*, 1977; 4:93-111.

36.  S.Y. Lu and I. Hamerton. Recent developments in the chemistry of flame retardant composite matrices. *Progress in Polymer Science*, 2002; 27:1661-1712.

37.  R. Walters and R. Lyon. *SAMPE Journal*, 1997; 42:1335.

38.  M. Bannister, P. Andrews, R. Rosu, R. Varley and W. Tian. Improved fire performance of carbon/epoxy composites through resin modification. In: *Proceedings of the 24th European SAMPE Conference*, 1-3 April 2003, Paris.

39.  C.A. Wilkie. Graft copolymerisation as a tool for flame retardancy. In: *Fire Retardant Materials*, ed. A.R. Horrocks and D. Price, Cambridge: Woodhead Publishing Limited, 2001, pp. 337-354.

40.  M. Susuki and C.A. Wilkie. The thermal degradation of acrylonitrile-butadiene-sytrene terpolymer grafted with methacrylic acid. *Polymer Degradation & Stability*, 1995; 47:223-228.

41.  J. Davidovits. GEOPOLYMERS: Inorganic polymeric new materials. *Journal of Thermal Analysis*, 1991; 37:1633-1756.

42.  J. Davidovits and M. Davidovits. GEOPOLYMER: Ultra-high temperature tooling material for the manufacture of advanced composites. In: *Proceedings of the 36th International SAMPE Symposium*, 1991, pp. 1939-1949.

43.  U. Sorathia. Improving the fire performance characteristics of composite materials for naval applications. *AMPTIAC Quarterly*, 2004; 7:49-54.

44.  J.W. Robinson, D.R. Hudson and A.M. Mazany. Fire testing inorganic composite structures. In: *Proceedings of the SAMPE Technical Conference*, 16-20 May 2004, Long Beach CA.

45.  R.E. Lyon, P. Balaguru, A.J. Foden, U. Sorathia and J. Davidovits. Fire resistant aluminosilicate composites. *Fire and Materials*, 1997; 21:61-73.

46.  T. Czigány. Basalt fiber reinforced hybrid polymer composites. *Materials Science Forum*, 2005; 473/474:59-66.

47.  S. Bourbigot and X. Flambard. Heat resistance and flammability of high performance fibres: A review. *Fire and Materials*, 2002; 26:155-168.

48.  U. Sorathia and C. Beck. Fire protection of glass/vinyl ester composites for structural applications. In: *Proceedings of the 41st International SAMPE Symposium*, March 1996, pp. 687-697.

49.  U. Sorathia, T. Gracik, J. Ness, A. Durkin, F. Williams, M. Hunstad and F. Berry. Evaluation of intumescent coatings for shipboard fire protection. *Journal of Fire Sciences*, 2003; 21:423-450.

50.  J.R. Brown, Z. Mathys, S.Z. Riddell and L.V. Wake. Fire-retardant performance of some surface coatings for naval ship interior applications. *Fire & Materials*, 1995; 19:109-118.

51.  J. Giancaspro, P. Balaguru and R. Lyon. Fire protection of flammable materials utilizing geopolymer. *SAMPE Journal*, 2004; 40:42-49.

52.  A. Tewarson and D.P. Macaione. Polymers and composites – an examination of fire spread and generation of heat and fire products. *Journal of Fire Sciences*, 1993; 11:421-441.

53.  A.P. Mouritz and Z. Mathys. Post-fire mechanical properties of glass-reinforced polyester composites. *Composites Science & Technology*, 2001; 61:475-490.

54. H.L. Vanderall. Intumescent coating systems, their development and chemistry. *Journal of Fire & Flammability*, 1971; 2:97-114.
55. M. Kay, A.F. Price and I. Lavery. A review of intumescent materials, with emphasis on melamine formulations. *Journal of Fire Retardant Chemistry*, 1979; 6:69-91.
56. G. Camino, L. Costa and G. Martinasso. Intumescent fire-retardant systems. *Polymer Degradation & Stability*, 1989; 23:359-376.
57. B.K. Kandola and A.R. Horrocks. Flame retardant composites, a review: The potential for use of intumescents. In: *Fire Retardancy of Polymers – The Use of Intumescence*, ed. R. Delobel, The Royal Society of Chemistry, 1998, pp. 395-417.
58. S.P. Quershi and D.W. Krassowski. Intumescent resin system for improving fire resistance of composites. In: *Proceedings of the 29th International SAMPE Technical Conference*, 28 Oct – 1 Nov 1997, pp. 625-629.
59. C.E. Anderson, J. Dziuk, W.A. Mallow and J. Buckmaster. Intumescent reaction mechanisms. *Journal of Fire Sciences*, 1985; 3:161-194.

# Chapter 9

# Fire Properties of Polymer Nanocomposites

D. Wang & C.A. Wilkie
*Department of Chemistry, Marquette University*

## 9.1     Introduction

The terminology polymer nanocomposite describes a composite in which one of the composite materials, the nano-material, has a minimum of one dimension which is on the nanoscale and it is completely dispersed throughout the polymer. The typical nano-material is a clay but graphite, single-wall and multiple-wall nanotubes and nanoscale spherical particles, such as polyhedral oligomeric silsesquioxane, POSS [1,2], silica [3-5] and titania [6,7], have also been used.

The bulk of this chapter will describe polymer-clay nanocomposites, because that is where the majority of the work has been carried out; only brief mention will be made of other nanocomposite systems. Attention will be directed towards the characterization of nanocomposite formation, evaluation of fire retardancy, the different types of clay modification and their effects, and the bulk of the chapter will address various examples of fire retardancy of polymer nanocomposites. The last sections of the chapter will provide the few details that are available on the mechanisms by which the presence of a nanomaterial can affect the fire retardancy and then a concluding section on future trends in this area.

Fire retardant fillers have been used with polymeric materials for many years, and these are described in chapter 8. In the traditional filled system - the microcomposite - a large loading is required which may lead to a significant decrease in the mechanical properties of the polymer. When nanophase particles are used, the situation is quite different. The reduction in size from the micro-scale to the nano-scale significantly enhances the surface area of the particles.  The increased surface area may lead to a reduction in the amount of material that is needed and one may also expect catalytic effects, due to the presence of the high surface area material, which may change degradation pathways and thus affect the heat release rate of the polymer.  Finally, the use of nano-scale materials can lead to the formation of a barrier which can prevent the evolution of volatiles during the degradation and thus increase the amount of char that is produced.

The interest in polymer-clay nanocomposites has developed due to the observation that the presence of layered silicate materials, such as montmorillonite, hectorite, bentonite, etc., at relatively low loading levels (typically 3% - 5%) can greatly improve the mechanical properties, enhance the barrier properties and improve the fire retardancy of polymers [8-13].  As early as the 1960's, Friedlander [14], Uskov [15] and Blumstein [16] began to study the thermal stability of polystyrene and poly(methyl methacrylate) that was prepared in the presence of clay.  They found that styrene and methyl methacrylate molecules can be adsorbed on the surface and interface of montmorillonite and form intercalated polymer-montmorillonite complexes. These complexes exhibited higher thermal stability and solvent resistance, presumably because the restriction of molecules in the confined environment retarded the movement of the polymer chain and delayed the degradation.  It was only after Toyota researchers [17,18] found that the addition of clay to polyamide-6, PA-6, (4.7%) led to superior mechanical properties and that the heat distortion temperature increased to 152°C, which is 87°C higher than that of virgin PA-6, that this area has received significant attention.

Clays are a family of layered silicate materials, known as 2:1 type phyllosilicate, consisting of montmorillonite, hectorite, saponite, fluoromica, fluorohectorite, vermiculite, kaolinite, magadiite, etc.  Montmorillonte has been the most used clay and is named after the city of Montmorillon in France, where this material was initially discovered in 1874 [19].  The montmorillonite clay structure can be viewed from two different points of view, microstructure and crystal structure. The microstructure of montmorillonite may be divided into three different categories, according to the particles studied: lamella structure, primary particle and aggregation state.  The lamella structure consists of a single layer, about 1 nm thick and 100-200 nm in length. The crystal structure of montmorillonite refers to the lamella structure.  Several lamellae associate and form the primary particle, which ranges from several nanometers to several tens of nanometers. Hundreds of thousands of primary particles aggregate and the size of the aggregate ranges from a tenth to several tens of micrometers.

From the point of view of the crystal structure, this mineral has a two-dimensional layered structure, consisting of a central octahedral sheet of Al or Mg fused by two tetrahedral sheets. The thickness of the sheets is 9.6 Å, and the space between the two layers is 2.0 Å; a view of this structure is shown in Fig. 9.1.  The gallery height in terms of d-spacing is 11.6 Å.  Within the layers the replacement of aluminum ions by magnesium or iron ions and magnesium ions by alkali metals, will generate a negative charge on the clay layers.  This negative charge requires positively-charged cations within the gallery space to balance the charge.  The cations, typically alkali metal and alkaline earths in the natural material, may be ion-exchanged with other cations, which then have the properties necessary for the formation of nanocomposites.  The capability for cation exchange is known as the cation exchange capacity, CEC, and is typically expressed in milliequivalents (meq)/100g.  Common values for montmorillonite clays are in the range of 100 meq/100g.  The cation exchange process is diffusion controlled, by Nernst diffusion, from the bulk to the clay interface, and by particle diffusion, the process within the clay particles.

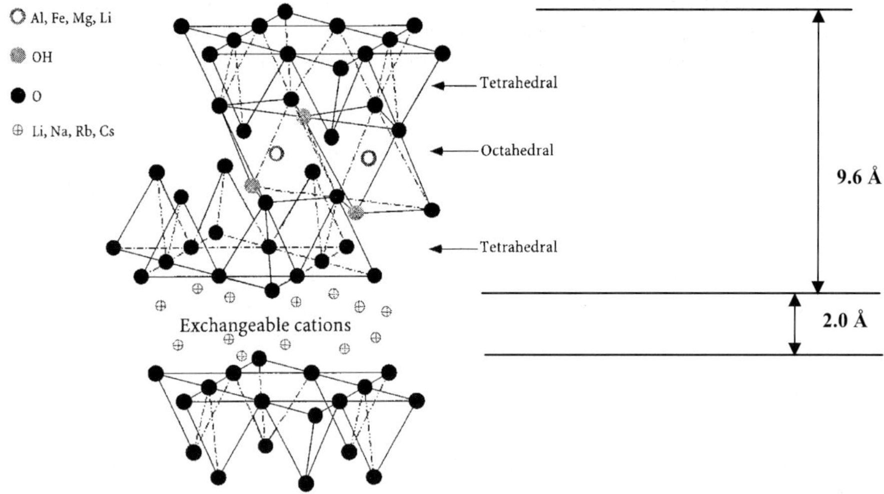


Figure 9.1.  Crystal structure of montmorillonite.

Since the gallery space is hydrophilic, materials which are also hydrophilic may be adsorbed into the gallery space of the naturally-occurring clay.  As early as 1940s and 1950s, Bradley [20], MacEwan [21] and Greene-Kelly [22,23] had shown that many types of organic molecules can intercalate into the gallery of clay and form montmorillonite-organic molecules complexes. Polyfunctional organic materials, including glycerols, resorcinol, catechol and pyrogallol, were introduced into the gallery-space of the clay.  The change in the spacing of the gallery depends on the size of the organic molecules and the number of layers of adsorbed molecules, ie. are mono- or di-molecular layers formed.  Lagaly [24] gave a summary of this research on the clay-organic interaction in 1987.

If one wishes to introduce organic polymers into the gallery space, one must perform ion-exchange to remove the hydrophilic sodium ions and replace them with an organophillic ion, the usual choice is an ammonium ion although phosphonium has also been used; in fact stibonium ions [25] and carbocations [26] have also been used as the counterions.  In order to make the gallery space sufficiently organophillic, there must be a least one long chain of 12 carbons or longer [27].  The gallery height in the pristine sodium clay is about 1 nm and this increases to 2 or more nm in the organically-modified clays.

Polymer-clay nanocomposites have been prepared both by *in-situ* polymerization and by blending processes.  The organophillic requirements are typically different for the two cases, with blending processes requiring more alkyl chains than does *in-situ* polymerization.  Intercalated PMMA and PS-sodium clay nanocomposites can be obtained using *in-situ* emulsion polymerization [28,29] while a partially exfoliated nanocomposite results when the organically-modified clay is utilized [30].

Three different types of nanocomposites may be described: (i) immiscible, also known as microcomposites, in which the clay is not nano-dispersed and is essentially behaving as a micron-sized filler; (ii) intercalated nanocomposites, in which the clay is completely nano-dispersed and the clay layers remain in registry; and (iii) exfoliated, also known as delaminated, nanocomposites, which also show good nano-dispersion and the registry between the clay layers has been lost.  These definitions are based on x-ray diffraction measurements; although as new techniques continue to be developed the definitions will probably undergo some change.  A schematic diagram showing these three types of materials is shown in Fig. 9.2.

The thermal stability of the clays has been studied by thermogravimetric analysis, TGA.  For pristine sodium montmorillonite, adsorbed water, in its many forms, is lost by 120°C while combined water, which is part of the montmorillonite structure, can begin at 250°C and continues to above 500°C.  The typical organically-modified clay, that which contains an ammonium ion, commences degradation at about 180-230°C by a Hofmann elimination process, which leads to a trialkylammonium cation and an alpha olefin.  This will lose trialkylamine at higher temperatures, about 400°C, so that the cation is now simply a proton.

Nanocomposites using graphite as the nano-material have been prepared in a number of cases [31-36].  In all cases the graphite must undergo some modification to permit the entry of the polymer between the graphite layers. This has either been accomplished by swelling the graphite through the preparation of potassium graphite, $KC_8$, by the preparation of expandable graphite through the intercalation of sulfuric acid between the graphite layers, or by ion-exchange on graphite oxides.  Nanotubes have also been used as the nano-material [37-39].  In some instances, one can simply disperse the nanotubes in a monomer and carry out a polymerization reaction, but better results appear to be obtained if an organic modification of the nanotube is first carried out in order to enhance compatibility between the nanotube and polymer.

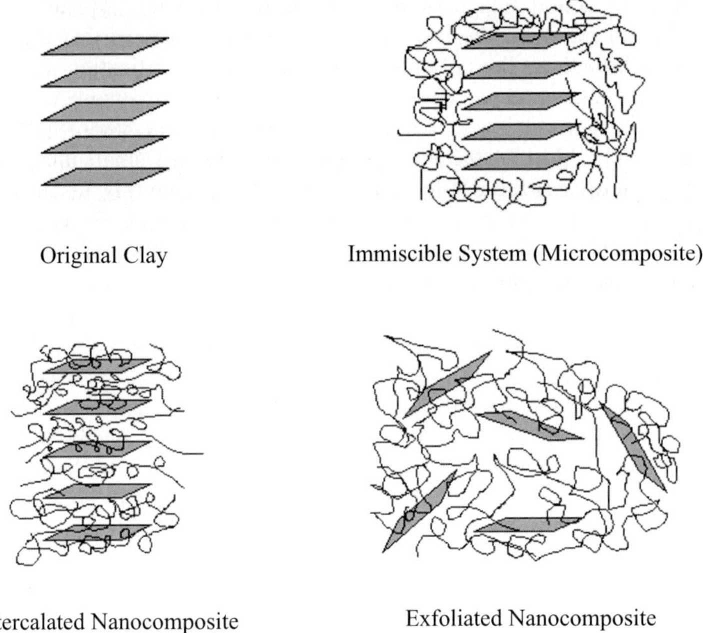



*Figure 9.2. The types of nanocomposites.*

## 9.2      Characterization of Nanocomposite Formation

Polymer-clay nanocomposites have not only the advantage of reduced flammability, but they also exhibit improved mechanical properties. This is a key advantage, because many fire retardants are used at relatively high loadings and this may lead to a significant reduction in the mechanical properties of the polymer. Characterization usually involves showing that the clay is well-dispersed throughout the polymer, ie. nano-dispersion, and establishing if an intercalated or exfoliated or a mixed material has been produced [40]. This usually involves a combination of x-ray diffraction, XRD, and transmission electron microscopy, TEM. XRD will provide the spacing of the gallery space, the d-spacing of the material; in an intercalated system, the d-spacing will increase from the value in the virgin organically-modified clay to some larger value and this is a good indicator of intercalation. When an exfoliated system is obtained, the registry between the clay layers is lost and no peak should be observed in the XRD. Unfortunately, in some instances disorder occurs in the process of combining a clay with a polymer and this also leads to the absence of a peak. Thus the observation of no peak in the XRD is ambiguous.

TEM provides an actual image of the clay in the polymer; typically at least two magnifications are required, the low magnification image shows that the clay is truly nano-dispersed throughout the polymer, while the high magnification image shows the actual clay layers and one can see if the registry has been maintained. TEM images present a problem in that the actual area that is imaged is very, very small compared to the whole material. Most of the time, investigators use this small image and purport that it is representative of the whole. To truly be able to state this, a statistical analysis of the entire material must be performed and enough images, focusing on different locations, must be obtained so that one can state with some degree of statistical certainty the state of the nanocomposite.

There are other techniques that are less-commonly used but should perhaps be more widely practiced, notably atomic force microscopy, AFM, nuclear magnetic resonance relaxation time, and cone calorimetry. AFM is an easier and quicker method but it is less direct than TEM; an example of AFM images of an intercalated, mixed intercalated-exfoliated, and an exfoliated structure are shown in Fig. 9.3. For the intercalated structure, the surface is quite smooth while for the exfoliated structure, very small pieces can be seen to be dispersed in the polymer matrix.

Van der Hart et al. [41-43] have shown in a series of publications that the proton relaxation time depends upon the state of the nanocomposite. As previously described, each plate-like layer of the layered structure of clay (montmorillonite) is about 1.0 nm thick and 50 to 100 nm in lateral dimension; paramagnetic iron ion ($Fe^{3+}$) can and does replace aluminium ion in the structure of clay. The typical concentration of $Fe^{3+}$ in naturally occurring montmorillonite produces nearest-neighbour Fe-Fe distances of about 1.0 - 1.4 nm. The relaxation time depends on the separation between nearest polymer-clay interfaces and the efficiency of direct paramagnetically induced relaxation [44]. In a microcomposite, since the clay is grouped in one region, the protons of the polymer will be relatively far removed from the interface and thus will be little affected. On the other hand, in an intercalated system the protons of the polymer will be much closer to the interface and an effect is to be expected. Finally, in an exfoliated system, since the clay layers now have a random orientation, the protons of the polymer will all be quite close to an interface and a larger reduction in relaxation time is expected. Since this experiment is conducted using one to two grams of nanocomposite, one is truly sampling the entire sample and this may be the best method for the evaluation of the extent of exfoliation. With the aid of a few assumptions, one can discuss the extent of exfoliation. The terminology exfoliation and intercalation are usually understood in terms of the registry between layers and this NMR technique offers a different method to examine the phenomena and this may eventually require the utilization of a new term.

In some of the earliest work on fire retardancy of polymer-clay nanocomposites, Gilman et al. [45,46] were able to show that cone calorimetry provides information on nanocomposite formation; microcomposites gave essentially no reduction in the peak heat release rate, PHRR, while nanocomposites, irrespective of whether they were intercalated or exfoliated, showed rather significant reductions. In work from these laboratories, we have also shown that there is a significant difference in reduction in PHRR of nanocomposites versus microcomposites.

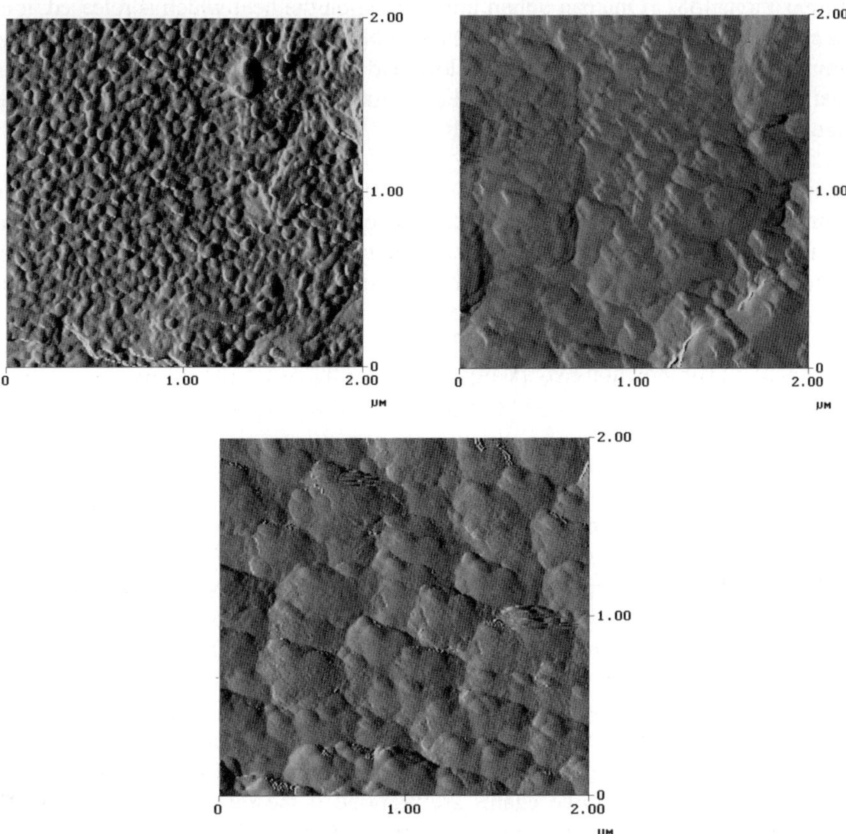

*Figure 9.3. AFM of polystyrene nanocomposites. The upper left is exfoliated, the upper right is mixed intercalated/exfoliated while the bottom is intercalated. The help of Kenneth A. Mauritz at the University of Southern Mississippi in obtaining these images is acknowledged.*

## 9.3    Evaluation of Fire Retardancy

The fire properties of materials are evaluated in many different ways: cone calorimetry (ASTM E 1354), radiative gasification [47], and Limiting Oxygen Index (LOI) (ASTM D2863, ISO 4589) are popularly used methods for the evaluation of fire retardancy of polymer materials.  For commercial products, the UL-94 (ISO 9772 and 9773, ASTM D635) test is frequently required to qualify a material as a fire retardant.  These techniques and others for characterizing the fire properties of materials are described in Chapter 11.

Cone calorimetry is widely used as a laboratory method to evaluate fire retardant compositions [48]. One can obtain information on the heat which is released, including time-to-ignition, the entire heat release rate as a function of time, the heat of combustion, the rate at which mass is lost, and the smoke which is produced. The entire heat release rate (HRR) curve is available, but attention is usually focused on its peak value, the peak heat release rate (PHRR).

Radiative gasification is a related technique in which the cone calorimetric experiment if performed in a nitrogen atmosphere. This eliminates the smoke so that one can take pictures of the material as it is heated and have actual pictorial evidence of the reaction. The limiting oxygen index measures the minimum oxygen concentration that is required to sustain burning of a sample. It is generally felt that if one can increase the LOI to a value that is much higher than about 20, close to the percentage of oxygen in air, that a fire retardant composition may be obtained.


## 9.4    Clay Modifications

In general, the gallery space in the clay is hydrophilic and the majority of polymers are hydrophobic. Thus, it is usually necessary to impart some organophilic character to this gallery space before the polymer may be inserted. This is most often accomplished by ion-exchanging the typical sodium or calcium cations for organophilic ammonium or other 'onium' ions. The most common ammonium ions contain at least one long alkyl chain, a minimum chain length of at least twelve carbons is required and frequently a chain length of sixteen is used. The identity of this organic modification is dependent both upon the polymer to be used and the mode of preparation. For instance, one sixteen carbon chain is required when a polystyrene nanocomposite is to be formed by bulk polymerization but two chains are required if the reaction is performed by melt blending. Likewise, if the polymer has some polarity, the preferred ammonium ion will likely have some polarity while a non-polar ammonium ion is preferred for non-polar polymers. Wang [49] has examined the effect of different preparative methods on the formation and type of nanocomposite that is produced.

The edges of the clay contain hydroxyl groups that provide another means to modify the clay. In addition to ammonium ions, phosphonium ions and stibonium ions [25] have both been used to modify the gallery space of clays. Very recently, Zhang [26] extended the possible organic modifications to include carbocations. Specifically, a styryltropylium ion has been used and ion-exchanged this into clay; the thermal stability, as measured by thermogravimetric analysis, is much higher than that observed from the typical 'onium' ions.

Xie et al. [50] studied the thermal degradation chemistry of alkyl quaternary ammonium modified montmorillonite. The degradation pathway is the loss of an olefin with the formation of a trialkylammonium ion; as the temperature is raised the trialkylamine is lost so that only a proton remains as the counter ion. The organically-modified clay undergoes degradation of the alkyl ammonium treatment 15-20°C earlier than does the parent alkyl ammonium salt. Phosphonium-modified clays show somewhat better thermal stability than ammonium-modified clays [51] and stibonium-modified clays show even higher thermal stability [25]. For the stibonium-modified clay the loss of the olefin still occurs but the stibine is never lost; indicating that a stibonium-modified clay may be better suited for melt blending with polymers that require higher temperatures for processing.

The 'onium' type modification is the only system that has been used for fire retardant compositions. However, the introduction of neutral organic molecules into the gallery space of clays had been studied for a long time. As early as 1945, Bradley [52] studied molecular association between montmorillonite and some polyfunctional organic liquids, such as amines, glycols, ethanol, 1,4-dioxane, pyridine, benzene and benzidene. Bradley found that distance between the layers ranges from about 3.4 up to 4.0 Å or more, and seems to increase with increasing chain length. This pioneering work does reveal the possibility of incorporation of simple molecules within a clay. MacEwan [21] prepared complexes of clays with organic compounds in 1948; markedly polar molecules would form a two-layer structure and non-polar molecules a one-layer structure. Very strongly polar, highly associated liquids link on even more than two layers. These results also showed that the clay sheet have an orienting effect on polar molecules. Greene-Kelly [23] studied the sorption behavior of aromatic organic compounds on montmorillonite. The results have shown that two orientations are common: at low surface concentrations, the molecules orient parallel to the silicate sheet and at higher surface concentrations the molecules reorient so that their planes are perpendicular to that of the silicate sheet.

Another interesting way to modify clay is through ion-dipole clay surface modification [53]. In clay, a charge imbalance arises in the oxygen layer shared by the aluminum and silicon and it is neutralized by exchangeable, surface cations. The exchanged cation cannot come closer than about 3.6 Å to the oxygen layer of the aluminosilicate, while in a normal ionically bonded system, the Na-O distance would be about 2.1 to 2.2 Å. The coulombic interaction decreases dramatically as the distance increases; at a distance of 3.6 Å approximately 60% of the bond strength has been lost. This renders a substantial partial positive charge on the gallery side of the atom so that montmorillonite has a very high hydration energy and water aids in neutralizing the partial positive charge by ion-dipole interaction.

Akelah et al. [54] have prepared montmorillonite in which the cation contains a polystyrene and they have used these as a support for organic reactions. Su [55] prepared an oligomeric styrene ammonium salt through the preparation of a copolymer of styrene and vinylbenzyl chloride, which was then used to quaternize an amine. The d-spacing of this clay is more than 6 nm; this clay was then used to prepare nanocomposites of polystyrene (PS), acrylonitrile-butadiene-styrene (ABS), high-impact polystyrene (HIPS), poly(methyl methacrylate) (PMMA), polypropylene (PP) and polyethylene (PE).

Biasci et al. [56] prepared poly(methyl methacrylate) montmorillonite adducts, in which the polymer was bound to the clay, by a radical polymerization of MMA in the presence of a clay on which the ammonium cation contained an acrylate. These materials were extracted to remove all non-bound polymer, leaving only that which was bound. The results showed that polymer was strongly fixed to the inorganic surfaces and enhanced the thermal stability. Su [57] synthesized an MMA-modified clay, again by the formation of a copolymer of MMA with vinylbenzyl chloride, and a polybutadiene-modified clay [58], and these clays was used to prepare polymer-clay nanocomposites of PS, PMMA, PP, PE, HIPS and ABS.

## 9.5     Examples of Fire Retardancy of Polymer Nanocomposites

In this section, previous work on polymer-clay and polymer-graphite nanocomposites that has relevance to fire retardancy will be addressed. The polymers that will be addressed include polystyrene (PS), acrylonitrile-butadiene-styrene (ABS), high-impact polystyrene (HIPS), polypropylene (PP), polyethylene (PE), poly(methyl methacrylate) (PMMA), poly(vinyl chloride) (PVC), ethylene-vinyl acetate (EVA), epoxy resins, polyamides (PA) and others. There is strong evidence that the fire retardant properties do not depend on the type of nanocomposite, ie. intercalated or exfoliated. In the majority of what follows, cone calorimetry is the technique that has been used to evaluate the fire retardancy and it is specifically the reduction in peak heat release rate (PHRR) that is used. As one will see in perusing this section, the reduction in PHRR is different for each polymer. A statement is made earlier in this chapter that cone calorimetry may be used to evaluate whether nano-dispersion of the clay throughout the polymer has occurred; this is only valid if one knows the magnitude of the reduction expected.

### 9.5.1   POLYSTYRENE NANOCOMPOSITE FLAMMABILITY

Zhu [59,60] studied the fire properties of polystyrene-clay nanocomposites prepared by bulk polymerization using both ammonium- and phosphonium-modified clays. The onset temperature of the degradation (TGA) is increased by about 50°C and the peak heat release rate (cone calorimetry) is reduced by 27-58%, depending upon the amount of clay that is present, and the mass loss rates are also significantly reduced in the presence of the clay. A typical plot of heat release rate is shown in Fig. 9.4 and clearly demonstrates the reduction in PHRR for nanocomposites.

Silicon-methoxide-modified clays and their polystyrene nanocomposites have also been studied [61]; the silicon-methoxide offers the possibility of reaction between the methoxide and a clay hydroxyl group to link together the cation and clay. Linkage between the silicon and clay apparently occurs in the clay but is not likely to occur in the nanocomposite, perhaps because of the presence of the polystyrene increases the distance between the reactive sites and the cation cannot span this distance. These systems are best described as intercalated nanocomposites, and the reduction in PHRR and TGA parameters are quite similar to other examples of PS nanocomposites.

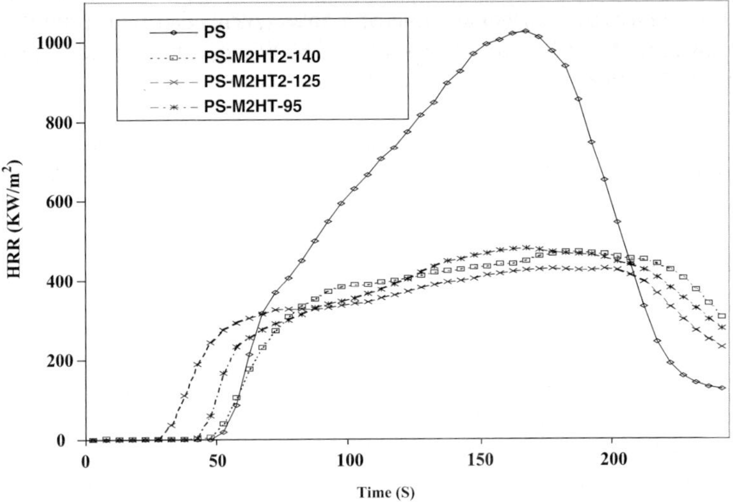


*Figure 9.4. Heat release rate for polystyrene and nanocomposites with different amount of ammonium salts (4 mass% clay). (Zhu, J.; Wilkie, C.A. Polym. Intl., 49, 2000, 1158-1163, reference 60. Reproduced with permission. Permission is granted by John Wiley & Sons Ltd on behalf of SCI.)*

Gilman [62] studied polystyrene nanocomposites using montmorillonite and fluorohectorite. It was shown that the type of layered silicate, nanodispersion, and processing conditions have an influence on the fire properties. Surprisingly, montmorillonite and fluorohectorite behave very differently; fluorohectorite has no effect on PHRR while montmorillonite shows a 60% reduction in PHRR compared to either virgin PS or an immiscible PS-NaMMT system.

Morgan [63,64] investigated the effect of the clay, the loading level and polymer melt viscosity on the flammability of polystyrene-clay nanocomposites. There were major reductions in peak heat release rates and increased carbonaceous char formation for these nanocomposites. It was detected that, while the viscosity of the PS nanocomposite played a role in lowering the PHRR, the clay loading level had the largest effect on PHRR. Finally, it was found that clay catalyzed carbonaceous char formation, and the reinforcement of the char by the clay was responsible for the lowered flammability of these nanocomposites.

Yao [65] used crown ethers to modify clays and obtained the corresponding polystyrene nanocomposites by bulk polymerization; in this instance, the sodium or potassium clay was used and rendered organophilic by the presence of the crown ether. Nanocomposites can be formed only from the potassium clays. The onset temperature of the degradation is higher for the nanocomposites and the PHRR is decreased by 25% - 30%. This is less than the normal value of a 50% - 60% reduction in PHRR which is commonly observed for PS nanocomposites, which suggests that there may also be some immiscible component present.

Su [55,57,58] has prepared styrene nanocomposites using clays that have oligomeric styrene, methacrylate and butadiene units on the cation of the clay. In the case of the styrene-containing clay, exfoliated systems are observed by melt blending in a Brabender mixer. The reduction in PHRR is comparable to what has been achieved with other organically-modified clays. Zhang [26] has prepared a styrene nanocomposite using a tropylium-substituted clay. The reduction in PHRR is less than that seen for the typical ammonium-substituted clays but is still substantial (30%).

A reactive blending method, combining polystyrene, maleic anhydride and a clay in a mixer, was also utilized in the preparation of PS clay nanocomposites [66]. In most cases, the reductions of PHRR are in the range of 20-30% and in only a few cases do they approach the 50% value. This must be a reflection of the somewhat immiscible nature of these systems. The time to PHRR is constant across the entire range of samples while the time-to-ignition does decrease and the greatest decrease in time-to-ignition occurs for the systems which give the greatest reduction in PHRR. As expected, the mass loss rate is decreased and the amount of smoke is constant or somewhat increased.

There has been some effort to look at synergy between nanocomposite formation and additive-type fire retardants. In one case a total of thirty-one phosphorus fire retardants were examined by high-throughput methods to ascertain which of the many phosphorus-based fire retardants would be most successful [67]. This led to the selection of three materials for further study. The clay alone gives a 56% reduction in PHRR while the phosphate alone gives a reduction of 20 to 40%. The synergistic combination gives a PHRR of 110 kW/m$^2$, a 92% reduction.

An alternative approach to this synergy has been to incorporate the phosphorus onto the ammonium cation. To this end a terpolymer was prepared containing styrene, *para*-vinylbenzyl chloride and a vinylphosphate. Because the phosphate is attached to the clay, it will be uniformly distributed throughout the polymer and it was hoped that this would enhance fire retardancy [68]. Since char formation is enhanced, it is suggested that there may be some condensed phase activity; the maximum reduction in PHRR by this approach was 81%.

Uhl [33,35,36] has examined nanocomposites of styrene in which graphite is the nano-dimensional material. When potassium graphite was used the reduction in PHRR was 43% while with sulfuric acid graphite the reduction was in the range of 50%; the use of a modified graphite oxide gives similar reductions in PHRR. These values are in the same range and indicate that nanocomposite formation has occurred and that graphite may be used in place of clays for fire retardancy.

## 9.5.2    ABS CLAY NANOCOMPOSITE FLAMMABILITY

Triphenylphosphate, TPP, nanocomposites (Nano TPP) were synthesized by intercalating TPP into the galleries of organically modified mica-type silicate (OMTS) and any retarding effect of nanocomposites on the evaporation of TPP was investigated [69]. It was found that Nano TPP has a higher evaporation temperature compared to TPP and the thermal stability is slightly enhanced by adding Nano TPP to acrylonitrile-butadiene-styrene copolymer (ABS). Epoxy resins and silane coupling agents were incorporated as co-fire retardants in this system. A very large increase in the limiting oxygen index (LOI) value was observed upon epoxy addition and further enhancement in thermal stability was obtained for the ABS compound containing a small amount of the coupling agent. It was also found that the enhancement is closely related to the morphologies of the chars formed after combustion. A synergistic effect of using the fire retardant nanocomposites and addition of epoxy resin and coupling agent as co-fire retardants was also confirmed for the compounds based on 2,6-dimethylphenol - resorcinol bis-(diphenyl phosphate) (DMP-RDP). LOI values as high as 44.8 were found for one particular formulation.

Su [55] has prepared ABS nanocomposites using his oligomeric styrene-substituted clay and reports a mixed intercalated-exfoliated system which gives a 37% reduction in PHRR. Graphite ABS nanocomposites give a slightly larger reduction in PHRR of 47% [36].

## 9.5.3    HIGH IMPACT POLYSTYRENE (HIPS)-CLAY NANOCOMPOSITES

HIPS nanocomposites have not been extensively studied; the only reports concern the use of the oligomeric-styrene-modified clay [55] and graphite [36]. Using the clay reduction of PHRR in the range of 40 to 50% are reported while with graphite the reductions are a little less, with the best reduction being 37%.

## 9.5.4    POLYPROPYLENE CLAY NANOCOMPOSITES

Poly(propylene-graft-maleic (anhydride))-layered silicate nanocomposites using montmorillonite and fluorohectorite have been studied [70]. The most important result from this work was the discovery that a clay-reinforced carbonaceous char forms during the combustion of the nanocomposites. This is particularly significant for systems whose base resin produces little or no char when burned alone. Further, reductions in heat release rates of 70%-80%, for these thermoplastic polymer nanocomposites were observed, as shown in Fig. 9.5.

A reactive blending approach, blending virgin PP, maleic anhydride and a clay, was also investigated by Wang [66]. Sodium clay and organically-modified clays were used, both in the presence and absence of maleic anhydride. The typical values of PHRR reduction ranges from 11% to 34%, much smaller than the values seen using PP-g-MA directly and likely indicating a substantial immiscible component. A PHRR greater than 15 or 20% is indicative of the presence of an intercalated or exfoliated nanocomposite. The mass loss rate does not change for all samples and the smoke is also relatively constant, but there is a variation in time-to-ignition. The greatest decrease in time-to-ignition occurs for those systems that show the greatest reduction in PHRR.

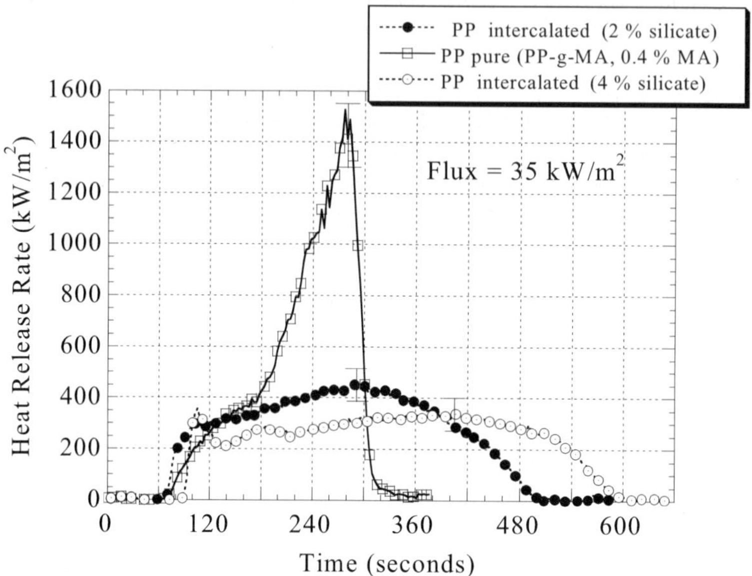


*Figure 9.5. Comparison of the heat release rate (HRR) plots for pure PP-g-MA and two PP-g-MA-layered-silicate nanocomposites, at 35kW/m² heat flux, showing a 70-80% reduction in peak HRR for the nanocomposites with a mass fraction of only 2% or 4% layered silicate, respectively. (Reprinted with permission from Chem. Mater. 12, **2000**, 1866-1873 copyright 2000 American Chemical Society).*

Synergy between conventional halogen-based fire retardants and clays has been studied [71]. An intercalated nanocomposite is formed when a simple organically-modified clay is melt blended with polypropylene-*graft*-maleic anhydride, and the addition of decabromdiphenyl oxide and antimony oxide does not change this structure. The nanocomposite shows a lower peak heat release rate than does the virgin polymer. The PHRR is reduced further when antimony oxide or decabromodiphenyloxide is present. When both additives are present, a synergy is observed and the reduction in PHRR is much larger than expected from the independent components.

Radiative gasification and vinyl polymer flammability were studied to assess the fire properties [72]. Radiative gasification experiments were carried out to detect flammability of intercalated polymer-clay nanocomposites prepared from polystyrene, and polypropylene-graft-maleic anhydride, PP-g-MA. The fire retardancy of PP-g-MA - clay nanocomposites is comparable to that of polyamide-6-clay nanocomposites and of the polymers containing fire retardants, eg. $Sb_2O_3$. The heat of combustion of the polymers is not changed by the presence of clay.

The use of the oligomeric styrene-modified clay enables the preparation of polypropylene nanocomposites in a Brabender mixer without the need for maleic anhydride [55]. The resulting material is mixed intercalated-exfoliated and the reduction in PHRR is 35%.

### 9.5.5    PE CLAY NANOCOMPOSITE FLAMMABILITY

Polyethylene clay nanocomposites have been prepared using melt blending in a Brabender mixer either in the presence or absence of maleic anhydride [73]. It was found that the PE-clay nanocomposites have a mixed immiscible-intercalated structure and there is better intercalation when maleic anhydride is combined with the polymer and clay to be melt blended. The reduction of PHRR is 30-40%. As is typically observed, the total heat released is the same for the nanocomposites as for virgin PE. The specific extinction area (SEA) (ie. smoke) is not changed from virgin PE, indicating that the presence of the clay does not cause increased smoke production. The reduction in PHRR and in mass loss rate is about the same whether maleic anhydride is present or absent. According to the cone criteria, the reduction in PHRR is about the same in all cases, so if nanocomposites are formed in any case, they are formed in all cases. Similar reductions in PHRR have been observed when the oligomeric styrene-containing clay is used [55].

Polyethylene/clay microcomposites and nanocomposites were prepared by the melt intercalation technique direct from sodium montmorillonite in the presence of a reactive compatibilizer, hexadecyltrimethylammonium bromide (C16). The loading level of C16 controls whether one obtains the microcomposite or the nanocomposite; the microcomposite showed the same combustion behavior as pure PE while the nanocomposite showed a 32% reduction in PHRR at 5 wt.% clay and 4 wt.% C16 loading [74].

### 9.5.6    PMMA CLAY NANOCOMPOSITE FLAMMABILITY

Zhu [75] reported the thermal stability and fire retardancy of poly(methyl methacrylate)-clay nanocomposites. Clays which contain a pendant double bond are more likely to give an exfoliated material while those which do not contain a double bond are likely to give intercalated materials. The PHRR decrease was in the range of 20% to 28% for PMMA nanocomposites, and there was no difference between intercalated and exfoliated systems. These nanocomposites also show lower mass loss rate than pure PMMA. For most polymer systems, the time-to-ignition of the nanocomposites are shorter than for the virgin polymer, but this is uniquely not true for PMMA.

Morgan [76] reported the thermal and flammability properties of a silica-PMMA nanocomposite. Silica nanoparticles are used as a fire retardant additive for PMMA. The thermal and physical properties of the composites were also investigated.

The flammability of a methacrylate copolymer, based upon exfoliated organophilic layered silicates, was investigated using a cone calorimeter [77]. The PHRR of these copolymer nanocomposites decreased by 27 to 35%; visual examination of the pyrolysis reaction revealed that at thirty seconds, when the mass loss rate for the nanocomposite slowed with respect to that of the neat copolymer, the surface of the nanocomposite was completely covered by char. Comparison of the residual yields for copolymers and nanocomposites indicated much higher char yields for nanocomposites.

### 9.5.7    PVC CLAY NANOCOMPOSITE FLAMMABILITY

Kalendova [78] intercalated PVC plasticizers into both a polar, organically-modified clay and into sodium montmorillonite; plate-like filler particles are generated by compounding the modified clay with PVC and plasticized PVC sheets without haze and these have been evaluated for fire resistance. Wang [79-81] prepared PVC clay nanocomposites both by melt blending organically-modified clay and sodium clay in the presence or absence of plasticizer. If one looks only at plasticized PVC, the presence of the plasticizer adds more fuel and the nanocomposites exhibit lower fire retardancy than the polymer. In the absence of the plasticizer, the time-to-ignition and time-to-PHRR both decrease, but there is little change in the PHRR, SEA and mass loss rate. Perhaps the most significant result is the specific extinction area. When PVC burns, smoke is a problem while fire retardancy *per se* is not. The PHRR values are all low and indicate that PVC does not present a large heat release problem. The observation that the presence of the clay does not increase the smoke and may in some cases cause smoke to decrease bodes well for the future of clays in the fire retardancy and smoke suppression of PVC.

## 9.5.8    EVA CLAY NANOCOMPOSITE FLAMMABILITY

Zanetti [47,82] studied the combustion behavior of exfoliated EVA/fluorohectorite nanocomposites. In the horizontal combustion mode of the mass loss calorimeter, the nanocomposite shows acceleration of EVA deacetylation and delayed volatilization of the resulting polyene. The overall heat release rate is much lower than in the case of EVA, as measured by the gas combustion temperature. Partial protection of EVA from the fire is also found in the case of the immiscible composite, although in this case the volatilization of the polymer during combustion occurs at a larger rate than for EVA alone. Accumulation of silicate on the surface of the burning specimen may create a protective barrier to heat and mass transfer which is, however, much more effective in the case of ablative reassembling of crystal layers of the nanocomposite than in that of the particles of the immiscible composite. Dripping of burning particles in vertical combustion is suppressed only in the case of the nanocomposite, which reduces the hazard of fire spread to surrounding flammable materials.

Beyer [83] combined a nano-filler with aluminum trihydrate to improve the fire retardancy properties of EVA-nanocomposites. Fire retardant nanocomposites were synthesized by melt-blending ethylene-vinyl acetate copolymers (EVA) with organically-modified clays; a mixed intercalated-exfoliated system was obtained. The use of the cone calorimeter to investigate the fire properties of the materials indicated that the nanocomposites showed a large decrease in heat release, and char-formation is believed to be the reason for the enhancement. Ethylene-vinyl acetate copolymer clay nanocomposite was also prepared by mechanical kneading [84].

Intumescent formulations, using a PA-6-clay nanocomposite as the carbonization agent, were studied in an EVA-based intumescent formulation [85]. Using PA-6 clay nanocomposite instead of pure PA-6 has been shown to improve the fire properties of the intumescent blend. The use of the PA-6 nanocomposite improved both the mechanical and fire properties of the EVA-based materials. The role played by the clay in the improvement of the FR performance was studied using FTIR and solid state NMR [86]. The clay allowed the thermal stabilization of a phosphorocarbonaceous structure in the intumescent char, which can act as a protective barrier.

Morgan [87] investigated the effects that clay dispersion has on flammability. Two polymer-clay nanocomposites made from polycaprolactam and poly(ethylene-co-vinyl acetate) were investigated. It is claimed that the type of dispersion of clay in the polymer (intercalated vs. delaminated) does not have a major effect on flammability.

Two different particle sizes of pristine montmorillonite (MMT) (38 μm and 48 μm) were used by Tang [88] to prepare EVA/MMT nanocomposites via direct melt intercalation in the presence of a compatibilizer (hexadecyltrimethylammonium bromide) in a 'one pot' reactive process. The combustion behavior of the nanocomposite and microcomposite in the cone calorimeter are remarkably different. The PHRR of the nanocomposite is 40% lower than pure EVA and 34% lower than the microcomposite. The PHRR of the nanocomposite loaded with 5% of the smaller size particles (38μm) is lower than that of the larger particle size (48μm). The initial heat release rate for the nanocomposites is higher at the beginning of combustion, probably because of decomposition of the organic alkylammonium cation and the accelerated evolution of acetic acid.

## 9.5.9   EPOXY CLAY NANOCOMPOSITE FLAMMABILITY

Phosphorus-containing epoxy-based hybrid nanocomposites were obtained from bis(3-glycidyloxy)phenylphosphine oxide, diaminodiphenylmethane, and tetraethoxysilane via an *in-situ* sol-gel process [89]. The nanometer-scale silica showed enhanced fire retardant properties, including an LOI value of 44.5.

Polyhedral oligomeric silsesquioxane, POSS, reinforced fire retarding epoxy vinyl ester resins have been investigated [90]. The POSS nanocomposites exhibit better fire retarding capability than the pure resin, with reduced smoke, heat release rate (HRR) and an increased time-to-ignition. The properties were found to be comparable or better than the halogenated version of the resin. The nano-reinforced composites acquire fire-retarding characteristics via two major mechanisms: (i) reduced volatilization of fuel and (ii) the formation of oxidatively stable, non-permeable surface chars.

## 9.5.10   PA CLAY NANOCOMPOSITE FLAMMABILITY

Polyamide-6-clay nanocomposites were prepared by Shu [91] via intercalation compounding. The peak value of heat release rate decreased by 32% and 63% respectively when the clay content was 2% and 5%, compared with virgin PA-6. An improvement in fire retardancy was achieved by preparation of polymer nanocomposites using intercalation of poly(vinyl alcohol) or PA-6 with 5% kaolinite [92]. This enabled the establishment of a concept to combine the intercalation of kaolinite with phosphorous fire retardants, trimethylphosphate, triphenylphosphate or triphenylphosphine in acrylonitrile-butadiene-styrene copolymer.

To evaluate the feasibility of controlling polymer flammability via a nanocomposite approach, the cone calorimeter was used by Gilman [93,94] to evaluate exfoliated PA-6-clay nanocomposites. The PHRR is reduced by 63% in a PA-6-clay nanocomposite containing a clay mass fraction of 5%. This is not only a very efficient FR system, but, in addition, the physical properties are not degraded by the presence of the additive, rather they are enhanced.

Uhl [34] has prepared nanocomposites of PA-6 in which graphite is the nano-dimensional material. Typical reductions in PHRR range from 50 to 65%, which is very similar to that seen for clay systems. The mass loss rate is also reduced and there is only a small increase in smoke production. These results suggest that graphite may replace clay for fire retardant purposes.

## 9.5.11   OTHER POLYMER CLAY NANOCOMPOSITE FLAMMABILITY

The addition of a melamine-treated clay in cyanate ester resins gives an exfoliated nanocomposite in which the PHRR is reduced by more than 50% [95]. It is anticipated that this nanocomposite approach would be especially useful for toughened cyanate ester resins, since the typical toughening agents often increase the flammability and lower the modulus.

Yeh [96] prepared a poly(o-methoxyaniline)-clay nanocomposite and the thermal stability and fire resistance were studied by TGA and limiting oxygen index (LOI). Biodegradable polyester-layered silicate nanocomposites based on poly($\varepsilon$-caprolactone) were prepared either by melt intercalation with PCL or by *in-situ* ring-opening polymerization of $\varepsilon$-caprolactone [97]. The PCL-layered silicate nanocomposites with 3 wt% of inorganic layered-silicate exhibited increased thermal stability as well as enhanced fire retardant characteristics as a result of a charring effect. Unsaturated polyester and phenolic resin nanocomposites were prepared by Lee [98], and these materials showed enhanced properties of thermal stability and fire retardancy, compared to the virgin polymers.

The cone calorimetric data for vinyl ester nanocomposites show a reduction in the order of 25 to 40% in the PHRR [99]. In addition to the change in PHRR, there is a change in the mass loss rate while the heat of combustion, soot and carbon monoxide yields are unchanged.

Vinyl ester/Nomex/clay nanocomposites were produced by the addition of both Nomex and clay in vinyl esters while the analogous fiberglass-containing systems were similarly produced [100]. (Nomex is the brand name for a family of high temperature resistant fibres invented and produced by DuPont.). All of the vinyl ester samples continued to burn and did not self-extinguish; even the glass fiber composite samples did not exhibit self-extinguishing properties. This particular vinyl ester composition did not appear to have enough inherent fire resistance. The fire retardant properties exhibited by the Nomex fibrid (fibers made of synthetic polymers) samples may be due to the inherent fire retardant properties of the Nomex fibrids themselves. There were differences in behavior between vertically suspended and horizontally mounted samples, with lower weight losses seen in the polymer/clay nanocomposites as compared to the neat resin. The use of polymer/clay nanocomposites may produce significant improvement in fire resistance only if the basic polymer system has a certain level of inherent fire resistance.

Poly($\varepsilon$-caprolactone)/clay nanocomposites were prepared by melt intercalation of the polymer with sodium montmorillonite and organically-modified montmorillonite [101,102]. Microcomposites or nanocomposites were prepared depending on the clay; the thermal stability improved with the filler loading up to 5 wt %. The addition of more clay resulted in a leveling of the effects and, perhaps, a decrease of these properties. A marked charring effect was observed upon exposure to a flame.

A new route for fire retardancy of fabrics was investigated by Bourbigot [103]. This work proposes two approaches for making fire retarded textiles using nanocomposite technology: nanocomposite yarn and nanocomposite coating. Polyamide-6 nanocomposites are exfoliated and can be processed via melt spinning to make multifilament yarns. Using montmorillonite clay and a new generation of organic-inorganic hybrid material - Polyhedral Oligomeric Silsesquioxanes (POSS) - polyurethane (PU) nanocomposites have been synthesized. These two types of nanocomposite were evaluated by cone calorimetry and the heat release rate is significantly reduced; these results offer a promising route to fire retarded textiles.

Montmorillonite clay was evaluated as an environmentally benign fire retardant for polycarbonate (PC) for use in computer housings [104]. PC was extruded at 220 °C with an organically-modified clay. The fire retardant properties were a complex function of two competing mechanisms: exfoliation vs. degradation of polymer molecular weight. For samples processed under high shear conditions, degradation dominated, and peak heat release rate (PHRR) increased with ash content. For samples processed at lower shear, a 30% reduction in PHRR was observed. Degradation of molecular weight and HRR were thus minimized by optimizing surface contact with MMT.

Poly(ether imide)-layered silicate nanocomposites with increased char yield and improved fire retardancy were prepared from 4,4'-(3,4-dicarboxyphenoxy) diphenylsulfide dianhydride and aliphatic diamines [105]. The layered silicates were prepared by cation exchange between lithium fluorohectorite or sodium montmorillonite and protonated primary amines.

Devaux et al. [106] compared montmorillonite clay and polyhedral oligomeric silsesquioxanes (POSS) as additives to process PU nanocomposites in order to provide fire retardancy to the coated textile structure. Results with polyurethane resins, (PU)/clay and PU/POSS coated on polyester or cotton fabrics, using cone calorimetry and thermogravimetric analysis, clearly highlights the great potential of POSS for fire retardant applications.


## 9.6      Mechanisms of Fire Retardancy in Nanocomposites

The mechanisms by which nanocomposite formation can enhance the thermal and fire stability of polymers have been of interest for some time. The first suggestion of a mechanism is due to Gilman and Kashiwagi [107], who suggested that the nanocomposite structure collapses during combustion and this forms a carbonaceous-silicate structure on the surface which can act as a barrier to mass transport and can also insulate the underlying polymer from the heat source. X-ray photoelectron spectroscopy, XPS, measurements have been carried out on polystyrene and poly(methyl methacrylate) nanocomposites and the accumulation of the aluminosilicate at the surface and the depletion of carbon has been confirmed [108,109]. When PVC systems were examined by XPS, carbon accumulation rather than aluminosilicate was observed; this is explained by the different degradation pathways for the three polymers in question [110].

A second mechanism, which appears to be only effective when the fraction of clay is quite low, is radical trapping by iron which is substitutionally present in the clay [111]. When clays which contain iron are compared with those in which iron is absent, there is a significant difference in the reduction in PHRR when the fraction of clay is less than 3%.

An explanation has been very recently offered for the great difference in reductions in PHRR for various polymers. The thermal degradation of polystyrene proceeds to give both monomer and oligomer while the degradation of polystyrene nanocomposites produces oligomer, but no monomer [112]. It is suggested that this is a catalytic effect, similar to that which was proposed by Blumstein for synthesis in the presence of a clay [16]. If this is correct, one may expect to see large reductions in the PHRR for polymers which degrade to give both monomer and oligomer while the reductions will be much smaller for polymers which degrade to give only monomer or only oligomer.

## 9.7    Future Trends in Fire Retardancy of Nanocomposites

It has been seen that the formation of nanocomposites brings about a large reduction in the rate of heat release. Polymer nanocomposites do provide an opportunity for the fire retardancy of polymeric materials. Much work has been done on many different polymers using a wide variety of organic modifications. The synthesis has been performed using virtually every method that is used for the preparation of polymers. The results show that in almost every system the PHRR and mass loss rate are reduced, but the total heat released is not affected and the time-to-ignition is almost always shorter. Even though there are these large effects on the cone calorimeter properties, the fact remains that the majority of nanocomposites do burn.

It is suggested that the synergy between nanocomposite formation and the use of fire retardants must be considered if one is to achieve fire retardancy through this technology. There are speculations on the process by which this occurs, but there is no agreement. More attention in the future must be paid to attempting to understand the process by which clays can affect the fire retardancy of polymers. We also think that more work must be performed using nanomaterials other than clays. It is quite possible, and there is evidence from work in this laboratory, that the mechanism for graphite nanocomposites is different from that for clay nanocomposites. Some of the questions that must be asked include: does the clay affect the degradation process? How does the clay effect the time-to-ignition? Can nanocomposite degradation be effectively modeled? How can one render the barrier more permanent?

## References


1.   A.R. Esker, B.A. Vastine, J. Deng, J.T. Polidan, B.D. Viers and S.K. Satija. Nanocomposite thin films based on POSS. In: *Abstracts of Papers, 225th ACS National* Meeting, New Orleans, LA, United States, 23-27 March 2003.

2.   H. Xu, S-W. Kuo, C-F. Huang, and F-C. Chang. Poly(acetoxystyrene-co-isobutylstyryl) POSS nanocomposites: characterization and molecular interaction. *Journal of Polymer Research*, 2002; 9:239-244.

3.   M. J. Percy, V. Michailidou, S.P. Armes, C. Perruchot, J.F. Watts and S.J. Greaves. Synthesis of vinyl polymer-silica colloidal nanocomposites via aqueous dispersion polymerization. *Langmuir*, 2003; 19:2072-2079.

4.   A.J. Waddon and Z.S. Petrovic. Spherulite crystallization in poly(ethylene oxide)-silica nanocomposites. retardation of growth rates through reduced molecular mobility. *Polymer Journal (Tokyo, Japan)*, 2002; 34:876-881.

5.   Q. Zhang and L.A. Archer. Structure and rheology of poly(ethylene oxide)/silica nanocomposites. *Langmuir,* 2002; 18:10435-10442.

6.   S.X. Wang, L.D. Zhang, H. Su, Z.P. Zhang, G.H. Li, G. W. Meng, J. Zhang, Y.W. Wang, J.C. Fan and T. Gao. Two-photon absorption and optical limiting in poly(styrene maleic anhydride)/$TiO_2$ nanocomposites. *Physics Letters A*, 2001; 281:59-63.

7.   N.R. de Tacconi, J. Carmona, W.L. Balsam and K. Rajeshwar. Photoelectrochromism in chemically modified nickel-titanium dioxide nanocomposite films. *Chemistry of Materials,* 1998; 10:25-26.

8.   M. Alexandre and P. Dubois. Polymer-layered silicate nanocomposites: preparation, properties and used of a new class of materials. *Materials Science & Engineering,* 2000; 28:1-63.

9.   E.P. Giannelis, R. Krishnamoorti and E. Manias. Polymer-silicate nanocomposites: model systems for confined polymers and polymer brushes. *Advanced Polymer Science*, 1999; 138:107-147.

10.  E.P. Giannelis. Polymer layered nanocomposite. *Advanced Materials*, 1996; 8:29-35.

11.  R.A.Vaia, K.D. Jandt, E.J. Kramer and E.P. Giannelis. Microstructural evolution of melt intercalated polymer-organically modified layered silicates nanocomposites. *Chemistry of Materials*, 1996; 8:2628-2635.

12.  D.A. Brune and J. Bicerano. Micromechanics of nanocomposites: comparison of tensile and compressive elastic moduli and prediction of effects of incomplete exfoliation and imperfect alignment on modulus. *Polymer*, 2002; 42:369-387.

13.  R.K. Bharadwaj. Modeling the barrier properties of polymer-layered silicate nanocomposites. *Macromolecules*, 2001; 34:9189-9192.

14.  H.Z. Friedlander. Organized polymerization. III. monomers intercalated in montmorillonite. *Journal of Polymer Science, Pt.B. Polymer Letters*, 1964; 2:475-479.

15.  I.A. Uskov. *High molecular weight compounds*, 1960; 6:926-930.

16.  A. Blumstein. Polymerization of adsorbed monolayers. 1: Preparation of the clay-polymer complex. *Journal of Polymer Science, Pt.A.,* 1965; 3:2653-64.

17.  Y. Kojima, A. Usuki, M. Kawasumi, A. Okada, Y. Fukushima, T. Karauchi and O. Kamigaito. Synthesis of nylon 6-clay hybrid. *Journal of Materials Research*, 1993; 8:1179-1184.

18.  Y. Kojima, A. Usuki, M. Kawasumi, A. Okada, Y. Fukushima, T. Karauchi and O. Kamigaito. Mechanical properties of nylon 6-clay hybrid. *Journal of Materials Research*, 1993; 8:1185-1189.

19.  A.B. Searle and R.W. Grimshaw. *The Chemistry and Physics of Clays*, 3rd ed. New York: Interscience Publishers Inc., 1959.

20.  W.F. Bradley. Molecular associations between montmorillonite and some polyfunctional organic liquids. *Journal of the American Chemical Society*, 1945; 67:975-981.

21.  D.M.C. MacEwan. Complexes of clays with organic compounds. I. complex formation between montmorillonite and halloysite and certain organic liquids. *Transactions of the Faraday Society*, 1948; 49:349-367.

22.  R. Greene-Kelly. Sorption of aromatic organic compounds by montmorillonite. I. orientation studies. *Transactions of the Faraday Society*, 1955; 51:412-430.

23.  R. Greene-Kelly. Sorption of aromatic organic compounds by montmorillonite. II. packing studies with pyridine. *Transactions of the Faraday Society,* 1955; 51:425-30.

24.  G. Lagaly. In: *Proceedings of the International Clay Conference. Denver, 1985*, eds. L.G. Schultz, H. van Olphen and F.A. Mumpton. Indiana: The Clay Minerals Society. Bloomington. 1987, pp. 343-351.

25.  D. Wang and C.A. Wilkie. A stibonium modified clay and its PS nanocomposite, *Polymer Degradation & Stability*, in press.

26.  J. Zhang and C.A. Wilkie. A carbocation substituted clay and its styrene nanocomposite, *Polymer Degradation & Stability*, 2004; 83:301-307.

27.  P. Reichert, H. Nitz, S. Klinke, R. Brandsch, R. Thomann, R. Mulhaupt. Poly(propylene)/organoclay nanocomposite formation: Influence of compatabilizer functionality and organoclay modification. *Macromolecular Materials Engineering*, 2000; 275:8-17.

28.  D.C. Lee and L.W. Jang. Preparation and characterization of PMMA-clay hybrid composite by emulsion polymerization. *Journal of Applied Polymer Science*, 1996; 61:1117-1122.

29.  M.W. Noh and D.C. Lee. Synthesis and characterization of PS-clay nanocomposite by emulsion polymerization. *Polymer Bulletin (Berlin)*, 1999; 42:619-626.

30.  S. Qutubuddin, X.-A. Fu and Y. Tajuddin. Synthesis of polystyrene-clay nanocomposites via emulsion polymerization using a reactive surfactant. *Polymer Bulletin (Berlin)*, 2002; 48:143-149.

31.  Y.-X. Pan, Z.-Z. Yu, Y.-C. Ou and G.-H. Hu. A new process of fabricating electrically conducting nylon 6/graphite nanocomposites via intercalation polymerization. *Journal of Polymer Science, Part B: Polymer Physics*, 2000; 38:1626-1633.

32.  G.-H. Chen, D.-J. Wu, W.-G. Weng, B. He, W.-L. Yan. Preparation of polystyrene-graphite conducting nanocomposites via intercalation polymerization. *Polymer International*, 2001; 50:980-985.

33.  F.M. Uhl and C.A. Wilkie. Polystyrene/graphite nanocomposites: effect on thermal stability. *Polymer Degradation & Stability*, 2002; 76:111-122.

34.  F.M. Uhl, Q. Yao and C.A. Wilkie. Expandable graphite/nylon 6 nanocomposites. in preparation.

35.  F.M. Uhl and C.A. Wilkie. Preparation of nanocomposites from styrene and modified graphite oxides. *Polymer Degradation & Stability*, in press.

36.  F.M. Uhl, Q. Yao and C.A. Wilkie. Formation of nanocomposites of styrene and its copolymers using graphite as the nanomaterial. Manuscript in preparation.

37.  C.A. Mitchell. Krishnamoorti, R. In: *Proceedings of Additives*. April 2003.

38.  H. Koerner, C.-S. Wang, R.A. Vaia, M.D. Alexander, N.A. and H. Bentley. In: *Proceedings of Additives*. April, 2003.

39.  C.A. Mitchell, J.L. Bahr, S. Arepalli, J.M. Tour and R. Krishnamoorti. Dispersion of functionalized carbon nanotubes in polystyrene. *Macromolecules*, 2002; 35:8825-8830.

40.  A.B. Morgan and J.W. Gilman. Characterization of polymer-layered silicate (clay) nanocomposites by transmission electron microscopy and x-ray diffraction: a comparative study. *Journal of Applied Polymer Science*, 2003; 87:1329-1338.

41.  D.L. Van der Hart, A. Asano and J.W. Gilman. NMR measurements related to clay-dispersion quality and organic-modifier stability in nylon-6/clay nanocomposites. *Macromolecules*, 2001; 34:3819-3822.

42.  D.L. Van der Hart, A. Asano and J.W. Gilman. Solid-state NMR investigation of paramagnetic nylon-6 clay nanocomposites. 1. crystallinity, morphology, and the direct influence of $Fe^{3+}$ on nuclear spins. *Chemistry of Materials*, 2001; 13:3781-3795.

43.  D.L. Van der Hart, A. Asano and J.W. Gilman. Solid-state NMR investigation of paramagnetic nylon-6 clay nanocomposites. 2. measurement of clay dispersion, crystal stratification, and stability of organic modifiers. *Chemistry of Materials*, 2001; 13:3796-3809.

44.  S. Bourbigot, J.W. Gilman, D.L. VanderHart, W.H. Awad, R.D. Davis, A.B. Morgan and C.A. Wilkie, C. A. Investigation of nanodispersion in polystyrene-montmorillonite nanocomposites by solid state NMR. *Journal of Polymer Science: Part B: Polymer Physics*, 2003; 41:3188-3213.

45.  J.W. Gilman, T. Kashiwagi, M. Nyden, J.E.T. Brown, C.L. Jackson, S. Lomakin, E.P. Gianelis and E. Manias. In: *Chemistry and Technology of Polymer Additives*. Eds. S. Al-Malaika, A. Golovoy and C.A. Wilkie, Blackwell Scientific, 1999, pp. 249-265.

46.  J.W. Gilman, T. Kashiwagi, E.P. Gianellis, E. Manias, S. Lomakin, J.D. Lichtenhan and P. Jones. In: *Fire Retardancy: The Use of Intumescence*, eds. M. Le Bras, G. Camino, S. Bourbigot and R. Delobel, Cambridge: Royal Society of Chemistry, 1998, pp. 203-221.

47.  M. Zanetti, T. Kashiwagi, L. Falqui and G. Camino. Cone calorimeter combustion and gasification studies of polymer layered silicate nanocomposites. *Chemistry of Materials*, 2002; 14:881-887.

48.  V. Babrauskas. Fire test methods for evaluation of fire-retardant efficacy in polymeric materials. In: *Fire Retardancy of Polymeric Materials*. eds. A.F. Grand and C.A. Wilkie, New York: Marcel Dekker, 2000, pp. 81-113.

49.    D. Wang, J. Zhu, Q. Yao and C.A. Wilkie. A comparison of various methods for the preparation of polystyrene and poly(methyl methacrylate) clay nanocomposites. *Chemistry of Materials*, 2002; 14: 3837-3843.

50.    W. Xie, Z. Gao, W.-P. Pan, D. Hunter, A. Singh and R. Vaia. Thermal degradation chemistry of alkyl quaternary ammonium montmorillonite. *Chemistry of Materials*, 2001; 13:2979-2990.

51.    W. Xie, R. Xie, W.-P. Pan, D. Hunter, B. Koene, L.-S. Tan and R. Vaia. Thermal stability of quaternary phosphonium modified montmorillonites. *Chemistry of Materials*, 2002; 14:4837-4845.

52.    W.F. Bradley. Molecular associations between montmorillonite and some polyfunctional organic liquids. *Journal of the American Chemical Society*, 1945; 67 I:975-981.

53.    G.W. Beall and S.J. Tsipursky. *Chemistry and Technology of Polymer Additives*, Oxford: Blackwell, 1999, pp. 266-280.

54.    A. Akelah, A. Rehab, A. Selim and T. Agag. Synthesis and application of organophilic polystyrene-montmorillonite supported onium salts in organic reactions. *Journal of Molecular Catalysts*, 1994; 94:311-322.

55.    S. Su, D.D. Jiang and C.A. Wilkie. Novel polymerically-modified clays permit the preparation of intercalated and exfoliated nanocomposites of styrene and its copolymers by melt blending, *Polymer Degradation & Stability*, 2004; 83:333-346..

56.    L. Biasci, M. Aglietto, G. Ruggeri and F. Ciardelli. Functionalization of montmorillonite by methyl methacrylate polymers containing side-chain ammonium cations. *Polymer*, 1994; 35:3296-3304.

57.    S. Su, D.D. Jiang and C.A. Wilkie. Methaxrylate modified cals and their polystyrene and poly(methyl methacrylate) nanocomposites, *Polymer Advances & Technology*, in press.

58.    S. Su, D.D. Jiang and C.A. Wilkie. Polybutadiene modified clay and its polystyrene nanocomposites, *Journal of Vinyl & Additive Technology*, 2004; 10:44-51.

59.    J. Zhu, A.B. Morgan, F.J. Lamelas and C.A. Wilkie. Fire properties of polystyrene-clay nanocomposites. *Chemistry of Materials*, 2001; 13:3774-3780.

60.    J. Zhu and C.A. Wilkie. Thermal and fire studies on polystyrene-clay nanocomposites. *Polymer International*, 2000; 49:1158-1163.

61.    J. Zhu, P. Start, K.A. Mauritz and C.A. Wilkie. Silicon-methoxide-modified clays and their polystyrene nanocomposites. *Journal of Polymer Science, Part A: Polymer Chemsitry*, 2002; 40:1498-1503.

62.    J.W. Gilman, C.L. Jackson, A.B. Morgan, R. Harris, E. Manias, E.P. Giannelis, M. Wuthenow, D. Hilton and S.H. Phillips. Flammability properties of polymer-layered-silicate nanocomposites. polypropylene and polystyrene nanocomposites. *Chemistry of Materials*, 2000; 12:1866-1873.

63.    A.B. Morgan, R.H. Harris, T. Kashiwagi, L.J. Chyall and J.W. Gilman. Flammability of polystyrene layered silicate (clay) nanocomposites: carbonaceous char formation. *Fire and Materials*, 2002; 26:247-253.

64.    A.B. Morgan, J.W. Gilman, R.H. Harris, C.L. Jackson, C.A. Wilkie and J. Zhu. In: *Abstracts of Papers - American Chemical Society*, 220th PMSE-064, 2000.

65.    H. Yao, J. Zhu, A.B. Morgan and C.A. Wilkie. Crown ether-modified clays and their polystyrene nanocomposites. *Polymer Engineering & Science*, 2002; 42:1808-1814.

66.    D. Wang and C.A. Wilkie. In-situ reactive blending to prepare polystyrene-clay and polypropylene-clay nanocomposites. *Polymer Degradation & Stability*, 2003; 80:171-182.

67.    G. Chigwada and C.A. Wilkie. Synergy between conventional phosphorus fire retardants and organically-modified clay can lead to fire retardancy of styrenics. *Polymer Degradation & Stability*, 2003; 81:551-557.

68.    X. Zheng and C.A. Wilkie. Flame retardancy of polystyrene nanocomposites based on an oligomeric organically-modified clay containing phosphorus. *Polymer Degradation & Stability*, 2003; 81:539-550.

69.    J. Kim, K. Lee, J. Bae, J. Yang and S. Hong. Studies on the thermal stabilization enhancement of ABS; synergistic effect of triphenyl phosphate nanocomposite, epoxy resin, and silane coupling agent mixtures. *Polymer Degradation & Stability*, 2003; 79:201-207.

70.    A.B. Morgan, T. Kashiwagi, R.H. Harris, J.R. Campbell, K. Shibayama, K. Iwasa and J.W. Gilman. Flammability properties of polymer-clay nanocomposites: polyamide-6 and polypropylene clay nanocomposites. In: *ACS Symposium Series*, 797 (Fire and Polymers), 2001, pp. 9-23.

71.    M. Zanetti, G. Camino, D. Canavese, A.B. Morgan, F.J. Lamelas and C.A. Wilkie. Fire retardant halogen-antimony-clay synergism in polypropylene layered silicate nanocomposites. *Chemistry of Materials*, 2002; 14:189-193.

72. J.W. Gilman, T. Kashiwagi, S. Lomakin, J.D. Lichtenhan, P. Jones, E.P. Giannelis and E. Manias. Nanocomposites: radiative gasification and vinyl polymer flammability. In: *Proceedings of the 46th International Wire & Cable Symposium*, 1997, pp. 761-774.

73. J. Zhang and C.A. Wilkie. Preparation and flammability of polyethylene-clay nanocomposites. *Polymer Degradation & Stability*, 2003; 80:163-169.

74. S. Wang, Y. Hu, Z. Qu, Z, Wang, Z. Chen and W. Fan. Preparation and flammability properties of polyethylene/clay nanocomposites by melt intercalation method from Na$^+$ montmorillonite. *Material Letters*, 2003; 57:2675-2678.

75. J. Zhu, P. Start, K.A. Mauritz and C.A. Wilkie. Thermal stability and flame retardancy of poly(methyl methacrylate)-clay nanocomposites. *Polymer Degradation & Stability*, 2002; 77:253-258.

76. A.B. Morgan, J.M. Antonucci, M.R. VanLandingham, R.H. Harris and T. Kashiwagi. Thermal and flammability properties of a silica-PMMA nanocomposite. *Polymer Materials Science & Engineering*, 2000; 83:57-58.

77. F. Dietsche and R. Mulhaupt. Thermal properties and flammability of acrylic nanocomposites based upon organophilic layered silicates. *Polymer Bulletin*, 1999; 43:395-402.

78. A. Kalendova, L. Kovarova, Z. Malac, J. Malac, J. Vaculik, J. Hrncirik, and J. Simonik. Modified montmorillonite clay nanocomposite in polyvinyl chloride. In *Proceedings of the 60th Annual Technical Conference of the Society of Plastic Engineers*, 2002, pp. 2250-2254.

79. D. Wang and C.A. Wilkie. PVC/clay nanocomposites: preparation, thermal and mechanical properties. *Journal of Vinyl & Additive Technology*, 2001; 7:203-213.

80. D. Wang, D. Parlow, Q. Yao and C.A. Wilkie. Melt blending preparation of PVC-sodium clay nanocomposites. *Journal of Vinyl & Additive Technology*, 2002; 8:139-150.

81. D. Wang and C.A. Wilkie. *Journal of Vinyl & Additive Technology*, 2002; 8:238-245.

82. M. Zanetti, G. Camino and R. Mulhaupt. Combustion behavior of EVA/fluorohectorite nanocomposites. *Polymer Degradation & Stability*, 2001; 74:413-417.

83. G. Beyer and E.A.G. Kabelwerk. Flame retardant properties of EVA-nanocomposites and improvements by combination of nano-fillers with aluminum trihydrate. *Fire & Materials*, 2002; 25: 193-197.

84. M. Alexandre, G. Beyer, C. Henrist, R. Cloots, A. Rulmont, R. Jerome and P. Dubois. Preparation and properties of layered silicate nanocomposites based on ethylene-vinyl acetate copolymers. *Macromolecules Rapid Communications*, 2001; 22:643-646.

85. F. Dabrowski, M. Le Bras, L. Cartier and S. Bourbigot. The use of clay in an EVA-based intumescent formulation. comparison with the intumescent formulation using polyamide-6 clay nanocomposite as carbonisation agent. *Journal of Fire Science*, 2001; 19:219-241.

86. S. Bourbigot, M. Le Bras, F. Dabrowski, J.W. Gilman and T. Kashiwagi. PA-6 clay nanocomposite hybrid as char forming agent in intumescent formulations. *Fire & Materials*, 2000; 24:201-208.

87. A.B. Morgan, J.W. Gilman, T. Kashiwagi and C.L. Jackson. In: *Fire Safety Developments: Emerging Needs, Product Developments, Non-Halogen FR's, Standards and Regulations, Papers, [Conference]*, Washington, DC, United States. Lancaster, PA: Fire Retardant Chemicals Association, 12-15 March 2000, pp. 25-41.

88. Y. Tang, Y. Hu, S. Wang, Z. Gui, Z. Chen and W. Fan. Preparation and flammability of ethylene-vinyl acetate copolymer/montmorillonite nanocomposites. *Polymer Degradation & Stability*, 2002; 78:555-559.

89. G.H. Hsiue, Y.L. Liu and H.H. Liao. Flame-retardant epoxy resins: an approach from organic-inorganic hybrid nanocomposites. *Journal of Polymer Science, Part A: Polymer Chemistry*, 2001; 39:986-996.

90. S.K. Gupta, J.J. Schwab, A. Lee, B.X. Fu and B.S. Hsiao. Polyhedral oligomeric silsesquioxane reinforced fire retarding epoxy vinyl ester resins. In: *Proceedings of the 47th International SAMPE Symposium & Exhibition*, 2002, pp. 1517-1526.

91. Z. Shu, G. Chen and Z. Qi. Polymer/clay nano-composite and its unique flame retardance. *Suliao Gongye*, 2000; 28:24-26.

92. G.E. Zaikov and S.M. Lomakin. New polymer flame retardant compositions for ABS and nylon 6. part 1. O*xidation Communications*, 1999; 22:556-558.

93. J.W. Gilman, T. Kashiwagi and J.D. Lichtenhan. Nanocomposites: a revolutionary new flame retardant approach. *SAMPE Journal*, 1997; 33:40-46.

94. J.W. Gilman, T. Kashiwagi and J.D. Lichtenhan. Nanocomposites: a revolutionary new flame retardant approach. *Proceedings of the 42th International SAMPE Symposium & Exhibition*, 1997, pp. 1078-1089.

95.   J.W. Gilman, R. Harris and D. Hunter. Cyanate ester clay nanocomposites: synthesis and flammability studies. *Proceedings of the 44th International SAMPE Symposium & Exhibition,* 1999, pp.1408-1423.

96.   J.-M. Yeh and C.-P. Chin. Structure and properties of poly(o-methoxyaniline)-clay nanocomposite materials. *Journal of Applied Polymer Science,* 2003; 88:1072-1080.

97.   N. Pantoustier, B. Lepoittevin, M. Alexandre, D. Kubies, C. Calberg, R. Jerome and P. Dubois. Biodegradable polyester layered silicate nanocomposites based on poly($\varepsilon$-caprolactone). *Polymer Engineering & Science.,* 2002; 42:1928-1937.

98.   J. Lee and E.P. Giannelis. Synthesis and characterization of unsaturated polyester and phenolic resin nanocomposites. *Polymer Prep. (American Chemical Society, Division of Polymer Chemistry),* 1997; 38:688-689.

99.   J.W. Gilman, T. Kashiwagi, M. Nyden, J.E.T. Brown, C.L. Jackson, S. Lomakin, E.P. Giannelis and E. Manias. Flammability studies of polymer layered silicate nanocomposites: polyolefin, epoxy, and vinyl ester resins. In: *Chemical Technology Polymer Additives,* Oxford: Blackwell, Oxford, 1999, pp. 249-265.

100.  A.P. Shah, R.K. Gupta, E.E. Smith and C.J. Hilado. Flammability and other characteristics of vinyl ester/clay, vinyl ester/nomex/clay, and vinyl ester/glass fiber/clay nanocomposites. In: *Proceedings of the 33rd International Conference on Fire Safety,* 2001, pp. 18-35.

101.  B. Lepoittevin, M. Devalckenaere, N. Pantoustier, M. Alexandre, D. Kubies, C. Calberg, R. Jerome and P. Dubois. Poly($\varepsilon$-caprolactone)/clay nanocomposites prepared by melt intercalation: mechanical, thermal and rheological properties., *Polymer,* 2002; 43:4017-4023.

102.  N. Pantoustier, M. Alexandre, P. Degree, C. Calberg, R. Jerome, C. Henrist, R. Cloots, A. Rulmont and P. Dubois. Poly($\varepsilon$ -caprolactone) layered silicate nanocomposites: effect of clay surface modifiers on the melt intercalation process., *e-Polymers [online computer file],* 2001, Paper No. 9.

103.  S. Bourbigot, E. Devaux, M. Rochery, X. Flambard and J.H. Koo. Nanocomposite textiles: new route for flame retardancy. In *Proceedings of the 47th International SAMPE Symposium & Exhibition,* 2002, pp. 1108-1118.

104.  H.A. Stretz, J.H. Koo, V.M. Dimas and Y. Zhang. Flame retardant properties of polycarbonate/montmorillonite clay nanocomposite blends. *Polymer Prep. (American Chemical Society, Division of Polymer Chemistry),* 2001; 42:50-51.

105.  J. Lee, T. Takekoshi and E.P. Giannelis. Fire retardant polyetherimide nanocomposites. In: *Proceedings of the Materials Research Society Symposium, 457(Nanophase and Nanocomposite Materials II),* 1997, pp. 513-518.

106.  E. Devaux, M. Rochery and S. Bourbigot. Polyurethane/clay and polyurethane/POSS nanocomposites as flame retarded coating for polyester and cotton fabrics. *Fire & Materials,* 2002; 26:149-154.

107.  J.W. Gilman and T. Kashiwagi. Polymer-layered silicate nanocomposites with conventional flame retardants. In: *Polymer-Clay Nanocomposites,* eds. T.J. Pinnavaia and G.W. Beall, New York: Wiley, 2000, pp. 193-206.

108.  J. Wang, J. Du, J. Zhu and C.A. Wilkie. An XPS study of the thermal degradation and flame retardancy of polystyrene-clay nanocomposites. *Polymer Degradation & Stability,* 2002; 77:249-252.

109.  J. Du, J. Zhu, C.A. Wilkie and J. Wang. An XPS study of thermal degradation and charring of PMMA clay nanocomposites. *Polymer Degradation & Stability,* 2002; 77:377-381.

110.  J. Du, D. Wang, C.A. Wilkie and J. Wang. An XPS investigation of thermal degradation and charring on poly(vinyl chloride)-clay nanocomposites. *Polymer Degradation & Stability,* 2003; 79:319-324.

111.  J. Zhu, F.M. Uhl, A.B. Morgan and C.A. Wilkie. Studies on the mechanism by which the formation of nanocomposites enhances thermal stability. *Chemistry of Materials,* 2001; 13:4649-4654.

112.  S. Su and C.A. Wilkie. The thermal degradation of nanocomposites that contain an oligomeric ammonium cation on the clay. *Polymer Degradation & Stability,* 2004; 83:347-362.

# Chapter 10

# Fire Safety Regulations

## 10.1    Introduction

An overview of the fire safety standards and regulations applied to polymer composite materials used in aircraft, ship, civil infrastructure and transport applications is given in this chapter. The fire standards are described in a general way, and it is not the purpose of this chapter to give an exhaustive description of all the regulations because it is outside the scope of this book. Regardless of the application, the fire safety standards applied to composites (and other combustible materials) should have the following characteristics: performance-based; accurately and realistically prescribe the fire performance limits of the composite component; only consider those fire reaction and fire resistive properties relevant to the application; consider the fire safety of the composite in the end-use condition rather than of the material itself; not require extensive and costly testing; internationally recognised; and provide a tool for the research and development of fire-safe materials.

Before outlining the fire safety standards, it is useful to consider several issues that are common to many of the regulations. Firstly, fire safety regulations often differ between nations, and in many cases these differences are substantial. For example, the regulations applied to the use of composites in rail carriages differ considerably between the United States, United Kingdom, Australasia, Japan and many other countries. As a result, certain types of composite may fulfil the fire safety requirements in one or several nations but fail in other countries. This difference arises because the international organisations needed to coordinate fire safety standards do not exist for many applications, including rail transportation, civil infrastructure and offshore platforms. The lack of uniform safety standards is considered one of the main impediments to the greater use of composites in many applications. A survey revealed that more than one-half of fire safety experts consider the lack of adequate and uniform regulations as the single greatest factor hindering the greater use of composites in many industries [1].

Some nations are forming collective bodies to establish and enforce common standards, and this is most notable in the countries of the European Union where an increasing number of fire test methods are being standardised by organisations such as ISO (International Organization for Standardisation) and CEN (European Committee for Standardisation). The most notable example is the fire safety standard applied to non-metallic materials used in ship construction, which is regulated globally by the International Maritime Organisation (IMO) but is enforced locally by the individual member countries of the IMO. In a few applications, the fire safety standards for composites are globally regulated. The best example is the Federal Aviation Administration (FAA) regulations related to cabin materials, which are applied to all large, wide-bodied civil aircraft.

Another important issue is that fire safety regulations often differ between the civilian and military sectors within a country. For example, the regulations applied to the use of composites in civilian aircraft may be diminished or waived when composites are used in military aircraft. This situation also exists for military land vehicles and ships, although some countries are applying the IMO regulations enforced for high speed craft to their naval vessels.

The regulations prescribed in many fire safety standards are continually evolving and changing, usually becoming more stringent, demanding and relevant. For example, an early version of FAA regulations for non-metallic materials used in the pressurised cabins of aircraft required a peak heat release below 100 kW/m$^2$ and a total heat release of 100 kW/m$^2$. In 1990 the regulations were tightened by reducing the peak and total heat release values to 65 kW/m$^2$ as well as enforcing an upper limit on smoke density. Traditionally, many standards required the fire testing of small composite specimens to determine specific fire reaction properties, usually flame spread rate and ignitability. The data generated by such tests was then used to rank the relative fire performance properties of composite materials. However, the test specimens were often not representative of their end-use condition, and therefore their fire performance could change in the actual application. As a result, there is increasingly a requirement of fire regulations to evaluate the performance of composites in their end-use condition, that includes geometry, surface coatings, attachments, and fire loads and scenarios expected for a given application.

## 10.2    Fire Safety Regulations for Rail

Polymer composite materials are used sparingly or not at all in passenger rail carriages due to concerns about their fire performance, in particular flammability and smoke toxicity, together with non-fire related concerns about their mechanical properties, processability and surface finish. Nevertheless, there is growing interest in the use of fire-safe composites in seating, panelling and roofing materials in carriages as a means to reduce weight. A survey sponsored by the Urban Mass Transportation Administration in the United States revealed that fire and smoke account for 1% to 5% of all safety incidents on US passenger trains [2]. Despite the rarity of fire, safety regulations on rail carriages in many countries are stringent and rigorously enforced because of the potential for large loss of life, especially if a fire were to occur in an underground tunnel where escape is difficult and the toxic smoke is trapped.

The fire safety requirements for passenger rail cars operated by Amtrak and other US inter-city/inter-state rail companies are mandated by the Federal Railroad Administration (FRA). The FRA is the regulatory Government agency responsible for US passenger train safety, and their standard Federal Register 49:192 (1984) gives the fire requirements for all materials used in US passenger carriages [3]. The fire performance limits set on combustible materials used in carriages and rolling stock are summarised in Table 10.1. The regulations have prescribed safety based on ignitability, flame spread, smoke density and fire endurance. It is seen that different limits are set on the various internal components of a carriage. The FRA guidelines presently require the fire testing of materials, rather than complete assemblies and components of a carriage. However, Amtrak has expanded the guidelines by stipulating that carriage components must be tested in their end-use condition, rather than as separate materials [4].

*Table 10.1. Summary of the fire test methods and performance limits prescribed for US railway vehicles relevant to use of composite materials*

| Material Application | Flammability Acceptance Limit | Smoke Emission Acceptance Limit | Flammability Test Method |
|---|---|---|---|
| Structural Flooring | Pass | None | ASTM E-119 |
| Seat Frame, Shroud, Exterior Shell, Wall and Ceiling Panels | Flame spread index $\leq 35$ | $D_s (90s) \leq 100$<br>$D_s (280\ s) \leq 200$ | ASTM E-162 |
| Insulation Materials | Flame spread index $\leq 25$ | $D_s (90s) \leq 100$<br>$D_s (280\ s) \leq 200$ | ASTM E-162 |

In Europe there is a lack of a common standard across the different countries. Each European country has its own regulations relating to flammability and smoke, and as a result there are currently thirty-five different standards applied across Europe to assess the fire safety of non-metallic materials for use in railway rolling stock. To overcome this situation, the European Union is presently developing a single fire standard, known as EN 45545 – "Fire Protection on Railway Vehicles", that will be enforced in all member nations. Most countries outside the European Union and United States have their own fire safety regulations, and at this time there is no push to develop an international standard.

## 10.3    Fire Safety Regulations for Automobiles, Buses and Trucks

Fire is not considered a major hazard for motor vehicles, despite there being over 400,000 car fires in the United States each year that regularly claim more than 700 lives and cause over 3,000 injuries. The growing use of moulded composites in place of sheet metal in body panels and interior components must be accompanied by improved fire safety regulations to ensure passenger safety. However, the need for regulations as stringent as for rail carriages, ships and aircraft is unnecessary because people can usually escape quickly from a burning vehicle.

The National Highway Transportation Safety Administration in the United States has implemented a fire safety standard for materials used in the interior of motor vehicles [5]. The standard is aimed at reducing the fire hazard caused by burning cigarettes and matches falling on the vehicle floor. The regulation requires that a 10 inch long test specimen of the floor material be exposed to the flame from a Bunsen burner to measure the rate of flame spread. Such a safety standard to assess the flammability and flame spread of automotive materials is currently not enforced in the European Union or most other countries [6].

## 10.4    Fire Safety Regulations for Civil Infrastructure

Composites are being used increasingly in infrastructure such as building façade panels, walkways and bridges. There is no international fire safety standard for infrastructure, and most countries have their own requirements for fire performance. However, the European Union has established fire safety regulations on composite materials used in construction products for buildings, including multi-occupancy dwellings, hospitals, schools, shops, clubs, leisure centres, stadiums, factories, stations, airports, tunnels and docks [7,8]. The following properties are evaluated in the European system: ignitability to a small flame, lateral flame spread, heat release, smoke, generation of flaming droplets, and fire resistance.

## 10.5    Fire Safety Regulations for Civilian Aircraft

The Federal Aviation Administration in the United States sets the fire safety regulations on the use of materials in wide-bodied passenger aircraft. These regulations are uniformly applied across the global aviation sector. Aircraft fires fall into three categories: ramp, in-flight and post-crash. Ramp fires occur when an aircraft is parked at the airport terminal, and the incident of such fires is very low. Fire is more likely to occur during flight or, in particular, post crash. Fire is ranked about the tenth greatest cause of aircraft crashes, and the fifth highest cause of aircraft fatalities (see Chapter 1). There were over 1,100 fatalities on US transport airliners between 1981 and 1990, and about 20% of these were caused by fire. The FAA believes that if aircraft accident rates continue at the current rate, then the number of deaths due to fire will increase at 4% per annum in-line with the growth in air passenger traffic.

Many fatalities occur in post-crash fires, when the flames, heat and toxic smoke generated by combustion of the aviation fuel and cabin materials can hinder escape from an aircraft and rapidly lead to death. For this reason, the FAA regulations consider the fire, smoke and toxicity properties of cabin materials for a post-crash fire scenario. The scenario dictates that passengers must be able to escape an aircraft within five minutes of a crash landing without being incapacitated, injured or hindered by heat, toxic fumes or smoke released from combustion of the cabin materials.

All non-metallic materials used inside the pressure vessel of commercial aircraft are subject to the FAA flammability regulations [9]. There are several fire tests mandated by the FAA to assess the flammability and fire performance of materials that are specified in FAR 25.853. The key fire properties considered by the FAA are total heat release, heat release rate and smoke emission. The FAA sets performance limits for heat and smoke on cabin materials to delay cabin flashover and provide increased escape time for passengers. Cabin flashover is a fire event characterised by ignition of the hot smoky layer below the cabin ceiling that contains incomplete combustion products released by the burning and smouldering cabin materials. When flashover occurs the cabin temperature rises suddenly, the flames spread rapidly, and the chances of survival for passengers and crew are virtually non-existent. Currently the FAA regulations are aimed at giving passengers an escape time of five minutes, but with the development of improved fire safe materials it is envisaged that this period could be increased to fifteen minutes within coming years.

The FAA regulations specify that the heat release properties of non-metallic materials must be measured using the Ohio State University (OSU) calorimeter test performed at a heat flux of 35 kW/m$^2$, as prescribed in ASTM E906. (The OSU calorimeter is described in the next chapter). As part of the safety regulations, the test material is required to have a total heat release of less than or equal to 65 kW/m$^2$ over two minutes and a peak heat release rate of less than or equal to 65 kW/m$^2$ over the five minute duration of the test. These specifications are used to ensure a cabin material does not contribute to the growth and spread of a fire during the first five minutes following a crash landing. The FAA requires that the smoke properties of non-metallic materials be measured using the NBS Smoke Chamber (which is described in Chapter 11) according to ASTM E662. This test is used to measure the smoke generating properties of cabin materials. Materials are required to have a smoke specific optical density ($D_s$) value that is less than or equal to 200 during the first four minutes. The FAA also specifies other fire regulations for ignition resistance and flame propagation using the Bunsen burner test. Further information on the fire safety regulations can be found in FAR 25.853, and McLean et al. [10] provide a review of the different test methods.

The fire performance requirements on composites used in military aircraft are different to those specified by the FAA. The military standards are application and aircraft specific. In general, the fire safety regulations for military aircraft, particularly fighters and assault helicopters, are less stringent than those mandated by the FAA.


## 10.6    Fire Safety Regulations for Ships and Submarines

The use of composites in the marine sector has greatly increased over the past thirty years, with these materials being used in a wide range of maritime vessels ranging in size from leisure craft and boats through to ships and submarines. The amount of composite material used in maritime vessels has increased on average by over 130% since 1970, and within the United States the annual consumption in the construction of boats and ships is currently about 190,000 tons. The growing use of composites has been accompanied by a deepening concern about their flammability and smoke toxicity. To minimise cost, the composites most often used maritime vessels have a polyester or vinyl ester matrix. These polymers are rarely used with a flame retardant additive or are chemically modified to reduce their flammability, and therefore the composites are highly combustible and produce copious amounts of thick smoke. The use of composites in large ships is a major concern because it is often difficult to contain and extinguish large fires. Furthermore, the heat and smoke fumes can remain trapped within the confines of the ship compartments.

The fire safety regulations for commercial ships licensed in the United States are specified in the Code of Federal Regulations, Title 46 [11]. This code covers small and large passenger ships, cargo vessels, tankers and shipbuilding materials. It also covers mobile offshore drilling platforms. The code makes it difficult to use composite materials in most ships due to the stringent requirements for no combustion or smoke. For most vessels the regulations require that the main ship structures (eg. hull, bulkheads, decks) be constructed using steel or an equivalent non-combustible material. The code also requires that the interior structures and fittings be made using non-combustible material. Consequently, composite materials cannot be used in ships, cargo vessels and tankers registered in the USA. The only vessels exempt from the code are small passenger vessels, lifeboats, various minor components, naval ships and submarines [4]. Certain small vessels (termed "T-vessels") can be constructed from composite material provided a low flame spread polymer approved by the United States Coast Guard is used.

Many nations have adopted or will soon adopt the safety regulations defined by the International Maritime Organisation (IMO). The IMO is an agency of the United Nations with responsibility for maintaining the "International Convention for the Safety of Life at Sea" (SOLAS), which are a series of international agreements designed to ensure the safety of passenger ships. The IMO regulations cover virtually every aspect of ship construction, including the fire safety of materials. While the IMO sets the regulations, the organisation has no power to enforce its rules; this is the responsibility of the individual member nations who are signatories to the agreement. There are over 150 members and several associate members of the IMO.

For many years the fire safety regulations required passenger ships to be constructed from non-combustible materials. The IMO defined a non-combustible material as a substance that "neither burns nor gives off flammable vapours in sufficient quantity for self-ignition when heated to approximately 750°C" in a vertical cylindrical chamber (ISO 1182). Organic matrix composites will decompose and release flammable vapours at 750°C, and therefore these materials were considered too unsafe to use in ships. The IMO introduced their High Speed Code (HSC) in 1996 to supersede earlier regulations covering high-speed craft, which includes catamarans, monohull passenger ships, hydrofoil passenger craft, and surface effect ships. The code is only applicable to passenger ships on voyages not further than four hours from a place of refuge in the event of fire or other major incident [12,13].

The HSC defines a new class of material known as 'fire-restricting material' that may be combustible but has low heat release rate, flame spread and smoke density properties. The code does not specify that composites can or cannot be used; it instead specifies a number of fire safety criteria that any material must fulfil to be considered safe to use in high-speed craft [14-16]. The regulations specify limits for peak and average heat release, peak and average smoke production, flame spread, and generation of flaming drops or debris. These fire properties must be measured on the ship materials in their end-use condition using the ISO 9705 room corner burn test (see Chapter 11).

A material will be accepted if it fulfils all of the following:
-   average heat release rate not exceeding 100 kW/m$^2$
-   maximum heat release rate not exceeding 500 kW/m$^2$
-   average smoke production rate not exceeding 1.4 m$^2$/s
-   maximum smoke production rate not exceeding 8.3 m$^2$/s (averaged over 60 seconds)
-   flame spread rate down a wall not exceeding 0.5 m from the floor
-   no flaming debris from the test material falling on the floor outside 1.2 m from the ignition source.

Furthermore, the material must pass a flame resistance test (ASTM E119). This test involves exposing a panel of the test material (in the end-use condition) to the SOLAS temperature-time condition for a specified period of time (up to one hour). The heating condition for the SOLAS test is shown in figure 11.10 (see the E119 cellulosic curve). In the flame resistance test, the average unexposed face of the panel must not exceed 139°C above the original temperature, and the temperature at any one point, including any joint, must not rise more than 180°C above the original temperature. These fire test conditions are stringent, and virtually no conventional composite material used in ship construction is able to fulfil all the criteria. Therefore, it is necessary to protect composites with a thermal barrier, intumescent coating or some other form of heat protection to meet the HSC definition of a fire restricting material.

The risk of fire on naval vessels is higher than for passenger ships because of the added dangers of weapons strikes in combat and accidental explosion of munitions. As discussed in chapter 1, in the period between 1983 and 1987 the United States Navy had

732 ship fires and between 1980 and 1985 had 164 submarine fires. Despite the higher fire risk with naval vessels, many navies do not have clearly defined fire safety regulations. Some navies are considering adopting the IMO code for high-speed craft to their surface vessels, while other navies – most notably the United States Navy (USN) – have established their own regulations to meet the unique operating conditions of warships and submarines [4,16-19].

The USN has established two guiding principles for the use of composite materials in their vessels: (i) the material itself must not be the fire source and (ii) the flammability resistance of the material must be sufficient to allow the crew to respond to the primary fire source [4,17-20]. Currently there is no fire safety standard applied to US naval ships, although several criteria have been set to reduce the hazard. There is a requirement that naval composite structures must be able to withstand a compartment fire for at least thirty minutes without flame or smoke penetrating the deck, ceiling or walls. There is a further requirement that composite structures exposed to fire must be able to support the ship loads for at least 30-60 minutes. The fire performance of candidate composite materials for US naval ships is based on a series of small-scale and full-scale fire tests, which are summarised in Table 10.2. Fire tests are performed to evaluate surface flammability; fire growth; fire resistance of bulkheads, decks, overheads, through penetrations, doors and joints; and structural integrity under load.

*Table 10.2. Fire Performance Goals and Fire Test Methods for US Navy Composite Topside Structures. Reproduced from Sorathia et al. [18].*

| Category | Test Method | Criteria |
|---|---|---|
| Surface Flammability | ASTM E-84 | <ul><li>Interior applications:<ul><li>Flame spread index : 25 max</li><li>Smoke developed index:15 max</li></ul></li><li>Exterior applications:<ul><li>Flame spread index : 25 max</li><li>Smoke data for review by NAVSEA 05P4</li></ul></li></ul> |
| Fire Growth | ISO 9705 "Fire Tests-Full-scale room test for surface products" Annex A standard ignition source file of 100 kW for 10 minutes and 300 kW for 10 minutes | <ul><li>Net Peak heat release rate over any 30 second period less than 500 kW</li><li>Net Average heat release rate for test less than 100 kW</li><li>Flame spread must not reach 0.5 m above the floor excluding the area 1.2 m from the corner with the ignition source</li><li>Not applicable to exterior weather surfaces</li></ul> |

| Category | Test Method | Criteria |
|---|---|---|
| Smoke Production | ISO 9705; 100 kW for 10 minutes and 300 kW for 10 minutes | • Peak smoke production rate less than $8.3\ m^2/s$ over any 60 second period<br>• Test average smoke production rate less than $1.4\ m^2/s$ |
| Smoke Toxicity | ASTM E662 (Flaming and non flaming mode) | • CO:350 ppm (max); HCl:30 ppm (max); HCN:30 ppm (max)<br>• Fire Gas IDLH Index, $I_{IDLH} < 1$; where $I_{IDLH} = (C_{co}/IDLH_{co} + C_{HCl}/IDLH_{HCl} + C_{HCN}/IDLH_{HCN})$ and $IDLH_{co} = 1200$ ppm, $IDLH_{HCl} = 50$ ppm and $IDLH_{HCN} = 50$ ppm |
| Fire Resistance and Structural Integrity Under Fire(1)<br><br>Bulkheads/ Overheads/ Decks/Doors Hatches/ Penetrations/ Composite Joints | • Navy modified: UL 1709 fire curve for 30 minutes using IMO A.754 (18) test procedures<br>• For composite joints, UL 1709 fire curve for 30 minutes using UL2079 "Tests for Fire Resistance of Building Joint Systems"<br>• Total number of thermocouples and their placement on unexposed side IAW MIL-PRF-XX381<br>• IMO App.A. III & A.IV apply<br>• Hose stream test (IMO) applies<br>• Maximum fire test load (2)<br>• Passive fire protection system shall be attached so as to survive the fire and other loads | • Average temperature rise on the unexposed surface not more than $250^\circ$ F ($139\,^\circ$C)<br>• Peak Temperature rise on the unexposed surface not more than $325^\circ$ F ($180\,^\circ$C)<br>• There should be no passage of flames, smoke, or hot gases on the unexposed face<br>• Approval of constructions restricted to the orientation in which they have been tested<br>• Structural Integrity Under Fire (under load):<br>  • No collapse or rupture of the structure for 30 minutes<br>  • The maximum average temperature on the unexposed side should not exceed the critical temperature of the composite where structural properties degrade rapidly. This applies if the critical temperature is less than the average temperature rise of $250\,^\circ$ F (3) |

Notes:

(1)  Fire resistance and structural integrity under fire is applicable only to those specific boundaries designated for each ship design. Fire resistance and structural integrity shall always be required for fire zone bulkheads. These requirements are normally not required for non-tight boundaries;

(2)  Tests shall be performed with maximum fire test load;

(3)  Critical temperature is defined as the temperature at which the rapid loss of modulus occurs when determined in accordance with DMTA.

The USN has defined a series of fire safety criteria on the use of composites inside the pressure hull of submarines, which are specified in the US military standard MIL-STD-2031: "Fire and Toxicity Test Methods and Qualification Procedure for Composite Material Systems Used in Hull, Machinery and Structure Applications inside Naval Submarines". The criteria are listed in Table 10.3, and are very stringent because death from heat or toxic smoke inhalation inside a submarine can occur within several minutes.

*Table 10.3. MIL-STD-2031 Submarine Composites Fire Performance Acceptance Criteria. Reproduced from Sorathia & Perez [19].*

| Fire Test/Characteristic | Requirement | Test Method |
|---|---|---|
| *Oxygen-Temperature Index (%)*<br>% oxygen at 25° C<br>% oxygen at 75° C<br>% oxygen at 300° C | <u>Minimum</u><br>35<br>30<br>21 | ASTM D-2863<br>(Modified) |
| *Flame Spread Index* | <u>Maximum</u><br>20 | ASTM E-162 |
| *Ignitability (sec)*<br>100 kW/m$^2$ irradiance<br>75 kW/m$^2$ irradiance<br>50 kW/m$^2$ irradiance<br>25 kW/m$^2$ irradiance | <u>Minimum</u><br>60<br>90<br>150<br>300 | ASTM E-1354 |
| *Heat Release Rate (kW/m$^2$)*<br>100 kW/m$^2$ irradiance, Peak/Average for 300 sec<br>75 kW/m$^2$ irradiance, Peak/Average for 300 sec<br>50 kW/m$^2$ irradiance, Peak/Average for 300 sec<br>25 kW/m$^2$ irradiance, Peak/Average for 300 sec | <u>Maximum</u><br>150/120<br>100/100<br>65/50<br>50/50 | ASTM E-1354 |
| *Smoke Obscuration*<br>Ds during 300 secs<br>D$_{max}$ | <u>Maximum</u><br>100<br>200 | ASTM E-662 |
| *Combustion Gas Generation*<br>(25 kW/m$^2$) | CO = 200 ppm<br>CO$_2$ = 4%<br>HCN = 30 ppm<br>HCl = 100 ppm | ASTM E-1354 |
| *Burn-Through Fire Test* | No burn-through in 30 minutes | Appendix B |
| *Quarter-Scale Fire Test* | No flashover in 10 minutes | Appendix C |
| *Large Scale open Environment Test* | Pass | Appendix D |
| *Large Scale Pressurizable Fire Test* | Pass | Appendix E |
| *N-Gas Model Smoke Toxicity Screening Test* | No deaths<br>Pass | Appendix F:<br>Modified<br>NBSTTM |

## References


1.    Anon. Survey results from the *3rd Conference on Composites in Fire*, Newcatle-upon-Tyne, September 2003.
2.    W.T. Hathaway. Fire safety in mass transit vehicle materials. In: *Proceedings of the 36th International SAMPE Symposium*, 15-18 April 1991.
3.    Federal Railroad Administration, Guidelines for Selecting Materials to Improve Their Fire Safety Characteristics, *Federal Register*, 49:162, 1984.
4.    U. Sorathia, R. Lyon, T. Ohlemiller and A. Grenier. A review of fire test methods and criteria for composites. *SAMPE Journal*, July/August 1997; 33:23-31.
5.    Urban Mass Transportation Administration, Recommended Fire Safety Practices for Rail Transit Materials Selection, *Federal Register*, 49: 158, 1984.
6.    Anon., 'The research requirements of the transport sectors to facilitate an increased usage of composite materials, Part II: The composite material research requirements of the automotive industry', www.compoitn.net.
7.    J. Murrell and P. Briggs. Developments in European and international fire test methods for composites used in building and transport applications. In: *Proceedings of the 2nd Conference on Composites in Fire*, Newcastle-upon-Tyne, September 2001.
8.    P. Briggs. Fire performance of composites in European construction applications. In: *Proceedings of the 3rd Conference on Composites in Fire*, Newcastle-upon-Tyne, September 2003.
9.    Federal Aviation Adminstration and Department of Transportation, *Aircraft Materials Fire Test Handbook*, Office of Aviation Research, Washington DC, 2000.
10.   B. McLean, S. Glicksberg and K. Coulliard. Qualification and certification of a new aerospace material with FAA fire property requirements. *SAMPE Journal*, Sept/Oct 2004; 40:6-12.
11.   P.J. Coxon and R.M. Letoureau. Development of fire safety standards for composite materials used in commercial marine applications. In: *Proceedings of the Marine Composites Symposium*, Savannah, GA, 8-10 November 1993, Paper F12.
12.   International Code for Application of Fire Test Procedures (FTP Code), Resolution MSC.61(67), International Maritime Organisation, London, 1997.
13.   International Code for High Speed Craft (HSC Code), Resolution MSC36(63), International Maritime Organisation, London, 1995
14.   A.T. Grenier and P.J. Maguire. Maritime fire safety standards – Some insight from an AHJ. www.uscg.mil.
15.   B. Høyning and J. Taby. Fire performance of composite marine structures in relation to the IMO high speed craft code. In: *Proceedings of the 2nd Conference of Composites in Fire*, Newcatle-upon-Tyne, September 2001.
16.   B. Høyning, 'Meeting commercial and military requirements for passive fire protection of composite high speed craft', In: *Proceedings of the 2nd Conference of Composites in Fire*, Newcatle-upon-Tyne, September 2003.
17.   E. Greene. *Marine Composites*. www.marinecomposites.com.
18.   U. Sorathia, G. Long, T. Gracik, M. Blum and J. Ness. Screening tests for fire safety of composites for marine applications. *Fire & Materials*, 2001; 25:215-222.
19.   U. Sorathia and I. Perez. Improving the fire safety of composite materials for naval applications. In: *Proceedings of the SAMPE Technical Conference & Exhibition*, Long Beach, CA, 16-20 May 2004.
20.   R.A. DeMarco. Composite applications at sea: fire-related issues. In: *Proceedings of the 36th International SAMPE Symposium*, 1991, pp. 1928–1937.

# Chapter 11

# Fire Tests for Composites

## 11.1　Introduction

The test methods most often used to measure the fire reaction and fire resistive properties of composites are outlined in this chapter. The majority of tests and standards described in this chapter are not specific to a certain class of material; although we seek here to examine them from the viewpoint of engineers interested in the use of composites. Determination of the fire reaction properties is important because of their strong influence on the early stages in the growth of fire. The reaction properties of composites that can affect the start-up and growth of fire include the time-to-ignition, oxygen index, peak and average heat release rates, and surface spread of flame. There are other reaction properties that may not influence the growth of fire but are critical to human survival, most notably smoke density and smoke toxicity. The test methods commonly used to measure these fire reaction properties are described.

Fire resistance is a general term that describes the ability of a structure to retain functionality in a fire and prevent the spread of fire. The capability of a structure to serve as a thermal barrier against the spread of fire to neighbouring rooms is an important aspect of fire resistance, which is determined by properties such as thermal insulation, burn-through rate, and flame-tightness of the structural material. Fire resistance also describes the changes to the load-bearing integrity of a structure during a fire, and here the key resistive properties are the retention of stiffness, strength and creep resistance during a fire and the residual mechanical properties following it. Fire resistance is critical to the safe use of composites in aircraft, ship, bridge and building structures. However, our understanding of fire resistance is limited by the need to characterise the properties using large composite structures in full-scale fires. Unlike the fire reaction properties, it is difficult to reliably determine the fire resistance of composite structures with bench-scale apparatus and small specimens. The high cost and complexity of many fire resistance tests has been a major impediment to characterising the fire behaviour of composite materials.

The first fire test was developed in the early twentieth century, and over the last one hundred years a large number of tests have been used to characterise the fire properties of combustible materials. Babrauskas and Peacock [1] and Babrauskas [2] give historical accounts of the development of fire reaction tests. The number of tests available to characterise fire reaction is now immense. For example, the American Society for Testing and Materials (ASTM) have over 80 fire test standards in the 1999 edition of their handbooks on testing standards. The standards are applicable to a wide range of materials, including building products, woods, plastics and other combustible materials. None of the ASTM fire standards are specific to polymer composites, but are instead generic to a wide range of materials. The ASTM is one of many international standardisation organisations that publish fire test methods. Babrauskas and Peacock [1] estimate there are thousands of fire test standards. Not all of the tests, of course, are relevant for testing composites; many tests are only suitable for evaluating building materials, woods or some other type of combustible material. Only a relatively small number of tests are suitable for determining the fire properties of composite materials or structures.

This chapter will not attempt to review every fire test method; the task would be too exhaustive and many tests are not directly relevant to composites. Instead, only those tests regularly used by research institutions, industry or fire safety authorities are described.

The fire reaction methods outlined here are the cone calorimeter (including the atmosphere controlled cone calorimeter), Ohio State University heat release rate calorimeter, limiting oxygen index test, flame spread tests, smoke density tests, single burning item method, and intermediate-scale to full-size room burn tests. The fire resistance tests described in this chapter include furnace and jet fire tests. In addition, a variety of unique, custom-designed tests that have been developed to measure fire resistance properties of specific composite structures, such as aircraft interiors, ship bulkheads or offshore pipes, are described.

The basic operating principles of the techniques are described, and the fire properties that can be measured with each method are identified. In addition, the capabilities and deficiencies of the tests for characterising the fire behaviour of composites are discussed. It is important to recognise that there is not a single, common approach to determine the fire response of composites [3], and often it is necessary to use a combination of techniques in order to completely characterise the behaviour of a composite material or structure in fire.

## 11.2     Scale of Fire Reaction Tests

Experimental techniques used to measure the fire properties of composites range in size from bench-top apparatus for testing small specimens weighing only a few grams up to full-scale tests for large structures. The scale of the fire tests reviewed in this chapter is shown in Fig. 11.1, where it can be seen that the sample size can vary from ~0.001 to 36 $m^2$. Regardless of scale, it is important that fire reaction tests are performed in conditions that closely replicate the type of fire to which the composite will be exposed. These fires range in heat flux from low intensity fires were the radiant heat flux is under 20 $kW/m^2$ to intense fires were the heat flux can exceed 150 $kW/m^2$. In addition, the material should be tested in the end-use or finished condition, for example covered with gel coat, paint or some other protective or decorative coating.

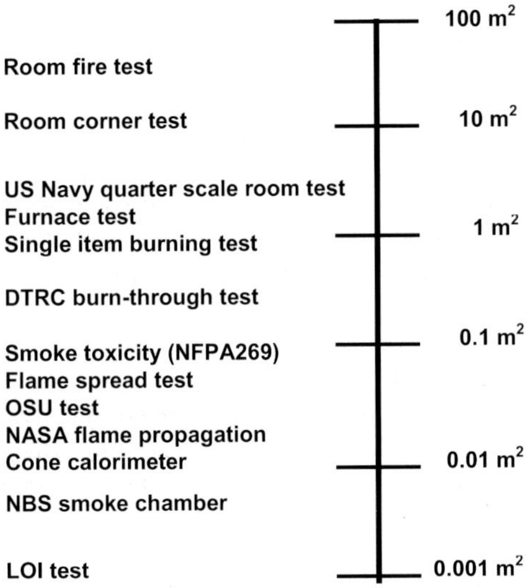


*Figure 11.1.  Range of specimen sizes of various fire test methods used to measure the fire reaction and fire resistive properties of materials.*

Of course, there is no single test for evaluating all the reaction and resistive properties of a composite for the many different fire scenarios. Instead, a broad range of techniques is available with each method relating to a specific type of fire threat. The most popular methods for measuring fire reaction properties are bench-scale tests, because they are quick, inexpensive and usually provide consistent, reproducible data. Bench-scale tests are often used to screen materials for their flammability and toxic smoke properties. A further use of such tests is to generate data that can be used to validate models for predicting the behaviour of materials in large fires [4]. The bench-scale tests described in this chapter include the cone calorimeter, Ohio State University heat release rate test, limiting oxygen index (LOI) test, N.B.S. smoke chamber, radiant panel, and smoke toxicity tests.

A limitation of many bench-scale tests (especially for heat release rate and smoke density) is they ignore the effects due to fire growth. Instead, they are said to relate "only to a snapshot of part of the overall fire behaviour" of the test material [5]. Another disadvantage is that it is difficult, if not impossible, to simulate actual fires using bench-scale techniques. For example, heat release rate, air movements and the oxygen/fuel ratio that exist in actual fires are often different to those in bench-scale fire tests, and this can affect substantially the measured fire reaction properties [6]. A further drawback of bench-scale tests are the entire sample is often completely consumed whereas, in real fires, this may not happen, due to the reduced oxygen levels encountered within enclosed, unventilated spaces [7]. The advantages and disadvantages of bench-scale fire tests are discussed at greater length by Babrauskas and Wickström [5].

There are a small number of intermediate-scale fire tests that can overcome some of the limitations of bench-scale tests. Intermediate-scale tests often involve a scale-model or part-section of a full-sized structure, although generally the surface area of the specimen is less than about 1-2 m$^2$. Such tests are often used to bridge the gap between bench-scale testing and full-scale testing, which is expensive. The intermediate-scale tests described are the single burning item (SBI) test, the furnace test, the US Navy quarter-scale room fire test, and the intermediate-scale cone calorimeter.

Large-scale tests are not usually performed to measure the fire reaction properties of composites because of the high cost and long set-up time. However, in certain cases, large-scale tests are unavoidable because of the uncertainty associated with scaling-up reaction data to large structures with a complex three-dimensional geometry. The large-scale tests most often used to measure the fire reaction properties of composite structures are room burn tests, which will be discussed. Finally, the commercial role of large scale tests as 'demonstrators' to show the viability of new materials and forms of construction should not be overlooked.


## 11.3    Cone Calorimeter

The most versatile bench-scale instrument for measuring the fire reaction properties of combustible materials is the oxygen-consumption cone calorimeter. The name 'cone calorimeter' comes from the truncated-shaped cone heater used to heat the specimen during fire testing. The instrument was developed by Dr Vytenis Babrauskas and colleagues at the National Institute for Standards and Testing (NIST) in the USA who were seeking improved methods to determine the flammability and heat release of combustible materials. Use of the cone calorimeter was first announced in 1982 [8].

From that time onwards the technique, and variants of it, have developed to become the most effective and widely used means of determining fire reaction properties. Presently over 100 instruments are in use in universities, research institutions and industries around the world.

The popularity of the cone calorimeter is due largely to its ability to determine a large number of fire reaction properties in a single test using a small specimen. The major capability of the instrument is that the heat release response of a burning material can be measured continuously to a high degree of accuracy, and values for peak and average heat release rate can be determined. The technique can also be used to measure time-to-ignition, time of sustained flaming, effective heat of combustion, smoke density, soot yield, mass loss rate, and yields of CO, $CO_2$ and other combustion gases. The only fire reaction property that cannot be directly determined is the flame spread rate.

Another reason for the success of the cone calorimeter is that the burning environment is considered a good representation of the majority of actual fire conditions, particularly for fire in a well-ventilated room [5, 9-11]. Materials can be fire tested at incident heat fluxes ranging from 0 to 100 kW/m$^2$, thereby simulating pre-flashover fires from low intensity up to high temperature fuel fires. However, the cone calorimeter is not able to represent flashover conditions when the heat flux may rise to about 125-150 kW/m$^2$ [12].

The cone calorimeter is a versatile technique that can be used to [10]:
- compare and rank the fire performance of different materials,
- pass or fail a material according to a certain fire criteria (eg. maximum heat release rate value or smoke obscuration level),
- assess the likely response of a material when exposed to a large fire,
- generate fire reaction data to validate fire models.

With regard to this last point, there are a number of models for predicting the behaviour of materials in full-size fires that are based on results from the cone calorimeter [11,13]. For example, the burning of furniture (eg. upholstered chairs) and wall linings in a large room fire can be accurately modelled using fire reaction data obtained using a cone calorimeter [eg. 14-16]. Furthermore, good agreement has been found between the peak heat release rate measured in a cone calorimeter for small representative samples of an article and the peak heat release rate of the whole article under realistic burning conditions. However, the capability of the cone calorimeter to replicate the fire response of composite structures in large fires has not yet been fully assessed. Greene [17] suggests that a disadvantage of the technique is that it is not very good for evaluating sandwich laminates or thermal barrier systems such as ceramic mats or intumescent coatings.

The ASTM first issued a full standard for cone calorimetry testing in 1990 under the designation 'Standard Test Method for Heat and Visible Smoke Release Rates for Materials and Products using an Oxygen Consumption Calorimeter' (ASTM E

1354-90). Since then a large number of standards for the cone calorimeter have been issued in North America (ASTM F 1550, ASTM D6113, NFPA 264, CAN ULC 135), mainland Europe (ISO5660), United Kingdom (BS 476 Part 15), Australia and New Zealand (AS/NZS 3837:1998), and elsewhere. The cone calorimeter has been described in numerous reviews by Babrauskas [1,4,7,10,18-23] and others [24-28], and therefore only the key features of the technique are outlined here. General and schematic views of a cone calorimeter are shown in Fig. 11.2.

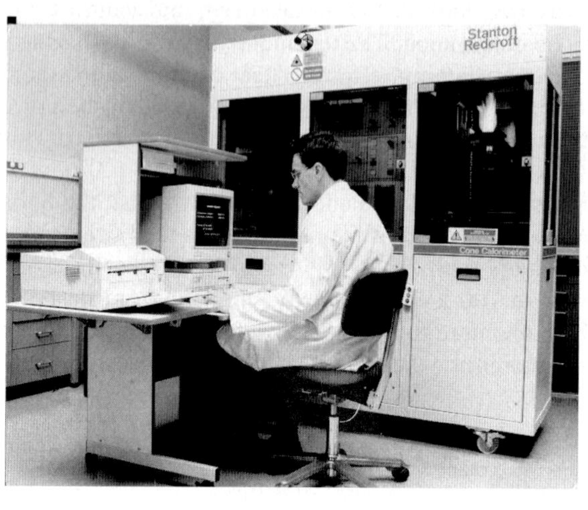


*(a)*

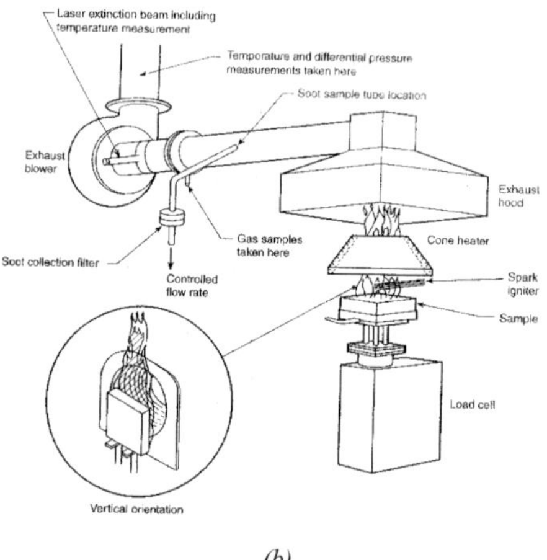


*(b)*

*Figure 11.2.  General view and schematic diagram of the cone calorimeter.  The diagram is reproduced with permission from Babrauskas and Peacock [1].*

Fire tests using the cone calorimeter involve exposing a flat specimen to an incident heat flux generated by a cone shaped heating element. The heater consists of an electric heater rod that is tightly wound into the shape of a truncated cone. Heating is precisely controlled using a thermostatically controlled radiant heater that is capable of subjecting a specimen to any incident heat flux up to 100 kW/m$^2$. Some instruments are able to generate a maximum heat flux of 110 kW/m$^2$.

Fire tests can be performed with the specimen in a horizontal or vertical direction, as illustrated in Fig. 11.3. Tests are usually performed in the horizontal orientation because in this mode the convective component of heat transfer is almost negligible. The heater and sample holder can be rotated through 90° to study the effect of vertical burning on the fire reaction properties. Vertical tests may of course be preferred if the actual fire scenario involves a vertical face or if it is required to study the effect of ply ablation on the fire behaviour of composites. Babrauskas [12] reports that the cone heater is capable of heating the specimen surface with a highly uniform heat flux. When fire tests are performed with the heater and specimen in the horizontal orientation, the peak variation in heating over the sample face is only about 2% when exposed to any heat flux between 25 and 100 kW/m$^2$. When the heater and specimen are in the vertical orientation, then the variation in surface heat flux is typically about 7%. Therefore, the cone calorimeter is able to produce a uniform, well-controlled heating condition, and this is a major advantage over many other fire test instruments.

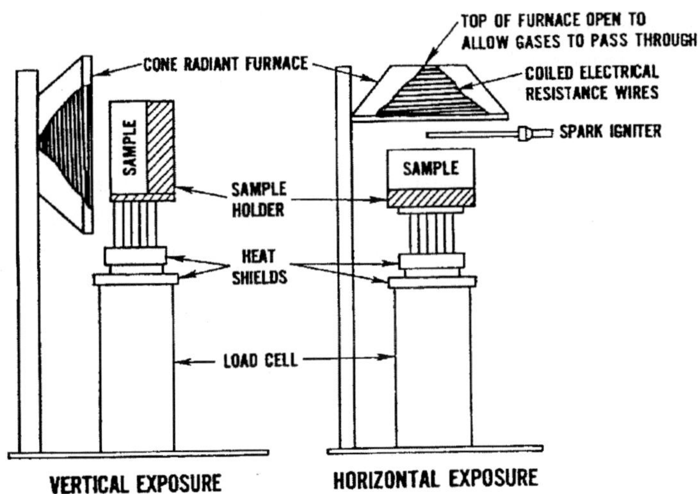


*Figure 11.3. Schematic showing the sample orientation in the vertical and horizontal directions in the cone calorimeter. The diagram is reproduced with permission from Greene [25].*

Specimens tested within the cone calorimeter are flat plates that are 100 mm long, 100 mm wide and up to 50 mm thick. The specimen is placed inside a sample holder lined with a low-density refractory material to minimise heat losses from the unexposed face. It is also necessary to insulate the specimen sides to minimise edge burning, which can give an artificially high heat release rate. The sides of composite specimens should also be sealed and insulated to avoid the escape of volatile gases from the edges, which can also affect the fire reaction properties [9]. During fire testing the mass loss of the specimen is recorded continuously using a load cell located beneath the sample holder.

The top surface of the specimen is positioned 25 mm from the cone heater. A spark igniter may be placed mid-way between the sample and heater, and is used to ignite combustible gases released from the thermally decomposing material when the concentration reaches the critical level needed to sustain flaming. Most fire tests on composites are performed using the igniter, although it is possible not to use the igniter to study the effect of auto-ignition on the reaction properties.

During fire testing a constant flow rate of air at 24 litres per second is maintained in the heating chamber. This air flow rate is needed to force all the gaseous combustion products released by the burning specimen up through the orifice of the cone heater and into the gas exhaust system above the heating chamber. As the entrained air and combustion products enter the exhaust system they are thoroughly mixed to avoid stratification of the gases that can cause inaccurate heat release, smoke or toxic gas measurements.

The most important capability of the cone calorimeter is the accurate measurement of heat release rate during combustion of a material. Heat is not measured directly in the cone calorimeter because of the difficulty in recording the thermal energy with a high degree of precision. Instead, the heat release rate is determined using the principle of oxygen consumption calorimetry [26,29]. This principle is based on the empirical observation by Huggett [30] that, for most organic materials, the heat evolved per unit mass of oxygen consumed is a near-constant value. Huggett found that this value ($\Delta h_c/r_o$) is 13.1 MJkg$^{-1}$ for most organic materials, including many polymers used in composite materials. ($\Delta h_c$ is the net heat of combustion and $r_o$ is the stoichiometric oxygen/fuel mass ratio). Huggett proposed that the heat release rate for a material can be determined from two simple measurements: the air flow rate through the fire environment and the residual oxygen concentration in the air after passing through a fire [29-31]. That is, the heat release rate can be calculated using the expression [31]:

$$ \dot{q} = \left( \frac{\Delta h_c}{r_o} \right) \left( \dot{m}_{O_{2,\infty}} - \dot{m}_{O_2} \right) \tag{11.1} $$

where $\dot{m}_{O_{2,\infty}}$ and $\dot{m}_{O_2}$ is the total mass flow of oxygen into and from the fire environment, respectively.

The oxygen content of the air entering the fire environment is a constant value (ie. 21%) while that in the exhaust stream can be measured with high precision using a paramagnetic oxygen analyser once water vapour has been removed using a cold trap. Parker [31] has shown that the heat release can be calculated with good accuracy using the expression:

$$\dot{q} = \left(\frac{\Delta h_c}{r_o}\right)\frac{M_{O_2}}{M_{air}} m_e \frac{(X^o_{O_2} - X_{O_2})}{\left[1-(b-1)X^o_{O_2}\right]-bX_{O_2}} \tag{11.2}$$

where $M$ is the molecular weight, $m_e$ is the mass flow rate of the exhaust stream, and $b$ is the stoichiometric factor. $X^o_{O_2}$ and $X_{O_2}$ are the mole fraction of oxygen before combustion and in the exhaust stream, respectively. This equation assumes the volume content of the combustion products (eg. CO, $CO_2$, HCl) to be negligible.

By considering combustion products in the analysis it is possible to calculate a more accurate heat release rate value. For example, when the mass fractions for both the oxygen and carbon dioxide are measured then a more precise heat release rate value can be calculated using:

$$\dot{q} = \left(\frac{\Delta h_c}{r_o}\right)\frac{M_{O_2}}{M_{air}} m_e \frac{1}{1-X_{O_2}-X_{CO_2}/[X^o_{O_2}(1-X_{CO_2})-X_{O_2}(1-X^o_{CO_2})]+(b+1)} \tag{11.3}$$

where $X^o_{CO_2}$ and $X_{CO_2}$ are the mole fraction of $CO_2$ in air before testing and in the exhaust stream, respectively.

Babrauskas [18], Janssens [26], and Janssens and Parker [29] give the equations for calculating the heat release rate when oxygen together with a variety of combustion products have been measured in the exhaust stream of a cone calorimeter. In most cases, however, Eqn. 11.2 is used because the concentrations of combustion gases are low compared to the $O_2$ content.

During fire testing the heat release of the specimen is recorded continuously by the cone calorimeter. It is common practice to express the heat release rate per unit area of the specimen, which is determined by:

$$q''(t) = \frac{\dot{q}(t)}{A_s} \tag{11.4}$$

where $A_s$ is the heat exposed surface area of the specimen. It is also usual to determine the total heat released during combustion (or up to a particular time), which is calculated by:

$$q'' = \sum_i \dot{q_i}''(t)\Delta t \tag{11.5}$$

where $\Delta t$ is the time period during which the heat release is measured.

To accurately determine the heat release rate it is essential to measure the oxygen concentration to a high degree of precision. As mentioned, a paramagnetic oxygen analyser within the exhaust section of the cone calorimeter is used to measure the oxygen content. The analyser is able to measure the oxygen level over the range of 0% to 25% to within an accuracy of 50 ppm.

The concentration of various combustion gases can also be measured. The carbon monoxide and carbon dioxide contents are measured using a real-time $CO/CO_2$ analyser connected to the same sampling line serving the oxygen analyser. However, the concentration of other combustion gases (HCl, HCN, $NO_x$) cannot be continuously measured in a standard cone calorimeter, although it is possible to take batch samples of the exhaust stream and then analyse for these gases using ion chromatography. A Fourier Transform InfraRed (FTIR) spectrometer attached to the cone calorimeter is being explored as a method for the real-time analysis of gases.

Smoke generation is another fire reaction property that can be measured in the cone calorimeter [32]. The smoke content in the exhaust stream is measured using a helium-neon laser photometer system with a split beam and silicon photodiode detectors. The sensing beam passes, over a known distance, through the smoke in the exhaust duct, while the reference beam travels direct to the detector. The intensity, $I$, of the sensing beam is then compared with $I_0$, that of the reference beam. The smoke obscuration is defined by the extinction coefficient ($k$):

$$k = \left(\frac{1}{L}\right) ln\frac{I_o}{I} \tag{11.6}$$

and the average specific extinction area $\sigma_{f(avg)}$:

$$\sigma_{f(avg)} = \frac{\sum_i V_i k_i \Delta t_i}{m_i - m_f} \tag{11.7}$$

where $L$ is the light path length, $V_i$ is the volume exhaust flow rate, $m_i$ is the initial specimen mass and $m_f$ is the final specimen mass. Smoke studies performed on various combustible materials (not including composites) by Marshall and Harrison [33] show that the cone calorimeter test gives a more reliable measurement of smoke density than many closed-box type smoke tests, including the N.B.S. Smoke Chamber described later.

In addition to optical smoke measurements, it is possible to equip the cone calorimeter with a device to measure the soot yield in smoke. This is a desirable test for composite materials because the release of ultra-small fragments of the reinforcing fibres during combustion could be a health concern [34,35]. The soot yield can be determined by placing a filter sampler into the exhaust stream of the cone calorimeter. A known mass fraction of the exhaust stream is passed through the filter that is weighed before and after the fire test to determine the soot yield. While the soot yield is relatively easy to determine by this method, most cone calorimeters are not fitted with a soot filter.


## 11.4    Atmosphere Controlled Cone Calorimeter

The standard cone calorimeter can only determine the fire reaction properties of combustible materials under normal atmospheric conditions (ie. 21% oxygen). During a fire, however, the oxygen content can decrease as it is consumed in the combustion reaction. This often happens when fires occur in enclosed spaces without ventilation, and the fire reaction properties of the combustible materials within the room can be altered by the depletion of oxygen. To study this, researchers at NIST developed the atmosphere controlled cone calorimeter for performing fire tests under different atmospheric conditions [36]. Atmosphere controlled calorimetry is not a widely used technique because of the high cost of the instrument. Furthermore, it is often unnecessary to determine the fire reaction properties under unusual atmospheric conditions. However, the technique is used occasionally to measure the fire behaviour of materials in a low oxygen atmosphere (such as at high altitude) or an oxygen-rich environment (such as inside manned space-craft) [37].

A schematic view of an atmosphere controlled cone calorimeter is given in Fig. 11.4. This instrument has the same capabilities as a conventional cone calorimeter, and is able to measure reaction properties such as time-to-ignition, mass loss, heat release rate, smoke density and combustion gases. The unique feature is that the heating chamber is enclosed and sealed from the outside environment. In this way the fire atmosphere can be controllably altered by the flow of gas into the test chamber. It is possible to vary the oxygen content between 0% and 50%, and to create any other atmosphere (eg. pure nitrogen) to study the effect on flammability.

Both standard and atmosphere controlled cone calorimeters currently operate under ambient pressure. Several research laboratories are interested in the possibility of constructing a cone calorimeter with a pressurised heating chamber.

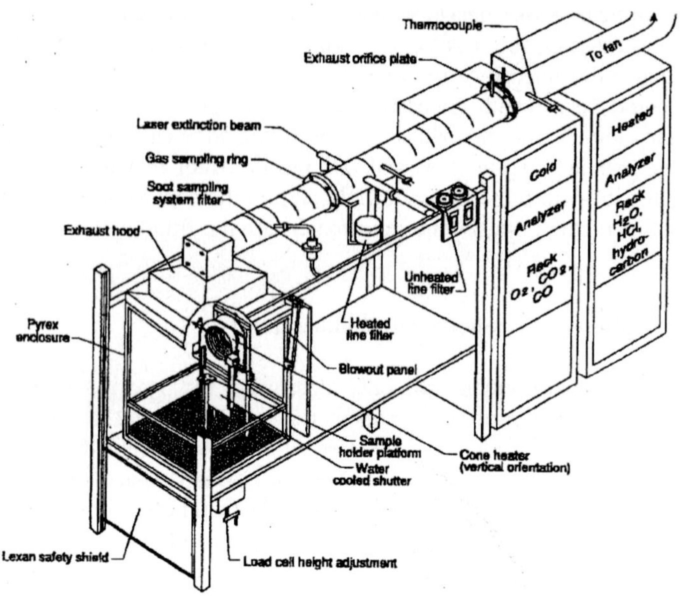


*Figure 11.4. Atmosphere controlled cone calorimeter. Reproduced with permission from Babrauskas et al. [33].*

## 11.5  Intermediate-Scale Cone Calorimeter

Despite the outstanding capability of the cone calorimeter to be able to determine many fire reaction properties in a single test, the apparatus is limited to testing small flat specimens. Tests on large complex structures cannot be performed using a conventional calorimeter, and this is a concern because such structures can behave in ways that can not be predicted using fire reaction data obtained from small samples. To overcome this limitation, Urbas and Luebbers [38] developed the intermediate-scale cone calorimeter shown schematically in Fig. 11.5. The operating principles of this apparatus are similar to a conventional cone calorimeter, and the testing procedure is outlined in the ASTM standard E1623. A feature of the intermediate-scale calorimeter is the wide fume exhaust hood and large fire chamber that can accommodate specimens up to 1.0 m x 1.0 m. This is sufficiently large to allow fire tests to be performed on complex structural sections and components. The intermediate-scale cone calorimeter has been used successfully to characterise the heat release properties of upholstered furniture and other household items [14-16].

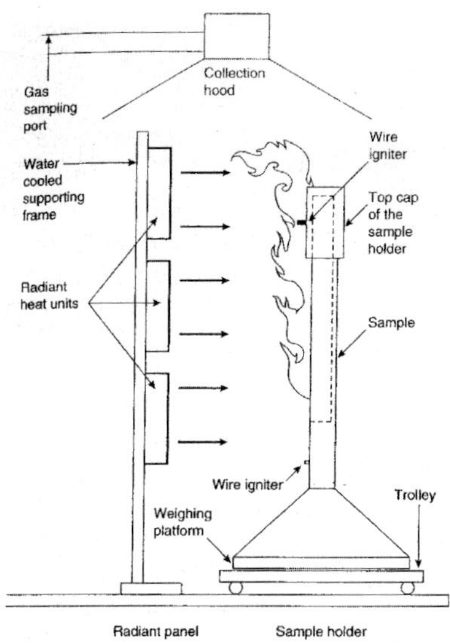


*Figure 11.5.  Schematic of the intermediate-scale cone calorimeter.  Reproduced with permission from Babrauskas [23].*

## 11.6     Ohio State University Calorimeter

The Ohio State University (OSU) calorimeter was developed in the early 1970s as a bench-scale technique for determining the flammability of building materials [39], and has since developed into a standard fire test for measuring the heat release rate of combustible materials, including polymers and polymer composites. The OSU calorimeter test is covered by ASTM E906 and is described by Babrauskas [6,7] and Sorathia et al. [3].

Figure 11.6 shows an OSU calorimeter, which basically consists of a small heating chamber that is designed to be adiabatic. Inside the chamber are four electric heating bars that directly face the specimen.  Fire tests can be performed with the specimen in a vertical or horizontal direction, although common practice is to test vertical samples. The vertical specimen size is 150 mm x 150 mm and up to 100 mm thick whereas the horizontal sample size is 110 mm x 150 mm and up to 45 mm thick.  Fire testing involves subjecting the specimen to a constant heat flux up to 80-100 kW/m$^2$, although many tests are performed at 35 kW/m$^2$.  The specimen is ignited during heating using a high temperature flame.  A problem with the flame igniter is that localised burning can occur, which affects the flammability results, particularly for horizontal samples that can be difficult to ignite.  Following ignition the heat release rate of the test material is measured continuously as it burns to completion. The smoke released by the burning material can be measured by a photometer located above the combustion chamber.

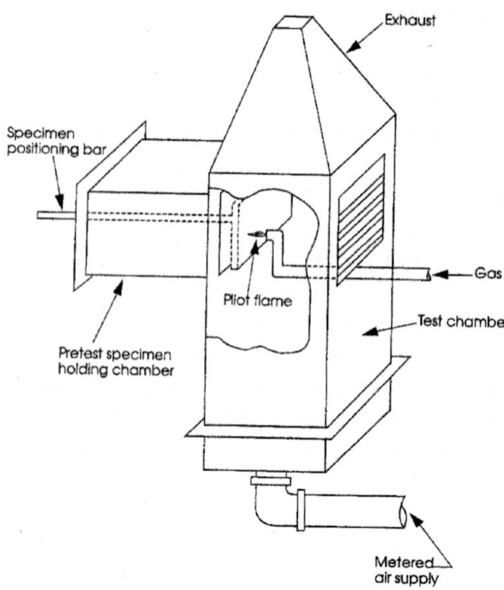


*Figure 11.6. Schematic of the Ohio State University calorimeter. Reproduced with permission from Babrauskas [7].*

The OSU technique is not as widely used as the cone calorimeter because the results are prone to greater error.  Babrauskas [2,40] and Hirschler [41] have listed the deficiencies of the OSU apparatus to include:

- the heating chamber is not perfectly adiabatic, and this can cause heat loss problems, resulting in errors in the determination of heat release rate,
- cannot measure mass loss during combustion,
- difficulties can be experienced when igniting the specimen that can cause severe localised burning,
- soot and smoke are forced to flow between the specimen and heating elements that can cause after-burn, and
- difficulties can occur in reaching and maintaining heat fluxes higher than ~50 kW/m$^2$.

Despite the problems, the OSU test is used to assess the fire performance of composite materials.  The Federal Aviation Administration (FAA), for instance, have adopted the technique as the standard method to characterise the fire safety of aircraft cabin materials at a heat flux of 35 kW/m$^2$.  It is also used by laboratories and a number of other industries to evaluate the heat release rate properties of new materials.

## 11.7    Limiting Oxygen Index Test

The ignition and flammability of combustible materials was evaluated for many years using a variety of small-scale tests, many of which involved burners. Because of reproducibility problems with these early methods, the Oxygen Index test was developed in the 1960s. The method was refined into what is now known as the 'Limiting Oxygen Index (LOI)' test, also known as the 'Critical Oxygen Index' test. The test measures the minimum percentage of oxygen in the test atmosphere needed to support ignition and flaming combustion of the test material.

The LOI method is described in various United States and European test standards, including ASTM D2863, ISO 4589-2 and NES 714. Figure 11.7 shows the LOI apparatus, which consists of a vertical chimney standing 450 or 500 mm high with an internal diameter of 75 or 100 mm. The chimney is made of a heat-resistant glass that allows the burning of the specimen to be observed. Oxygen and nitrogen are pumped in at the base of the chimney where they pass through a layer of glass beads that ensures even mixing before entering the main test chamber. A needle valve or paramagnetic oxygen cell inside the chimney is used to measure the oxygen concentration to better than 0.1%. The gas mixture flows upwards to a vertical test specimen that is between 6.3 to 12.7 mm wide, up to 10.5 mm thick and 152 mm high. A small gas flame is used to ignite the upper end of the specimen, and the subsequent burning behaviour is monitored.

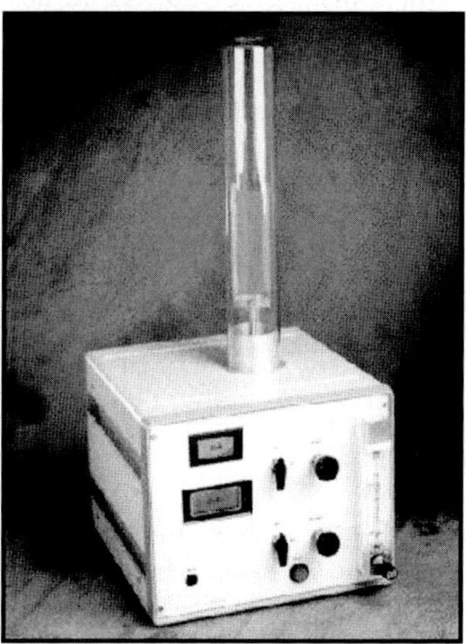

*Figure 11.7.  General view of the limiting oxygen index apparatus.  Photograph used with the permission of Fire Testing Technology Ltd.*

A series of tests must be performed at increasing oxygen concentration, with the LOI being determined by the minimum percentage of oxygen needed by the sample to burn with a candle-like flame for exactly three minutes. The LOI can also be defined as the minimum oxygen concentration required for the flame to creep down the sample for 50 mm. Whichever definition is used, the LOI value is not a fundamental fire reaction property of a material. Therefore it should only be used to rank the relative flammability of different materials.

The LOI is usually measured at room temperature, although it is possible to heat the chimney to conduct tests at elevated temperature. LOI equipment is available to determine the oxygen index between room temperature and 125°C, while custom-built devices can be operated at much higher temperatures. A limitation of the test, however, is that the oxygen index value can change with temperature, often changing the relative flammability ranking of some materials [42-45]. Therefore, care must be exercised when selecting the test temperature. In addition to temperature, other test variables can affect the oxygen index value. The effects of gas flow pressure and velocity, sample dimensions, moisture content, and equipment dimensions on the index value have been investigated.

Despite its widespread use, the LOI method has several deficiencies. Most notably, the test condition has never been correlated to any aspect of full-size fires [1]. Therefore, it is not possible to imply the burning conditions of a material observed in the LOI test with behaviour in an actual fire. For this reason, several fire researchers suggest that the significance of the test is questionable [1,46]. Another problem with the test is the use of downward flame spread for a distance of 50 mm to define the oxygen index of a material. In real fires the downward spread of flame is of little importance, and the sense of measuring this property is also questionable. Indeed, Weil et al. [46] makes the criticism that the characteristics of the heat transfer and rate of burning under downward flame spread conditions are fundamentally different from those in the much more important upward-burning configuration that dominates real fires. Furthermore, the LOI test is conducted at oxygen concentrations usually above the normal oxygen content of air, which again does not occur in most fires. Because of these problems, the LOI test does not, except in unusual circumstances, predict the real fire performance of materials.

## 11.8    Flame Spread Tests

An important fire safety consideration when using polymer composites is the rate at which flames can spread over the surface. The theory of surface flame spread is reviewed by Quintiere [47], and despite considerable progress in recent years the ability of theory to predict flame spread over composite materials is limited by a lack of material data and phenomenological information about fire conditions. Therefore, the most reliable approach to determine the flame spread properties is by testing.

The radiant panel flame spread test is arguably the most widely used method. The test has been standardised by the ASTM for measuring the surface flammability of building products (ASTM E 162) and cellular plastics (ASTM D3675), and it can also be used on polymer laminates and sandwich composites. The purpose of the test is to determine the rate at which a flame front spreads down the surface of a burning specimen [3,7,24,48]. The apparatus is shown in Fig. 11.8, and testing involves exposing a flat specimen panel to a heat flux produced by a 0.30 m x 0.61 m gas-fired heater. The test panel is 0.15 m wide x 0.46 m high, and is inclined at an angle of 45° to the heater. The heater is positioned so that ignition occurs at the upper edge of the specimen, and during a test the rate at which the flame front travels down the panel is measured. Also measured during testing is the temperature rise in the stack. From these two measurements the Flame Spread Index, $I_s$, can be calculated using the equation:

$$I_s = F_s Q \qquad (11.8)$$

where $Q$ is the heat evolution factor.

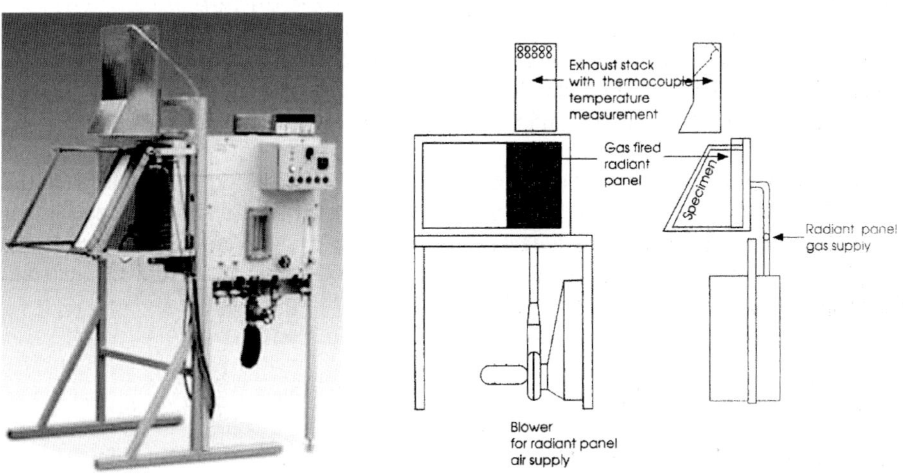


*Figure 11.8. General view and schematic diagram of the radiant panel flame spread test apparatus. The photograph and diagram is reproduced with permission from Fire Testing Technology Ltd. and Greene [25], respectively.*

Both the flame spread speed and heat liberated from the panel change over the course of a test, however the flame spread index is formulated to be a constant value in order to provide a common scale for ranking different materials. The flame spread index is also an indirect measure of the heat release response of certain types of flammable materials.

The radiant panel flame spread test is often used to determine the fire behaviour of composite materials [49,50]. However, the downward movement of flame is a slow and unrealistic mode for fire to spread. Of course, upward flame spread, which is much faster, is the more likely mode in an actual fire. A material that exhibits good flame spread resistance when tested in the downward mode may not necessarily have good performance in upward spread. Therefore, the practical relevance of the flame spread index is questionable.

The radiant panel flame test apparatus (ASTM E162) can also be used to measure smoke density. An exhaust hood above the panel is used to collect smoke released by the burning material. A filter in the exhaust stream is used to measure the smoke density. However, this test is rarely used to measure smoke because other techniques such as the cone calorimeter and N.B.S. Smoke Chamber are more reliable.

Several fire tests have been developed to create more realistic modes of flame spread [39,51,52], although they are not as widely used as the radiant panel flame spread technique. NASA developed the Upward Flame Propagation Test to evaluate the flame spread behaviour of materials used in spacecraft, including polymer composites [53]. In this test a small specimen (30.5 cm long x 64 cm wide) is exposed to an incident heat flux of 75 kW/m$^2$ while an igniter is used to ignite the bottom edge. The test lasts for only 25 seconds, and the sample is deemed to have passed if the flame spread length is under 15.3 cm (or half the specimen length). The average flame spread rate is calculated by dividing the flame spread length by the burn time.

Other tests include the lateral flame spread method (ASTM E1321) and the fire tunnel test (ASTM E84). The BS 476 range of procedures, used in the United Kingdom and Europe, contains a flame spread technique that employs a vertical 1.2 metre wide sample placed at right angles to a vertical gas matrix burner. The rate of horizontal flame spread is measured.

## 11.9    Smoke Density Tests

Smoke is defined by the ASTM Fire Standards Committee as a visible 'airborne suspension of solid and liquid particles evolved when a material undergoes pyrolysis and combustion'. The particles suspended in smoke are ultra-fine, and typically have an average particle size distribution of 0.3 to 3 μm. The quantity and size of smoke particles is determined by the chemical composition and char yield of the material. The smoke is also affected by the nature of the combustion process, with flaming, pyrolysis and smouldering conditions affecting smoke in different ways [54]. One of the main concerns with smoke is it greatly limits visibility and hinders the ability of people to escape and of fire fighters to locate and suppress the fire.

Smoke formation is not an inherent fire property of a material. The level of smoke measured in a test depends on the burning condition (eg. heat flux, oxygen level, the presence or absence of flame) as well as the test apparatus (eg. test chamber volume, ventilation, specimen geometry, etc). As a result, no single smoke test or even a set of smoke measurements from different tests is likely to provide a comprehensive definition of smoke behaviour in a real fire. A variety of experimental techniques for measuring the smoke density of burning materials have been developed, and these are reviewed by Hirschler [41] and others [eg. 55,56]. No single smoke test is universally recognised for its predictive ability or correlation with real fire situations. Many of the techniques have inherent problems, the most frequent one being that the smoke in the test is produced under unrealistic combustion conditions that do not occur in actual fires. Care therefore needs to be exercised when selecting a method for measuring smoke properties.

As mentioned, the cone calorimeter, the Ohio State University calorimeter, and the radiant panel flame techniques can be used to measure the smoke density of burning materials, in addition to other fire properties. Although, the single most widely used test is the N.B.S. smoke chamber, shown in Fig. 11.9 [3,24,57]. The technique determines the specific optical density of smoke generated by materials, and the test procedure is outlined in several international standards including ASTM E662, ASTM F814, NFPA 258 and BS 6401. The apparatus consists of a sealed chamber that holds a specimen measuring 76 mm x 76 mm that can be up to 25.4 mm thick. The specimen is mounted in the vertical position facing an electric radiant heater that generates an incident heat flux of 25 kW/m$^2$. The heater to some N.B.S. Smoke Chambers is modified so tests can be performed at variable heat fluxes up to 50 kW/m$^2$. Tests can be performed with the specimen in the flaming or non-flaming (smouldering) modes.

Smoke released by the burning material is measured in the smoke chamber using a monochromatic light beam. A photometric system with a vertical light path is used to measure the light transmission as smoke accumulates during combustion. The light transmittance measurements are used to calculate the specific optical density of the smoke. The specific optical density ($D_s$) is a dimensionless number that is inversely related to the visibility through smoke, and is calculated using the equation:

$$D_s = (V / AL)log_{10}(100 / T)$$
(11.9)

where $V$ is the volume of the smoke chamber, $A$ is the exposed surface area of the sample, $L$ is the length of the light path, and $T$ is the percentage light transmittance. The two parameters often measured in a smoke test are the maximum specific optical density ($D_{max}$) and specific optical density at 300 seconds ($D_{300s}$). As well as smoke density, the N.B.S. Smoke Chamber can be instrumented to measure the presence of combustion gases (eg. CO, $CO_2$, HCN, HCl, reduced $O_2$) within the smoke.

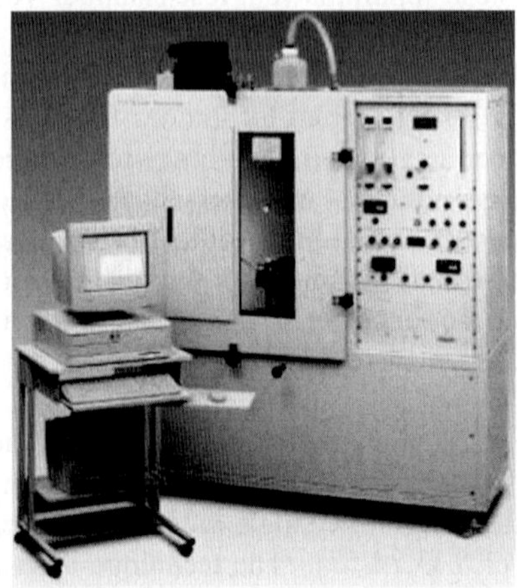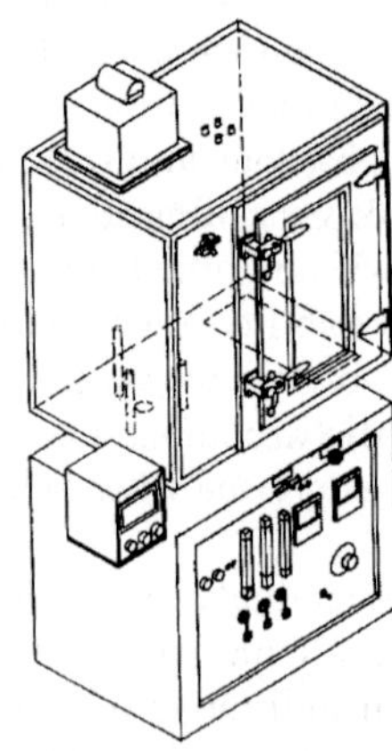

*Figure 11.9. General view and schematic diagram of the N.B.S. smoke chamber. The photograph and diagram is reproduced with permission from Fire Testing Technology Ltd. and Greene [25], respectively.*

While the N.B.S. Smoke Chamber is a popular technique, the test has several deficiencies that include [41,58,59]:

- The results do not correlate with full-scale fires.
- The sample weight loss is not measured.
- The fire self-extinguishes when the oxygen level drops below ~14%.
- The chamber is not adiabatic, with heat loss from the walls being significant.
- Soot gets deposited on the optical system, which can produce misleadingly high specific optical density values.
- Smoke is recirculated through the heater and through the flame, thereby being recombusted.
- Some chambers can only test specimens in the vertical position at a single heat flux (25 kW/m$^2$).
- The smoke distribution is not uniform through the chamber, but can concentrate as layers near the chamber ceiling.


## 11.10    Furnace Tests

The ability of composites to provide thermal insulation, smoke tightness and burn-through resistance is essential to slow the spread of fire between rooms. Fire scientists have yet to adopt a single method to determine these fire resistant properties. Instead several tests are used, most notably furnace and burn-through fire tests. Furnace testing is often used to certify the fire resistance of large composite structures used in ships, such as decks, bulkheads and panels for superstructures. This testing is also used to evaluate the fire performance of composites to be used in buildings and offshore platforms.

The furnace method involves heating a large composite panel attached to the open side of a gas or electric furnace. Depending on the furnace size, composite panels between 1 $m^2$ and 10 $m^2$ can be tested. The panel is tested in the end-use condition with the appropriate surface coatings, joints and other structural details to ensure a realistic test for the fire behaviour of a full-size composite structure. The panel is instrumented with thermocouples and, if desired, heat flux gauges to measure its thermal response in the fire.

The furnace method can be used to test the fire resistance of composites under a diverse range of heating conditions. Although it is more meaningful to perform fire tests under controlled heat flux rather than controlled temperature conditions, the majority of furnace tests involve subjecting the panel to controlled temperature-time conditions. Figure 11.10 shows the temperature-time heating curves most often used to determine the fire resistance of composites. The lower temperature curve is the cellulosic fire (ASTM E119), and this is used to generate a thermal environment similar to that experienced when materials such as woods and fabrics are the fuel source. The more severe fire test is performed using the hydrocarbon (UL1709) heating condition. The hydrocarbon curve is intended to simulate a fire with petroleum or oil is the fuel source. The UL1709 curve was devised after the fire incident aboard the *USS Stark* when it was realised that the cellulosic/E119 heating condition did not accurately replicate munition fires on warships. Likewise, there is concern that the cellulosic/E119 condition does not reliability represent the fire resistance required for building materials.

During a furnace test the thermal response of the composite is monitored using thermocouples buried within and attached to the exposed and unexposed surfaces of the panel. The fire resistance is usually defined by the time taken for the unexposed panel surface to rise to 160$^{o}$C or for a hot spot to reach 180$^{o}$C above ambient temperature. Thermal cameras are useful for real-time monitoring of the temperature profile over the unexposed surface of the panel.

Furnace tests are also used to determine the structural integrity of loaded composite panels when exposed to fire. During fire testing the panel is subjected to a compressive and/or bending load. The change in the structural response of the panel is recorded using strain gauges and displacement transducers placed on the back-face to measure the change in stiffness and strength. These furnace tests have been used to assess the structural integrity of composite ship panels in fire for the United States Navy. The British Standard BS476 contains furnace procedures for testing three metre square vertical panels, for beams and columns under load.

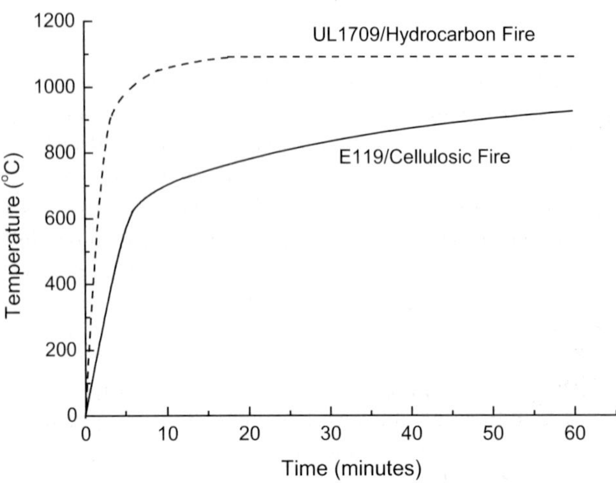



*Figure 11.10. Temperature-time heating curves commonly used in furnace testing of composites.*

The furnace method has several advantages when evaluating the fire resistance of composite structures, such as controlled heating and realistic fire conditions. However, the method has some deficiencies that can affect the reliability of the test results, most notably variable results between different furnaces and testing organisations, even though all technically comply with the requirements of the standards [60]. Gibson [61] attributes these discrepancies to several factors, including different levels of vitiation (ie. oxygen depletion near the panel surface), different emissivity values of the furnace linings, and interactions between the furnace control thermocouple and the hot surface of the composite panel when flaming combustion occurs. The flaming effect with composites can produce a rapid temperature rise, of the order of 100°C or more, in the near vicinity of the specimen hot face. Some progress has been made recently in the improvement of furnace control through the use of plate thermocouples in the furnace, facing the hot side of the panel. A plate thermocouple comprises a flat plate, often insulated at the rear, with the thermocouple junction on the front face. Plate thermocouples are designed to sense temperatures in front of the plate, without being affected by those behind it. Although their use arguably improves the ability of the furnace control loop to follow the required temperature curve, this results in practice in the gas supply being significantly reduced whenever flaming takes place at the hot surface. Since there is no equivalent of this effect in a real fire this raises the question of whether such a procedure is appropriate. The debate concerning the control of testing furnaces is likely to continue for some time and it is probable that special procedures will eventually be introduced for materials that show significant surface flaming.

## 11.11    Burn-Through and Jet Fire Tests

Burn-through resistance is an important, although not well understood, fire resistive property of composites. Burn-though tests are designed to simulate a jet flame burning through a panel, which is a danger on ships and offshore platforms when the rupture of a fuel line or gas pipe can produce a high-pressure flame. Jet fires are potentially more damaging than pool fires because the heat flux is greater and because the high gas velocity can erode the burning article. Various techniques have been developed to evaluate the burn-through resistance of composites, and all methods involve directing a jet flame onto the test piece and measuring either the burn-through time, the time to loss of integrity under load or, in the case of composite passive fire protection, the temperature rise of the substrate surface being protected [25,48,62-65].

The United States Navy widespread use of the 'DTRC Burn-Through Test' [25,48]. This test was developed at the former David Taylor Research Center (USA) to measure the burn-through rate of composite materials to be used in warships, although it can be used to assess the burn-though resistance of any combustible material.  The test is shown schematically in Fig. 11.11, and involves directing the flame from a propane burner on to a composite panel that is about 0.6 m x 0.6 m in size. The flame temperature is ~1100°C (2000°F), and this is produces the jet fire conditions similar to a burst hydrocarbon fuel line on a ship. The test specimen is exposed to the jet flame until the back-face reaches a certain temperature, and this is used to define the burn-through resistance.  The United States Navy considers the test material to have adequate fire resistance when the back-face temperature remains below 120°C after thirty minutes exposure.

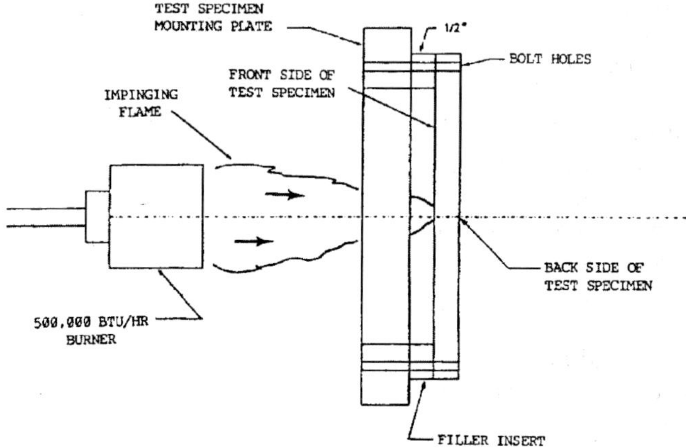


*Figure 11.11.  Schematic of the DTRC Burn-Through test.  Reproduced with permission from Greene [25].*

A small number of jet fire rigs exist in the United Kingdom, United States and Norway. These were designed principally for testing pipes, vessels and structures with and without passive fire protection under conditions relevant to the oil and gas industries. These rigs are used to determine the fire resistance of fuel and gas pipes under jet fire attack on an offshore structure [63,64,66]. The largest jet fire rig is operated by British Gas, and is capable of directing a 20 meter long horizontal flame of burning natural gas onto the component under test.  The test article is located about half-way down the flame length where it is subjected to a combination of high heat flux (~300 kW/m$^2$) and high gas jet velocity (~50 ms$^{-1}$). These test conditions are extremely severe due to the combination of high temperature and the erosion effects of the gas jet. The specimen is exposed to the jet fire for a fixed time, and after testing the functionality and burn-through of the test article is assessed. While the British Gas rig produces jet fire conditions that can occur on offshore platforms, the rig is extremely expensive to operate. Therefore, small to medium-sized jet fire resistance tests have been developed that are capable of subjecting test articles to high heat flux and gas velocity, but without the high cost [63-65].

## 11.12    Single Burning Item Test

The 'Single Burning Item (SBI) Test' simulates a practical hazard: a burning waste receptacle (the burning item) in a room corner. It is a relatively new procedure for determining fire reaction and arose as a result of the European Construction Products Directive, which sought to harmonize test procedures across the European Union. It is envisaged that most construction products, including composite items, will be tested and classified according to the SBI protocol specified in EN 13823 [77].  The SBI procedure was chosen because several fire reaction and resistive properties can be measured in a single test, including the time-to-ignition, heat release rate, smoke production, flame spread rate and fire growth, under realistic fire conditions. In addition, the generation of flaming droplets and particles produced during thermal decomposition of the test structure can be evaluated.

The SBI apparatus is illustrated in Fig. 11.12. It is an intermediate-scale corner fire test consisting of two wall panels made of the test material. One wall is 1.5 m high x 1.0 m wide while a second narrower wall is 1.5 m high x 0.5 m wide. A 0.25 m-sided triangular propane gas burner located in the corner generates a heat flux of ~50 kW/m$^2$, and this is intended to simulate a fire in a waste-paper bin. The SBI test is performed inside a fire room that has a fume extraction system in the ceiling.

The time-to-ignition and flame spread rate are determined by observing the response of the wall panels to the fire. The heat, smoke and gases released by the burning wall materials are extracted from the room through the exhaust hood in a manner similar to the cone calorimeter. The temperature, air-flow rate, smoke density, $O_2$ and $CO_2$ concentrations are measured continuously using sensors inside the exhaust duct. The heat release rate is calculated from the oxygen consumed during burning of the wall materials using the oxygen consumption principle. It is also possible to measure changes to the structural capacity of the wall panels when exposed to fire. By applying compressive dead-loads to the panels during testing it is possible to determine the loss in stiffness and the time-to-failure of the walls.

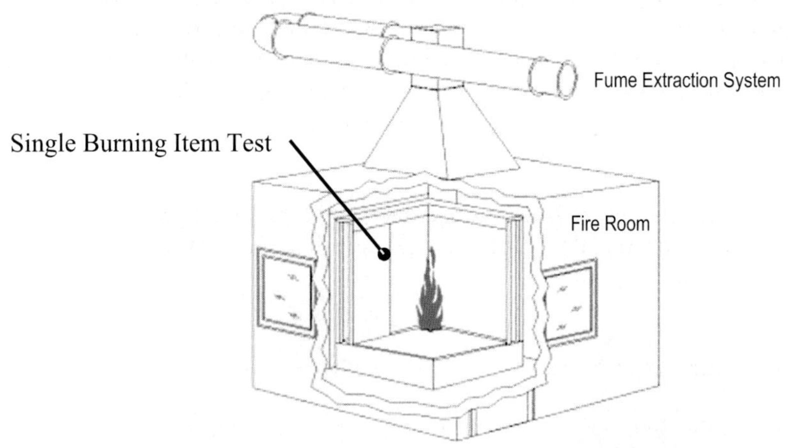


*Figure 11.12. Schematic of the single burning item test.*

While the SBI procedure offers a credible European alternative to a wide range of nationally-based test protocols, a number of drawbacks have been noted; the most significance being the scale of the test, which results in significantly higher costs than many other fire reaction tests, such as the cone calorimeter. The other problem arises from the requirement that the specimens be provided in the form of flat sheets. This is appropriate for cladding or panels, but the majority of composite products, including mouldings, pultruded sections and pipes, are not available in this form. Interim solutions have involved the making up of flat sections from strips of products but it is probable that an alternative procedure may eventually be needed. Some success has been achieved in predicting product performance in the SBI test from the results of cone calorimetry.

## 11.13    Room Fire Tests

Several intermediate to full-scale room fire tests can be used to determine the fire behaviour of composites for use in buildings and ship compartments. The smallest and simplest of these is the room corner test, which consists of two wall panels that are 2.1 m (7 ft) high and 1.2 m (4 ft) wide, and this geometry is similar to that employed in the

SBI test.  A 30 litre (8 gallon) pan of hexane fuel placed in the corner is used as the fire source. The response of the wall panels to flame spread and fire extinguishment can be observed directly during testing, as seen in Fig. 11.13. It is also possible to apply compressive dead-loads to the walls to determine their structural response to fire. The room corner test is popular for several reasons, most notably its simplicity and moderate cost. However, it is limited in the number of fire resistive properties that can be measured. Furthermore, the test is very severe because the geometry causes the draft from each wall to converge at the inside corner where the heat flux is extremely high and localised. For this reason most conventional composite materials fail the room corner test, and only pass with highly effective thermal barrier protection.

*Figure 11.13.  Room corner test.  Photograph supplied courtesy of the CRC for Advanced Composite Structures Ltd.*

A larger fire test is the 'U.S. Navy Quarter Scale Room Fire Test' [25,48].  This test was developed at NIST to determine the flashover potential of composite compartments in United States Naval ships. Testing is performed inside a room measuring 3.0 m (10 ft) long by 3.0 m (10 ft) wide by 2.4 (8 ft) high, with one wall having an open doorway. Inside the room is placed three test panels of composite material that are 0.91 m (3 ft) x 0.76 m (30 in) and one composite panel that is 0.91 m (3 ft) x 0.91 m (3 ft), which are joined into a square room. The fire is started using a burning fuel source located at the room corner. The key advantage of this test is it provides realistic data on the flashover behaviour of composite structures without incurring the high cost and long testing time experienced with full-scale room burn tests.

The largest of the fire tests to determine the fire reaction and fire resistive properties of composite materials is the room calorimeter test. Full-size room fire tests were originally developed in the 1920s, although a test to measure the heat release rate inside a burning room was not devised until the 1980s at the University of California [68] and NIST [69] and later refined by researchers at Weyerhaeuser and N.B.S. The room calorimeter test can provide information on a number of fire reaction properties, including the time-to-ignition, heat release rate, smoke density, toxic gas emission and flame spread [7]. In addition, fire resistive properties can be measured, including heat penetration through the walls, burn-through rate, and structural response of the test article to fire.

A schematic diagram of the room fire calorimeter is shown in Fig. 11.14. The procedure for conducting the test is outlined in ISO 9705 standard 'Fire Tests – Full-Scale Room Test for Surface Products' and ASTM E603-98 'Standard Guide for Room Fire Experiments', and is also described by Babrauskas [7], Høyning and Taby [11] and Greene [25]. The ISO 9705 test represents a full-size room that is 3.66 m (12 ft) long, 2.44 m (8 ft) wide and 2.44 m (8 ft) high. An open doorway is located at one end of the room, and immediately outside the doorway is a fume extraction hood. Panels of the test material are installed over the ceiling, two side walls and back wall. The only surfaces not covered with the test material are the floor and wall with the doorway. The material must be mounted on the walls and ceilings in the end-use condition, with the same joints, fixtures and surface coatings used in the actual construction. A propane burner is placed flush against the corner of the back and a side wall, and this simulates the effect of a burning item in a room corner.

A room fire test is usually recorded with video and thermal imaging cameras to observe ignition and flame spread. In addition, thermocouples located throughout the room are used to monitor the temperature rise and flashover conditions. The heat, smoke and fumes produced during combustion escape through the open door, where they enter the fume extraction hood. The fume exhaust duct is instrumented with an oxygen analyser to determine the change in oxygen content of the room fire atmosphere, and from this the heat release rate can be determined. The exhaust system is also instrumented with $CO/CO_2$ and smoke analysers. While the room calorimeter is the most common room fire method for testing composite materials, a variety of other room fire tests exist including the 'Monsanto Room Calorimeter' and the 'HORDTEST/ISO room fire test' [70].

The room fire calorimeter is arguably the most reliable method to evaluate the fire reaction behaviour of large composite panels. Høyning and Taby [11] report that values for the fire reaction properties of composites measured in a room fire test are often different to the results of bench-scale tests. The thermal environment for many small-scale tests, such as the OSU calorimeter and LOI test, is less severe than the fire condition imposed in the room calorimeter test. Consequently, fire reaction properties such as peak heat release rate and smoke obscuration are generally higher when measured in the room test. Other bench-scale tests such as the cone calorimeter have comparable heat flux conditions to the room test but cannot cause flashover at which point the fire reaction property values for composites can increase dramatically.

An advantage of the room test is that the effects of delamination damage, failure of joints and fixtures, ablation of charred plies, and falling debris on the reaction properties can be realistically studied, which is not possible with many small-scale tests. However, while the room fire calorimeter has many important advantages, it is used sparingly to assess the fire reaction properties of composites because of the high cost involved in testing.

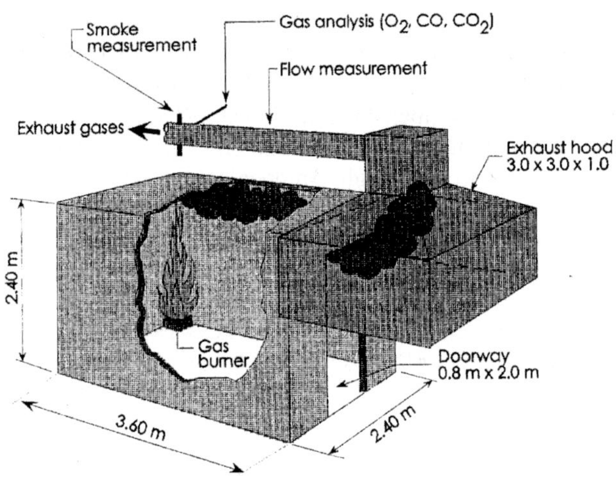


*Figure 11.14. Schematic of the room fire calorimeter.*

## 11.14    Structural Integrity in Fire Tests

Degradation of the mechanical properties of composite materials due to fire has been a subject of intensive research and analysis since the 1980s. Resin degradation, delamination damage and thermal softening of composites can cause significant losses in stiffness, strength, creep resistance and other mechanical properties, as discussed in chapters 6 and 7. Models for predicting the degradation to the load-bearing properties of composites are established for simple structures, such as flat panels, however accurately determining the properties of complex structural assemblies is not possible with existing models. Assessing the structural integrity of composites in a fire is still largely reliant on experimental testing. Various methods have been developed to evaluate the loading capacity of composite structures in fire, although it is important to recognise that no method has been standardised or is universally accepted. The tests range in scale from small coupons through to large panels, and generally the larger the test specimen the better the correlation with the fire integrity of full-scale composite structures under load [71-76].

One of the simplest methods is known as the '3-foot E 119 test with multiplane load', which was developed by Dao and Asaro [77,78]. The test can be used to assess the structural behaviour of any type of composite material under load, and has proven especially useful for determining the fire resistance of candidate materials for ship panel structures such as decking, bulkheads and hull plating that experience combined compression and bending loads. The test involves subjecting a composite panel to compression and bending loads when exposed to fire [25,77,78]. The panel is 0.91 m (3 ft) long, 0.71 m (2.3 ft) wide and 12.2 mm (0.5 in) thick, and therefore is limited to representing a section from a much larger composite structure. The structural integrity is determined by measuring the reduction to the stiffness and strength of the panel exposed to the fire. Only flat panels can be tested; it is not possible to test complex structural sections because of restrictions imposed by the apparatus on size and shape of the test article. Therefore, the test can not be used to determine the fire resistance of large composite assemblies.

## 11.15    Aircraft Fire Tests

The fire resistance of composites used in aircraft structures and cabin furnishings has been a safety concern since these materials were first used nearly forty years ago. While in-flight and post-crash fires are extremely rare, when they do occur they are responsible for a large number of aircraft fatalities (as described in Chapter 1), and therefore numerous tests are used to assess the fire safety performance of aircraft composites. Bench-scale fire reaction tests such as the cone calorimeter and OSU calorimeter are used to evaluate the fire behaviour of aircraft materials. However, a major deficiency of small-scale tests is their failure to replicate the in-flight and post-crash fire environments involving large composite structures and cabin furnishings.

The Federal Aviation Administration devised full-scale in-flight and post-crash fire tests to assess the fire behaviour of composites used in the cabin of wide-body aircraft [79-83]. The post-crash test involves subjecting a C-133 aircraft fuselage to an external jet-fuel fire. The FAA enlarged a C-133 fuselage to an internal diameter of 5 meter to resemble a wide-body passenger aircraft. Prior to fire testing, the C-133 fuselage is fitted with candidate composite materials in furnishing such as sidewalls, partitions, the ceilings, and storage bins. A pool of burning jet-fuel is located outside an open forward cabin door to replicate the fire caused by the rupture of a wing fuel tank in a crash landing. During the ensuing fire the cabin environment is monitored for temperature, heat flux, smoke density and gases ($O_2$, CO, $CO_2$, HCl, HF and HCN). The fire reaction of composites used in interior furnishings in aircraft can also be studied for an in-flight fire scenario by igniting a passenger seat doused with one quart of fuel. This test method has proven reliable in assessing the fire resistance of cabin composite materials, although it is an extremely expensive test to perform.

## 11.16  Concluding Remarks

The range of test methods for determining the fire reaction and fire resistive properties of composites is clearly very diverse, although some attempts are now being made to rationalize this situation. The tests vary in scale and complexity, and considerable care must be exercised when selecting the most appropriate procedure. Most fire reaction testing is performed using bench-scale apparatus because of the lower cost and greater simplicity. Test methods such as the cone calorimeter, OSU calorimeter and radiant flame spread technique have an excellent track record and are widely used for determining the fire behaviour of composites. However, the problem of scaling-up from small-scale test results to fire performance of large structures in real fires is a major difficulty.

Problems common to bench-scale tests are that the ignition, heating and atmospheric (oxygen content) conditions are not representative of an actual fire. The growth and spread of fire cannot be accurately replicated using these tests, and it is usually not possible to achieve flashover. Further, the specimens are rarely tested in the end-use condition.

Large-scale fire tests, such as the room calorimeter, avoid many of the problems associated with small size, and allow the composite article to be tested in the end-use condition. However, large fire tests are generally slow and expensive to perform and may only produce data relevant to one particular situation.

It is also important to recognise that no single test method is adequate to evaluate all the fire properties of a composite.  Table 11.1 lists the properties that can be determined using the test methods described in this chapter.  Of these, it is those methods capable of determining the heat release rate that are most important when characterising the fire reaction of composites.  While a few methods have the capability to determine the heat release rate together with several other properties, it is necessary to use two or more methods to obtain a complete understanding of the fire behaviour of a composite material.

*Table 11.1. Fire reaction properties measured using test methods.*

| Test Method | Heat Release Rate | Ignition Time | Flame-Out Time | Heat of Combustion | Mass Loss Rate | Flame Spread | Smoke Density | Soot Yield | Gas Emissions |
|---|---|---|---|---|---|---|---|---|---|
| Cone calorimeter | ✔ | ✔ | ✔ | ✔ | ✔ | | ✔ | ✔ | ✔ |
| OSU calorimeter | ✔ | | | | | | ✔ | | |
| Bomb calorimeter | | | | ✔ | | | | | |
| LOI test | | ✔ | ✔ | | | | | | |
| Radiant panel flame spread | | ✔ | ✔ | | | ✔ | ✔ | | |
| NBS smoke chamber | | | | | | | ✔ | | ✔ |
| NFPA 269 toxicity test | | | | | | | | | ✔ |
| Room calorimeter | ✔ | ✔ | ✔ | | | ✔ | ✔ | | ✔ |

# References


1. V. Babrauskas and R.D. Peacock. Heat release rate: the single most important variable in fire hazard. *Fire Safety Journal*, 1992; 18:255-272.
2. V. Babrauskas. From Bunsen burner to heat release rate calorimeter. In: *Heat Release in Fires*, ed. V. Babrauskas and S.J. Grayson, London: Elsevier Applied Science, 1992, pp. 7-29.
3. U. Sorathia, R. Lyon, T. Ohlemiller and A. Grenier. A review of fire test methods and criteria for composites. *SAMPE Journal*, July/August 1997; 33:23-31.
4. V. Babrauskas. Designing products for fire performance: the state of the art of test methods and fire models. *Fire Safety Journal*, 1995; 24:299-312.
5. V. Babrauskas and U. Wickström. The rational development of bench-scale fire tests for full-scale fire prediction. In: *Proceedings of the 2ⁿᵈ International Fire Safety Science Symposium.* 1989, pp. 813-822.
6. V. Babrauskas. Effective measurement techniques for heat, smoke, and toxic fire gases. *Fire Safety Journal*, 1991; 17:13-26.
7. V. Babrauskas. Fire test methods for evaluation of fire-retardant efficacy in polymeric materials. In: *Fire Retardancy of Polymeric Materials*, ed. A.F. Grand and C.A. Wilkie, New York: Marcel Dekker, Inc., 2000, pp. 81-113.
8. V. Babrauskas. Ten years of heat release research with the cone calorimeter. In: *Heat Release and Fire Hazard*, Vol. 1., ed. Y. Hasemi, Building Research Institute, Japan: Tsukuba, 1993, pp. III-1 to III-8.
9. J. Hume. Assessing the fire performance characteristics of GRP composites. In *Proceedings of the International Conference on Materials & Design Against Fire*, 27-28 October 1992, London, pp. 11-15.
10. V. Babrauskas. Why was the fire so big? HHR: The role of heat release rate in described fires. *Fire & Arson Investigator*, 1997; 47:54-57.
11. B. Høyning and J. Taby. Fire protection of composite vessels: fire protection and structural integrity – an integrated approach. In: *Proceedings of the Fourth International Conference on Fast Sea Transportation*, 21-23 July 1997, Sydney, South Yarra (Victoria): Baird Publications, pp. 811-816.
12. V. Babrauskas. The cone calorimeter. In: *Heat Release in Fires*, ed. V. Babrauskas and S.J. Grayson, London: Elsevier Applied Science, 1992, pp. 61-91.
13. U. Wickström and U. Göransson. Full-scale/bench-scale correlation of wall and ceiling linings. *Fire & Materials*, 1992; 16:15-22.
14. V. Babrauskas, D. Baroudi, J. Myllyniäki and M. Kokkala. The cone calorimeter used for predictions of the full-scale burning behaviour of upholstered furniture. *Fire & Materials*, 1997; 21:95-105.

15. V. Babrauskas and I. Wetterlund. Testing of furniture composites in the cone calorimeter: A new specimen preparation method and round robin tests. *Journal of Fire Safety*, 1998; 30:179-194.

16. J.F. Krasny, W.J. Parker and V. Babrauskas. *Fire Behaviour of Upholstered Furniture and Mattresses*,. Norwich NY: William Andrew Publishing, 2001.

17. E. Greene. Putting out the fire in marine composite construction. *Composites Fabrication*, July 2001, pp. 20-24.

18. V. Babrauskas. Development of the cone calorimeter – a bench-scale heat release rate apparatus based on oxygen consumption. *Fire & Materials*, 1994; 8:81-95.

19. S.J. Greyson, V. Babrauskas & M.M. Hirschler. A new international standard for flammability testing. *Plastics Engineering*, April 1994; 29-31.

20. V. Babrauskas. The cone calorimeter – a versatile bench-scale tool for the evaluation for fire properties. In: *New Technology to Reduce Fire Losses & Costs*, eds. S.J. Grayson & D.A. Smith, 1986, pp. 78-87.

21. V. Babrauskas and W.J. Parker. Ignitability measurements with the cone calorimeter. *Fire & Materials*, 1987; 11:31-43.

22. V. Babrauskas. The development and evolution of the cone calorimeter: a review of 12 years of research and standardization. In: *Fire Standards in the International Marketplace, ASTM STP 1163*, ed. A.F. Grand, Philadelphia: American Society for Testing and Materials, 1995, pp. 3-22.

23. V. Babrauskas. The cone calorimeter. In: *SFPE Handbook of Fire Protection Engineering*, 3rd ed., ed. P.J. DiNenno, D. Drysdale, C.L. Beyler, W.D. Walton, R.L.P. Custer, J.R. Hall and J.M. Watts, Bethesda, Maryland: Society of Fire Protection Engineers, 2002, pp. 3-63 to 3-81.

24. U. Sorathia, T. Dapp and J. Kerr. Flammability characteristics of composites for shipboard and submarine internal applications. In: *Proceedings of the 36th International SAMPE Symposium*, 15-18 April 1991, pp. 1868-1878.

25. E. Greene, *Marine Composites*, www.marinecomposites.com.

26. M. Janssens. Calorimetry. In: *SFPE Handbook of Fire Protection Engineering*, 3rd ed., eds. P.J. DiNenno, D. Drysdale, C.L. Beyler, W.D. Walton, R.L.P. Custer, J.R. Hall and J.M. Watts, Bethesda, Maryland: Society of Fire Protection Engineers, 2002, pp. 3-38 to 3-62.

27. M.J. Scudamore, P.J. Briggs and F.H. Prager. Cone calorimetry – a review of tests carried out on plastics for the association of plastics manufacturers in Europe. *Fire & Materials*, 1991; 15:65-84.

28. J.P. Redfern. Rate of heat release measurement using the cone calorimeter. *Journal of Thermal Analysis*, 1989; 35:1861-1877.

29. M. Janssens and W.J. Parker. Oxygen consumption calorimetry. In: *Heat Release in Fires*, ed. V. Babrauskas and S.J. Grayson, London: Elsevier Applied Science, pp. 31-59.

30. C. Huggett. Estimation of rate of heat release by means of oxygen consumption measurements. *Fire & Materials*, 1980; 4:61-65.

31. W.J. Parker. Calculations of the heat release rate by oxygen consumption for various applications. *National Bureau of Standards (US)*, NBSIR 81-2427, 1982.

32. M. Foley and D. Drysdale. Note: smoke measurements and the cone calorimeter. *Fire & Materials*, 1994; 18:385-387.

33. N.R. Marshall and R. Harrison. Comparison of smoke particles generated within a small scale hood and duct smoke test apparatus with those in a cumulative apparatus. Fire Research Station, Borehamwood, 1991.

34. F. Barthorpe. Danger – fibres on fire. *Professional Engineering*, June 1995; 10-11.

35. S. Gandhi, R. Lyon and L. Speitel. Potential health hazards from burning aircraft composites. *Journal of Fire Sciences*, 1999; 17:20-41.

36. V. Babrauskas, W.H. Twilley, M. Janssens and S. Yusa. A cone calorimeter for controlled-atmosphere studies. *Fire & Materials*, 1992; 16:37-43.

37. J.E. Leonard, P.A. Bowditch and V.P. Dowling. Development of a controlled-atmosphere cone calorimeter. *Fire & Materials*, 2000; 24:143-150.

38. J. Urbas and G.E. Luebbers. The intermediate scale calorimeter development. *Fire & Materials*, 1995; 19:65-70.

39. E.E. Smith. Heat release rate of building materials. In: *Ignition, Heat Release and Noncombustibility of Materials (ASTM STP 502)*, American Society for Testing and Materials, 1972, pp. 119-134.

40. V. Babrauskas. Performance of the Ohio State University rate of heat release apparatus using polymethlymethacrylate and gaseous fuels. *Fire Safety Journal*, 1982; 5:9-20.

41. M.M. Hirschler. How to measure smoke obscuration in a manner relevant to fire hazard assessment: Use of heat release calorimetry test equipment. *Journal of Fire Sciences*, 1991; 9:183-222.

42. D.P. Macaione, R.P. Dowling and P.R. Berguist. *Summary Report AMMRC TR 83-53*, Watertown, MA, 1983.

43. J. DiPietro and H. Stepniczka. Factors affecting the oxygen index flammability ratings. *Journal of the Society of Plastics Engineers*, 1971; 27:23-31.

44. C. Abbott and R. Chalabi. Heated oxygen index test. In: *Proceedings of the International Symposium on Fire Safety of Combustible Materials*, Edinburgh, Scotland, October 1975, pp. 296-303.

45. G. Rietz. Influences on the determination of oxygen index. II. Influence of temperature on the oxygen index. *Plaste und Kautschuk*, 1988; 35:386-388.

46. E.D. Weil, M.M. Hirschler, N.G. Patel, M.M. Said and S. Shakir. Oxygen index: correlations to other fire tests. *Fire & Materials*, 1992; 16:159-167.

47. J.G. Quintiere. Surface flame spread. In: *SFPE Handbook of Fire Protection Engineering*, ed. P.J. DiNenno, D. Drysdale, C.L. Beyler, W.D. Walton, R.L.P. Custer, J.R. Hall and J.M. Watts, Society of Fire Protection Engineers, 2000, pp. 2-246 to 2-257.

48. R.A. De Marco. Composite applications at sea: fire-related issues. In: *Proceedings of the 36$^{th}$ International SAMPE Symposium*, 15-18 April 1991, Anaheim, CA, pp. 1928-1938.

49. R.M. Morchat and J.A. Hiltz. Fire-safe composites for marine applications. In: *Proceedings of the 24$^{th}$ International SAMPE Technical Conference*, 20-22 October 1992, pp. 153- 163.

50. A.G. Gibson and J. Hume. Fire performance of composite panels for large marine structures. *Plastics, Rubbers & Composites: Processes & Applications*, 1995; 23:175-183.

51. T.J. Ohlemiller and T.G. Cleary. Upward flame spread on composite materials. In: *Fire and Polymers II*, pp. 422-434.

52. V. Babrauskas, J.A. White and J. Urbas. Testing for surface spread of flame: New tests to come into use. *Building Standards*, March/April 1997; 13-18.

53. NHB 8060.1C. *Flammability, Odor, Offgassing, and Compatability Requirements and Test Procedures for Materials in Environments That Support Combustion*, NASA Office of Safety and Mission Quality, 1991, pp. 4-1.

54. G.W. Mulholland. Smoke production and properties. In: *SFPE Handbook of Fire Protection Engineering*, ed. P.J. DiNenno, D. Drysdale, C.L. Beyler, W.D. Walton, R.L.P. Custer, J.R. Hall and J.M. Watts, Society of Fire Protection Engineers, 2002, pp. 2-258 to 2-268.

55. F.J. Rarig and A.J. Bartosic. Evaluation of the XP2 smoke density chamber. In: *Symposium on Fire Test Methods – Restraint & Smoke 1966, ASTM STP 422*, American Society for Testing and Materials, 1967, pp. 106- 124.

56. D. Gross, J.J. Loftus and A.F. Robertson. Method for measuring smoke from burning materials. *Symposium on Fire Test Methods – Restraint and Smoke, 1966, ASTM STP 422*, American Society of Testing and Materials, 1967, p. 166-204.

57. W.P. Chien and J.D. Seader. Prediction of specific optical density for smoke obscuration in an NBS smoke density chamber. *Fire Technology*, 1975; 11:206-217.

58. V. Babrauskas. Applications of predictive smoke measurements. *Journal of Fire & Flammability*, 1981; 12:51-64.

59. M.M. Hirschler. Smoke and heat release and ignitability as measures of fire hazard from burning of carpet tiles. *Fire Safety Journal*, 1992; 18:305-324.

60. M.A. Sultan. The effect of furnace parameters on fire severity in standard fire resistance tests. *Fire & Materials*, 1996; 20:245-252.

61. A.G. Gibson. Fundamentals of composite fire behaviour. In: *Proceedings of Composites in Fire 2*, 12-13 September 2001, Newcastle-upon-Tyne, UK.

62. Anon. Interim jet fire test for determining the effectiveness of passive fire protection materials. *Offshore Technology Report OTO 93028*, H.S.E., 1993.

63. P.S. Hill and G.C. White. Jet fire testing and performance of composite materials. In: *Proceedings of Composites in Fire*, Newcastle-upon-Tyne, 1999.

64. J. Folkers. Fire testing and performance of fiber glass pipe. In: *Proceedings of Composites in Fire*, Newcastle-upon-Tyne, 1999.

65. A.G. Gibson, Y.-S. Wu, A.L. Nass and R.J. McNaught. A low cost burner technique for the development and modelling of laminates in fire. In: *Proceedings of Composites in Fire 3*, 9-10 September 2003, Newcastle-upon-Tyne.

66. G.C. Grim. Fire endurance of glass fibre reinforced plastic pipes onboard ships. In: *Proceedings of Polymers in a Marine Environment*, 23-24 October 1991, paper 10.

67. J. Murrell and P. Briggs. Developments in European and international fire test methods for composites used in building and transport applications. In: *Proceedings of Composites in Fire 2*, 12-13 September 2001, Newcastle-upon-Tyne.

68. F.L. Fisher and R.B. Williamson. Intralaboratory Evaluation of a Room Fire Test Method. *National Bureau of Standards, NBS-GCR-83-421*, 1983.

69. B.T. Lee. Standard room fire test development at the national bureau of standards, in *Fire Safety: Science and Engineering, ASTM STP 882*, ed. T.Z. Harmathy, Philadelphia: ASTM, 1985, pp. 29-44.

70. V. Babrauskas. Full-scale heat release rate measurements. In: *Heat Release in Fires*, ed. V. Babrauskas and S.J. Grayson, London: Elsevier Applied Science, pp. 93-111.

71. J.K. Chen, C.T. Sun and C.I. Chang. Failure analysis of a graphite/epoxy laminate subjected to combined thermal and mechanical loading. *Journal of Composite Materials*, 1985; 19:408-423.

72. C.A. Griffiths, J.A. Nemes, F.R. Stonesifer and C.I. Chang. Degradation in strength of laminated composites subjected to intense heating and mechanical loading. *Journal of Composite Materials*, 1986; 20:216-235.

73. J.A. Milke and A.J. Vizzini. The effects of simulated fire exposure on glass-reinforced thermoplastic materials. *Journal of Fire Protection Engineering,* 1993; 5:113-124.

74. A.G. Gibson, P.N.H. Wright, Y.-S. Wu, A.P. Mouritz, Z. Mathys and C.P. Gardiner. The integrity of polymer composites during and after fire. *Journal of Composite Materials*, 2004; 38:1283-1307.

75. J. Bausano, S. Boyd, J. Lesko and S. Case. Composite life under sustained compression and one sided simulated fire exposure: characterisation and prediction. In: *Proceedings of Composites in Fire 3*, 9-10 September 2003, Newcastle-upon-Tyne.

76. B.Y. Lattimer, J. Ouelette and U. Sorathia. Large-scale fire resistance tests on sandwich composites. In: *Proceedings of the SAMPE Technical Conference & Exhibition*, 16-20 May 2004, Long Beach, CA.

77. R.J. Asaro and M. Dao, 'Fire degradation of fiber composites', *Marine Technology*, 34, (1997), 197-210.

78. M. Dao and R.J. Asaro. A study on failure prediction and design criteria for fiber composites under fire degradation. *Composites*, 1999; 30A:123-131.

79. L.J. Brown. Cabin Hazards from a Large External Fuel Fire Adjacent to an Aircraft Fuselage. *FAA-RD-79-65*, August 1979.

80. R.G. Hill, G.R. Johnson and C.P. Sarkos. Postcrash fuel fire hazard measurements in wide-body aircraft cabin. *FAA-NA-79-42*, December 1979.

81. C.P. Sarkos, R.G. Hill and W.D. Howell. The development and application of a full-scale wide body test article to study the behaviour of interior materials during a postcrash fuel fire. *North Atlantic Treaty Organisation (NATO), Advisory Group for Aerospace Research and Development (AGARD) Lecture Series No 123 – Aircraft Fire Safety*, Washington DC, 15-16 June 1982.

82. C.P. Sarkos and R.G. Hill. Evaluation of aircraft interior panels under full-scale cabin fire test conditions. In: *Proceedings of the AIAA 23rd Aerospace Sciences Meeting*, 14-17 January 1985, Reno, NV.

83. R.E. Lyon. Fire-safe aircraft cabin materials. in *Fire and Polymers, ACS Symposium Series No. 599.* ed. G.L. Nelson, Washington DC: American Chemical Society, 1995, pp. 618-638.

# Chapter 12

# Health Hazards of Composites in Fire

## 12.1    Introduction

Fibre reinforced polymer composite materials can pose a serious health hazard in fire. Smouldering or flaming composites can produce copious amounts of dense smoke consisting of a potentially toxic mix of combustion gases, soot particles and fibres. These combustion products can cause acute and delayed health problems and, in the worst case, cause death. The short-term effects of inhaling toxic smoke include impaired judgement and decision-making capacity which can jeopardise the safety of a person attempting to escape from a burning composite structure, such as a building, aircraft, ship or rail carriage. The irritants in smoke, which include combustion gases (eg. HCl, HBr and $NO_2$), soot particles and fibre fragments, can also delay escape by causing severe bouts of coughing and choking as well as extreme eye irritation that prevents a person from keeping their eyes open long enough to find an exit. The delayed, long-term health problems that result from inhaling smoke may include damage to tissues and organs, possibly leading to cancers and tumours.

Several aircraft fires have revealed the types of health problems experienced by people when exposed to burning or burnt composite material. In October 1990 an accident investigation team from the Royal Air Force attended the crash site in Denmark of a burnt-out Harrier GR5, which had contained carbon/epoxy composite in the fuselage and wings. Many of the team suffered a variety of health problems when exposed to the burnt composite material that varied in severity from eye and skin irritations to severe breathing difficulties [1-4]. These health problems were experienced despite the fire being extinguished and the accident investigation crew wearing standard protective clothing, masks and goggles. As another example, in 1997 twenty-two fire-fighters were hospitalised after attending a fire on a United States Air Force F-117A 'Night Hawk' stealth fighter, which contains a large amount of composite material. The fire-fighters suffered from nausea, headache, eye soreness, skin irritation and laboured breathing when exposed to the smoke from the burning composites on the aircraft.

The health problems have been attributed largely to the release of ultra-small fragments of fibre, which are believed to pierce the skin causing irritation, adhere to the eyes causing soreness, and cause breathing difficulties when inhaled [4]. The smoke and fumes released during combustion of the organic matrix also contribute to the health complaints. Other health hazards have been attributed to composites in fire, including airborne fibres released from burning composites behaving as microscopic "poison darts" that deposit toxins in the respiratory system [5]. While the short-term health problems due to smoke inhalation are well documented, the delayed and long-term hazards are not as well understood. There are no reported cases of tissue and organ injury, cancers or other serious health problems to humans due to smoke inhalation from composite materials, although much remains unknown.

A review of the health hazards associated with the combustion gases, fibre fragments and char (soot) particles within the smoke plume of burning composite materials are given in this chapter. A description of the experimental test protocols used to assess the toxic potency of smoke is presented, that involve chemical analysis of the combustion products and animal exposure lethality experiments. An overview of chemical and toxicological studies into the health hazards of the combustion gases released by composites is given, and gaps in our understanding of their health effects are identified. Following this, the N-Gas model for calculating the toxicity of a single combustion gas or combinations of different gases is described as a method for assessing the smoke lethality of composites. Toxicology and epidemiology studies of the hazards associated with inhaling fibre fragments and soot particles are reviewed, with most attention given to the health problems experienced with inhaling glass and carbon fibres. At the end of the chapter is a brief description of the types of safety clothing and equipment that should be worn when exposed to burning composite materials.


## 12.2    Smoke Toxicity Test Methods

The majority of people who die in fire are killed by inhaling toxic gases present in the smoke rather than from the heat or burns. It is estimated that 65% to 80% of deaths in residential fires in the United States occur due to smoke poisoning. For this reason, the main concern with using polymer composite materials within enclosed spaces should be the generation of toxic smoke in the event of fire. It is often assumed that incapacitation and death is due primarily to carbon monoxide in smoke. However, epidemiological studies have revealed that deaths are rarely due to poisoning from CO alone, and it is other toxic compounds in the smoke together with CO that cause most deaths. In addition to CO, composites can release other toxic gases, such as HCN and HCl, which contribute to death during or soon after the fire. Certain polymers used in composites also yield mutagenic compounds (eg. quinoline) or gases that induce acute symptoms such as shock or convulsions (eg. phenol).

Unfortunately, thermal degradation models for polymers cannot accurately predict the types and amounts of combustion gases released from burning composites, and therefore the smoke toxicity must be determined by experimentation. There are essentially two experimental approaches used to assess smoke toxicity: the analytical chemical method and the animal exposure test [eg. 6-9]. The chemical approach involves using analytical techniques such as gas chromatography/mass spectrometry (GC/MS) and Fourier Transform Infrared Spectroscopy (FTIR) to identify all the compounds and their concentrations in the smoke. Based on this information, an attempt is made to predict the smoke toxicity. However, this approach has many problems and limitations that make it difficult to accurately predict the toxic potency. A major problem is that the number of compounds released by a burning composite material is often extremely large. For example, Vogt [10] identified over 100 different gases in the smoke produced by a carbon/epoxy composite. As another example, Levin [11] conducted a survey of the chemicals found in the combustion gases to seven plastics – ABS, nylon, polyester, polyethylene, polystyrene, poly(vinyl chloride), rigid polyurethane foam – and identified a combined total of over 400 compounds. It is extremely difficult and time-consuming to determine the composition and concentration of every gas species, particularly for different fire scenarios. Furthermore, the toxic potency and health effects of many compounds are not known, and the approach of examining the toxicity of all the various combinations of compounds is an impossible task for most materials.

Animal tests are the most common approach for determining smoke toxicity. The tests basically involve exposing animals (usually Fischer 344 male rats) inside a closed chamber to smoke produced by the thermal decomposition of a composite material. The acute responses (eg. laboured breathing, incapacitation, asphyxia) of the animals during testing are monitored, and after testing the animals can be studied over an extended period to monitor the development of long-term health problems (eg. tumours, cancers). A series of experiments are usually performed at different smoke concentrations to determine the incapacitation index value ($EC_{50}$) or lethality index value ($LC_{50}$) for a specific set of combustion conditions. The $EC_{50}$ and $LC_{50}$ values respectively define the smoke conditions under which 50% of the animals have been incapacitated or killed. The lower the $EC_{50}$ or $LC_{50}$ values the more toxic are the combustion products from the material. After testing, the animals are usually dissected to determine the lethality mechanism, such as tissue or organ damage. Genetic tests can also be conducted to determine whether the DNA has been damaged and if this may lead to mutations, tumours, cancers or some other genetically-controlled health problem.

There are many animal test methods to determine smoke toxic potency, most notably the NIST radiant panel test, NIBS toxic hazard test, NBS cup furnace method, SwRI/NIST method, and University of Pittsburgh test [6,12-15]. The NIST radiant panel test uses an apparatus that basically consists of three components: a radiant furnace, a chemical analysis system, and an animal exposure chamber, as shown schematically in Fig. 12.1. The radiant furnace thermally decomposes a specific amount of test material, and then the smoke is transferred to a 200 litre closed chamber where it is cooled to ambient temperature. The chamber is instrumented with various chemical analysers to determine the composition and concentration of the combustion gases. Alternatively, a small sample of the smoke can be extracted from the chamber for separate analysis. However, instead of determining the composition of every combustion species, analysis is usually restricted to the most common combustion products that are CO, $CO_2$, HCN, HCl, $NO_2$ as well as reduced level of $O_2$. Six rodents are placed in restrainers and then inserted into portholes located along one wall to the chamber. The test protocol has been carefully planned to minimise the number of animals, and it is considered that six is the minimum number needed to determine the smoke toxicity. The animals are placed facing into the chamber to ensure they are directly exposed to the smoke for 30 minutes, during which time their acute responses are monitored. Blood samples can be taken during the test to measure the rise in carboxyhemoglobin loading with time. Several tests are conducted at increasing smoke concentrations using different rodents in each test to determine the mass of combustible material needed to cause 50% (three) of the animals to die within the 30 minute exposure period and/or during the 14 day post-exposure period. This mass of material is used to define the approximate $LC_{50}$ value that is a quantitative measure of the smoke toxicity. The test procedure is described in ASTM E1678 (Test Method for Measuring Smoke Toxicity for Use in Fire Hazard Analysis), NFPA 269 (Standard Test Method for Developing Toxic Potency Data for Use in Fire Hazard Modelling), and ISO 13344 (Determination of the Lethal Toxic Potency of Fire Effluents) standards as the accepted animal test method for determining smoke toxicity.

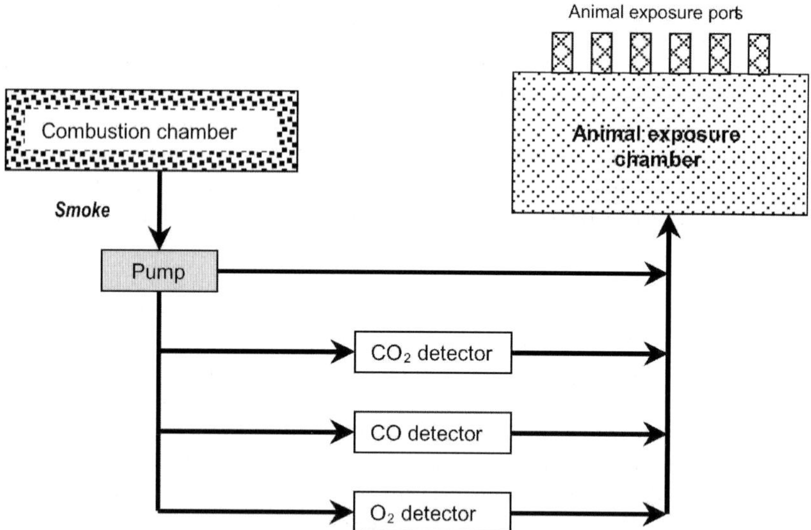


*Figure 12.1. NIST radiant panel test apparatus for smoke toxicity testing. Adapted from Levin et al. [16].*

The NIBS smoke toxicity test has many features common with the NIST radiant panel test. The heater used to thermally decompose the test material in order to produce the smoke, the instruments used to chemically analyse the smoke, and the smoke chamber containing the animals are similar for both test methods. The most significant difference is that with the NIBS test the toxic potency ($IT_{50}$) is defined by the time taken for the smoke to kill 50% of the animals during a 30 minute exposure or 14-day post-exposure period. As mentioned, with the NIST test the toxic potency of the smoke is defined by the mass of combustible material.

The University of Pittsburgh Toxicity Test is another animal test commonly used to determine the toxic potency of smoke [17]. The test involves heating a sample at a rate of 20°C/min until it is completely decomposed. The smoke is chemically analysed for the main combustion gases, and then mixed with cold room air to raise the oxygen to the normal level (ie. 21%) and to minimise the heat stress to the animals. The diluted smoke is then circulated through a test chamber at a flow rate of 20 litres/min. The chamber contains four port-holes that each hold a laboratory rat. The animals are exposed in a head-only position to the smoke for 30 minutes during which time their acute responses are monitored. In a similar manner to the NIST test, a series of experiments must be performed with increasing sample weight to raise the smoke density. The sample mass that causes 50% of the animals to die is used to define the toxic potency of the test material, $LC_{50}$.

The incapacitation of people due to smoke is a major hazard, and various experimental tests based on physical motor skills are used to determine the types and concentration of combustion products that cause incapacitation. Most tests involve an animal performing a task when exposed to smoke, such as running in a motorised wheel, running in a maze, or pushing the correct lever to open a door to escape a smoky atmosphere [18]. The tests are repeated at different smoke densities to determine the critical concentration at which the motor skills become affected and when the animal is incapacitated. Although, Levin [18] argues that the value of such tests is limited because the smoke concentration that causes incapacitation is usually only slightly less than the lethal amount, and therefore the tests does not add much extra information about toxic potency.

While experimental animal tests are the most common method to assess the toxic potency of smoke, these tests have several limitations. In toxicity tests it is essential that the test material is thermally decomposed under experimental conditions that simulate the realistic fire scenarios of concern. However, many fire scenarios (particularly flashover) are difficult, if not, impossible to replicate with bench-scale toxicity tests. Furthermore, with many tests the combustion environment is different from a real fire, and consequently differences can occur in the concentration of toxic gases in the smoke [15,19,20]. In particular, the CO level in a small-scale test can be substantially different to the level in a real fire because of differences in the amount of oxygen available for combustion. Babrauskas [20] reports that the CO yield measured in the various bench-scale tests are up to an order of magnitude different to those in real fires. However, the levels of certain other combustion gases including $CO_2$, HCl, HBr and HCN are roughly independent of the size of the test apparatus, and therefore the smoke toxicity tests generate approximately the same concentrations of these gases as an actual fire. Another problem is that toxicity data measured using animal tests is usually dependent on the apparatus, and quite often two toxicity test methods will provide different results about the smoke lethality of the same material [21]. Other problems with toxicity tests are they are expensive and time-consuming to perform. Also, while the tests are designed to use a minimum number of animals, the use of any animal for experimental purposes is always controversial.

A factor complicating animal toxicity tests is that the many gas species within smoke have different potency levels and affect the tissues and organs in different ways. Mixtures of certain gases in smoke can have a synergistic effect by interacting in a way that raises their toxic potency above the level due simply by the addition of the toxicity of the individual gases. Often a mixture of gases at a low concentration can be more hazardous than a single gas at a much higher concentration because a combination of toxicological reactions can occur. For example, Levin et al. [22] found a synergistic effect when CO and $CO_2$ are present together in smoke. CO becomes increasingly more potent when the concentration of $CO_2$ in smoke increases up to 5%. Above a $CO_2$ content of 5%, the toxicity of CO reverts back towards the toxicity of CO by itself. As another example, the toxic potency of $NO_2$ is greater if CO, $CO_2$ or $O_2$ are also present in smoke [23]. In other cases, an antagonistic toxicological effect can occur when one gas species reduces the toxic potency of another gas. For example, the toxicity of $NO_2$ and HCN are reduced when both gases are present in smoke. It is believed that $NO_2$ reacts with $H_2O$ in the lungs to produce nitrous acid in the blood that dissociates into nitrite ions. These ions aid in the production of methamoglobin that is an antidote to cyanide poisoning [22]. The many complex interactions between gases are extremely arduous and technically difficult to quantify by experimental toxicity testing on animals.

## 12.3    Health Hazards of Combustion Gases

Despite the numerous incidents of health problems experienced by people exposed to burning composites [1-4], medical researchers and toxicologists have paid scant attention to the toxicity of smoke produced by composite materials. At this time, the toxicity and health problems caused by the inhalation of most types of compounds released by thermally decomposing polymers and polymer composites are not known [24-26]. Understanding the health effects of the smoke is complicated by the fact that the types and amounts of gases released by composites depend on many factors. The properties of the composite obviously have a large influence, in particular the volume content, chemical composition and formulation of the polymer matrix, which may contain plasticisers, stabilisers, flame retardant additives, fillers, and excess catalysts and cross-linking agents that have not reacted with the resin. The nature of the fire also has a significant influence, most notably the flame temperature, the oxygen content of the fire atmosphere, and the mode of combustion. For example, the decomposition reaction rate of the polymer matrix increases with the fire temperature, and this results in higher yield rates of CO, $CO_2$, HCN and $NO_2$ gases in the smoke at the expense of heavier organic volatiles. Likewise, the yield of low molecular weight gases such as CO and $CO_2$ is dependent on the type of combustion. For example, Hunter and Forsdyke [27] determined that a glass fibre/phenolic composite released about 50 ppm CO and 300 ppm $CO_2$ when decomposition occurred by smouldering combustion. However, when the same material was burnt in a flaming mode the gas concentrations increased substantially to 100 ppm CO and 5000 ppm $CO_2$ due to the higher decomposition reaction rate of the phenolic matrix. Due to the many factors that influence the types and amounts of gas compounds released by thermally decomposing composites it is difficult to specify exactly the chemical nature and toxicity of the smoke, although some general comments about the potential health hazards can be made.

Composite materials release a large number of different volatiles when the polymer matrix decomposes, as described in Chapter 2. For example, Lipscomb [24] analysed the combustion gases released by a burning carbon/bismaleimide composite using GC/MS and identified 90 compounds, several of which are mutagenic or carcinogenic in laboratory mice. Quinoline was one of the main combustion gases, and this compound is known to damage DNA and alter the functions of the retina, optic nerves, cardiovascular system and central nervous system. The smoke also contained toluidine that induces numerous health problems including vertigo, headaches and methemoglobinemia. In addition, the smoke contained N-hydroxymethylcarbazole that is a mutagen, and phenol that may cause acute symptoms including shock. As mentioned, Vogt [10] identified over 100 different volatile compounds produced in the thermal decomposition of a carbon fibre/epoxy composite. As another example, Tewarson & Macaione [28] analysed the smoke from glass fibre/phenolic composite and detected a large variety of gases including CO, $CO_2$, and a mixture of organic compounds including toluene, methane, acetone, propanol, propane benzene and low molecular weight aromatic compounds.

While a large number of different compounds are found in the smoke produced by burning composites, most fire reaction studies have focussed only on the presence of CO and $CO_2$. This is because most fire studies are performed using the oxygen consumption cone calorimeter technique, and usually the only gas analysis system on this instrument is a $CO/CO_2$ analyser. It is possible to equip a cone calorimeter with other types of gas analysis instruments, although this is rarely done. The yields of CO and $CO_2$ from a wide variety of composite materials with thermoset (eg. epoxy, polyester, phenolic, BMI, phthalonitrile) or thermoplastic matrix (eg. PEEK, PPS) composites have been determined for a range of fire test conditions [29-37]. The formation of these gases during the decomposition of composites containing combustible organic fibres has also been studied [32]. An analysis of the gas data determined from these studies reveal, not surprisingly, that the amount CO drops with an increase in the char yield of the polymer matrix. Figure 12.2 shows plots of CO yield against mass loss for several thermoset polymers and their fibreglass composites. The CO yield is defined by the mass yield of CO normalised to the mass of material

consumed in the fire. The mass loss is inversely related to the char yield. It is seen that
the CO yield increases rapidly with the mass loss, and this is due to a greater mass of
polymer being decomposed into volatile gases (including CO). This clearly indicates
that minimising the release of toxic CO gas from a composite is dependent on the use of
a high char-yielding polymer that has a high resistance to volatilisation in fire.

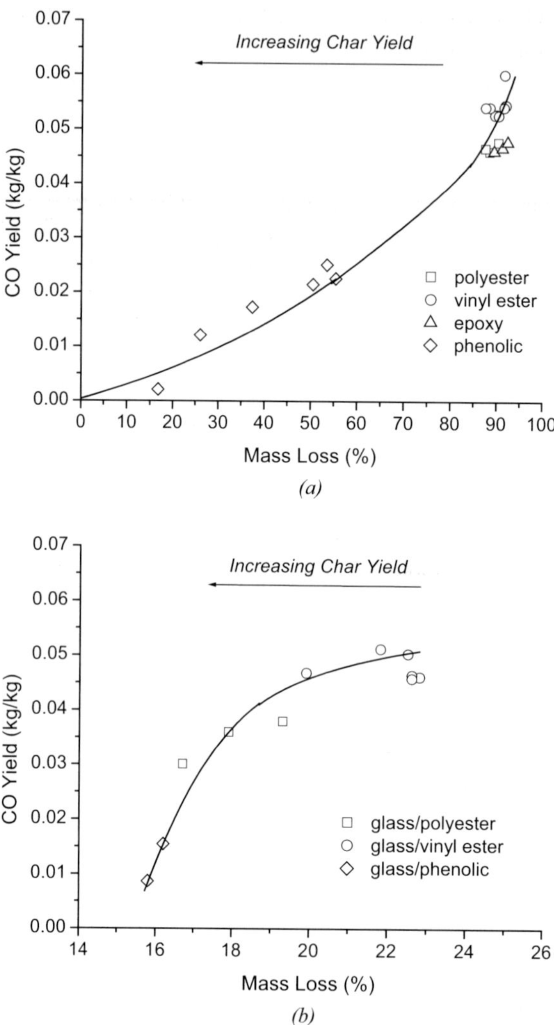


*Figure 12.2. Relationship between mass loss and CO yield for several (b) thermoset polymers
and (b) woven glass/thermoset polymer composites. Data from Brown et al. [33].*

Table 12.1 gives the average yields of CO, $CO_2$, HCl and HCN released during the flaming combustion of various types of fibreglass composite materials [34]. For comparison, Table 12.2 gives the exposure limits to toxic gases above which death will occur within ten minutes. It is seen in Table 12.1 that the concentration of CO gas for most of the materials is within the range of 200 to 300 ppm. Several of the composites also release HCl gas that can form a corrosive acid in the moist air environment of the lungs. While the concentration of HCl produced by many composite materials is low, it can reach high levels when sandwich composite materials with a poly(vinyl chloride) foam core are decomposed in fire.

*Table 12.1. Gas yields from various fibreglass composites. Data from Sastri et al. [34].*

| Composite | Carbon dioxide (ppm) | Carbon dioxide (vol%) | Hydrogen cyanide (ppm) | Hydrogen chloride (ppm) |
|---|---|---|---|---|
| Glass/vinyl ester | 230 | 0.3 | 0 | 0 |
| Glass/epoxy | 283 | 1.5 | 5 | 0 |
| Glass/bismaleimide | 300 | 0.1 | 7 | trace amount |
| Glass/phenolic | 300 | 1.0 | 1 | 1 |
| Glass/polyimide | 200 | 1.0 | trace amount | 2 |
| Glass/phthalonitrile | 40 | 0.5 | trace amount | 0 |

*Table 12.2. Maximum exposure limits of toxic gases. Above these limits death occurs within 10 minutes.*

| Toxic Gas Species | Exposure Limit |
|---|---|
| Carbon monoxide (CO) | 1500 ppm |
| Carbon dioxide ($CO_2$) | 50000 ppm |
| Hydrogen cyanide (HCN) | 50 ppm |
| Hydrogen chloride (HCl) | 30 ppm |
| Sulfur dioxide ($SO_2$) | 30 ppm |
| Nitrogen oxides ($NO_x$) | 30 ppm |

Dailey and Shuff [38] measured the smoke toxicity values for several polymers that are commonly used in composites, including polymers containing flame retardant additives, and the results are given in Fig. 12.3. These values were measured using the University of Pittsburgh toxicity test. Smoke toxicity ($LC_{50}$) was defined by the mass of polymer needed to produce smoke that was lethal to 50% of the test animals within 30 minutes. It was found that the toxic potency of the smoke varied greatly within a single class of polymer. For example, the $LC_{50}$ value for the unsaturated polyester was about four times higher than the general purpose polyester. Phenolic resin was found to have the lowest toxicity of the group of polymers studied, presumably because of the lower yield of combustion gases. Despite this work, toxicity tests on the large number of other polymers used in composites have not been performed, and much remains unknown about the smoke toxicity of most materials.

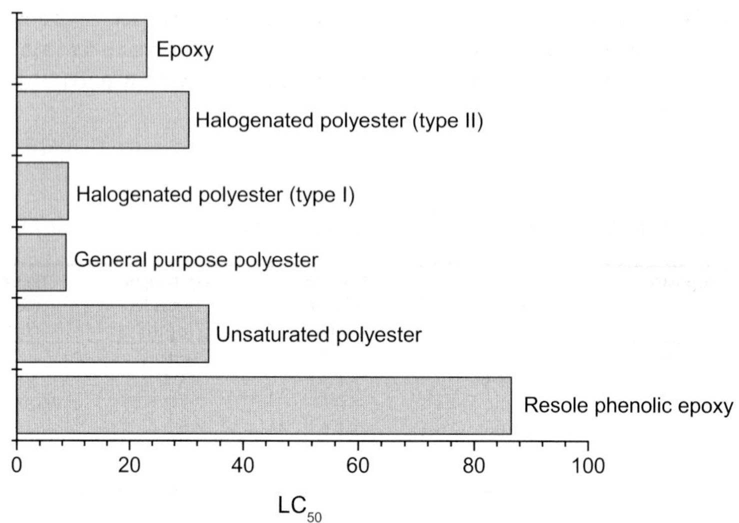


*Figure 12.3. Smoke toxicity (LC$_{50}$) values for various polymers. Data from Dailey and Shuff [38].*

Composite materials that are used in high fire risk applications can have a polymer matrix that has been chemical modified (eg. halogenated) or blended with flame retardant additives (eg. antimony trioxide) to delay ignition and reduce the heat release rate and smoke density, as described in chapter 8. The yields of toxic products from flame retardant composites can be much less than the untreated material, particularly when the additive acts in the solid phase to promote char formation or the formation of other barriers to decomposition. These additives slow or prevent the thermal decomposition of the composite material and thereby reduce the yield of smoke and toxic gases. However, other types of flame retardants can increase greatly the smoke toxicity of the polymer matrix. Brominated polymers release dioxins and dibenzofurans while chlorinated polymers produce hydrogen chloride, dibenzo-p-dioxins and related dioxin volatile compounds that are extremely hazardous [39]. Human studies have shown that exposure to high levels of dioxins can cause chloracne. Dioxins are also potent modulators of cellular growth and differentiation, particularly of epithelial tissues, that can promote cancer. Figure 12.4 shows the maximum HCl concentration in the smoke of a halogenated polyester resin containing different fire retardant additives [40]. For comparison, unmodified (non-halogenated) polyesters do not produce HCl. It is seen that the standard halogenated polyester yields about 800 ppm HCl, and this drops slightly with the addition of antimony trioxide or Nyacol. However, zinc borate and alumina trihydrate increase greatly the HCl gas concentration. The inhalation of HCl causes sensory irritation at low doses and pulmonary irritation and then bronchoconstriction at higher concentrations [7]. The tendency for severe bronchoconstriction and laryngeal spasms eventually leads to death during or shortly after exposure. The dioxins, hydrogen chloride and other toxic gases released by burning halogenated polymers have led to these materials being banned in many countries, particularly the member nations of the European Union.

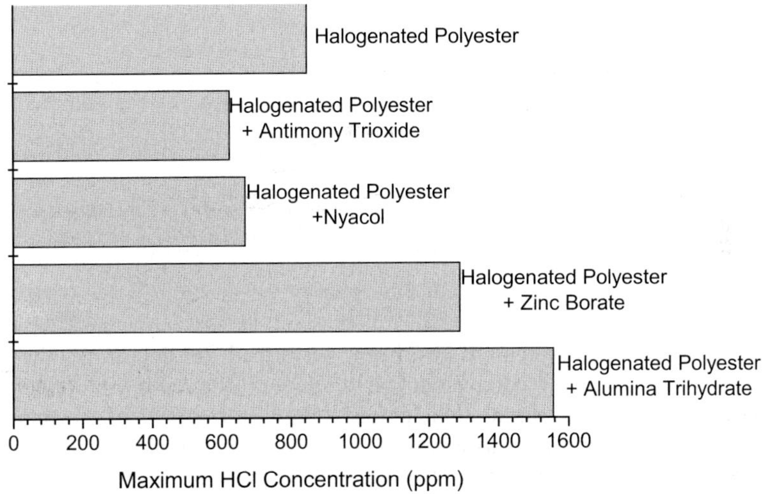


*Figure 12.4. Effect of fire retardant additives on the maximum yield of HCl gas from a halogenated polyester resin (Hetron 27196). Data from Morchat and Hiltz [40].*

Smoke toxicity can be increased considerably when composites contain organic fibres, such as aramid or polyethylene. Aramid fibres release CO, $CO_2$, HCN, nitrogen oxides and various organic compounds whereas polyethylene fibres yield high amounts of CO, $CO_2$, $n$-alkanes and $n$-alkenes [41-45]. The combination of gases released by both the organic fibres and polymer matrix can increase considerably the smoke toxicity of a composite material. For example, Fig. 12.5 compares the CO yield from aramid/phenolic and polyethylene/phenolic composites against a variety of phenolic composites reinforced with glass or carbon fibres, which are not combustible [46]. It is seen that the CO concentration from the composites containing the organic fibres is much higher, demonstrating the large influence fibre combustion can have on smoke toxicity.

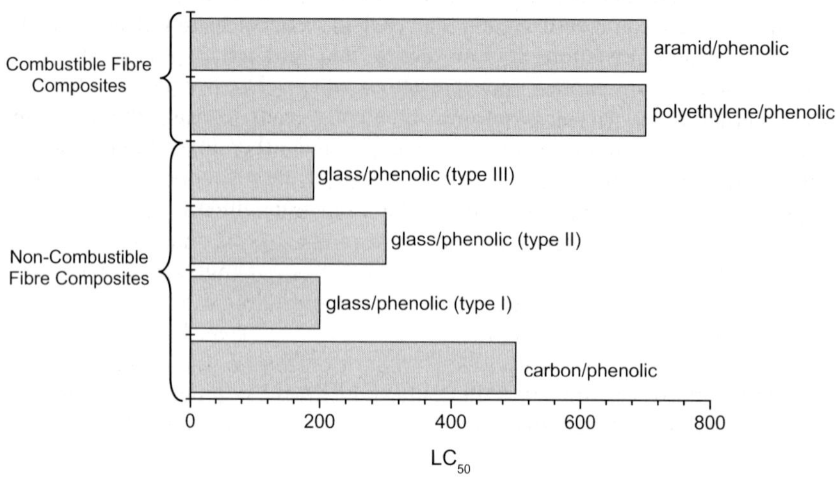


*Figure 12.5. Comparison of the CO yield from phenolic matrix composites containing combustible organic or non-combustible inorganic fibres. Data from Sorathia et al. [46].*

While a great deal of research has been performed to determine the chemical composition of the principal combustion gases released from composite materials, little research has been reported on the toxic potency and health effects of the smoke. In one of the few reported studies, Castiostro [47] assessed the smoke toxicity of burning carbon/epoxy materials to determine the conditions that cause incapacitation and death inside an aircraft during a fire. The toxicity test was designed to replicate the smoke conditions inside an aircraft by having the ratio of the composite sample weight to smoke chamber volume scaled to a similar ratio as the composite panel weight to cabin volume in a wide-bodied passenger aircraft. The respiratory and cardiac responses of several rats were monitored during testing and the concentration of their blood enzymes was measured to provide an indication of tissue necrosis in the respiratory, neural, cardiovascular, liver and kidney systems. In addition, the delay in reaction time of the animals due to smoke inhalation was studied. Rats were conditioned to jump on a pole to avoid a mild electric shock passing through a metal grid on the cage floor to the smoke chamber. Prior to testing, the rats had been trained to jump given a light or sound signal warning them that the gird was about to be electrified. The time delay after the warning was used in the smoke toxicity tests as a measure of the loss of avoidance response of the animals. When the rats were exposed to the smoke from a burning carbon/epoxy composite, their avoidance and escape responses were seriously affected after 6 and 12 minutes, respectively. Most of the rats died after about 16 minutes, and an autopsy revealed extensive pulmonary edema and kidney damage consistent with the inhalation of a lethal dose of toxic smoke. However, the testing did not identify the combustion gases responsible for the delayed responses and death of the animals.

Kimmel et al. [25] used guinea pigs to determine the toxic potency of smoke released by a smouldering carbon/epoxy composite. The smoke consisted mostly of $CO_2$ and CO, with small amounts of $NO_2$ and $SO_2$. Below a critical concentration, the breathing response of the guinea pigs was not affected by the smoke, but above this level the animals experienced restricted breathing similar to acute asthma. The difficulty in breathing increased with the smoke dosage, and at the highest concentration the animals suffered convulsions. Kimmel and colleagues observed that filtering the smoke of soot and airborne fibre particles before exposing the animals moderated these reactions, but did not eliminate the breathing problems. It was also found that breathing quickly returned to normal when the animals were given fresh air, however the recovery was short-lived for several guinea pigs which suffered recurring bouts of laboured breathing. The implications of this study for humans are not conclusive, but obviously severe breathing problems will be experienced when exposed to the smoke of burning carbon/epoxy composites. However, the toxicity of smoke produced by other types of composites, including materials that release HCl, HCN and other highly toxic gases, has not been studied.

## 12.4    N-Gas Model for Smoke Toxic Potency

Levin and colleagues at NIST have developed a mathematical approach to calculate smoke toxicity called the N-gas model [16,22,23,48-51]. The model is based on an assumption that a small number (N) of gases in smoke account for a large percentage of the toxic potency. That is, the model assumes that only a few gases released from a burning material are toxic, and the smoke toxicity is dependent on the total concentration of these gases. The toxic potency of the large number of different organic compounds that occur in moderate or trace amounts within the smoke of burning composites are not considered important, and are ignored in the N-gas model. To use the model, the test material must first be thermally decomposed using a bench-scale smoke method, such as NIST radiant panel or NIBS toxic hazard tests. The concentrations of the primary fire gases in the smoke are measured by chemical analysis techniques, and the data is then used to calculate the smoke toxicity using the N-gas model.

When originally developed in the 1980s, the N-gas model only considered the affect of two gases: the residual $O_2$ content in the smoke and the CO concentration. The model was soon expanded to a 3-gas model that considered the toxic interactions of $O_2$, CO and $CO_2$. Later, the 6-gas model that considered $O_2$, CO, $CO_2$, HCN, HCl and HBr was developed, which is expressed by the empirical equation:

$$N-GasValue = \frac{m[CO]}{[CO_2]-b} + \frac{21-[O_2]}{21-LC_{50}(O_2)} + \frac{[HCN]}{LC_{50}HCN} + \frac{[HCl]}{LC_{50}HCl} + \frac{[HBr]}{LC_{50}(HBr)} \quad (12.1)$$

where $LC_{50}$ is the toxic potency that is a experimentally measured value based on animal exposure tests to the smoke. The numbers in brackets indicate the measured concentrations of the CO, $CO_2$, HCN, HCl and HBr gases in ppm and $O_2$ percent present in the smoke. m and b are empirical constants, where m = -18 and b = 122,000 when the $CO_2$ concentration is below 5%, and m = 22.7 and b = -39,000 at $CO_2$ levels above 5%. The $LC_{50}$ value for HCN is 200 ppm for 30 minute exposure or 150 ppm for 30 minute exposure plus the 14 day post-exposure observation period. The $LC_{50}$ value of $O_2$ is 5.4%, which is substracted from the normal concentration for $O_2$ in air (ie. 21%). The $LC_{50}$ values for HCl and HCN are 3700 and 3000 ppm, respectively. It is worth noting that the $LC_{50}$ values are dependent on several factors, most notably the smoke test method, the source of test animals, and the strain on the animals during testing, although the values given provide a good estimation of toxic potency.

When the mass of burnt material generates a sufficient amount of combustion gases to produce an N-gas value of about 1, then the toxic potency is high enough to cause some animals to die. Above an N-gas value of about 1.3 it is predicted that all the animals will die, whereas when the value is below 0.8 it is expected all the animals will survive. The N-gas models have been shown to correctly predict the level of smoke toxicity for a variety of materials (eg. wood, plastics). However, the accuracy of the N-gas model in the prediction of the smoke toxicity for polymer composite materials has not been assessed.

The 6-gas model expressed in equation 12.1 is the accepted model for calculating the smoke toxicity of combustible materials, and is included in the toxicity test standards ASTM E1678, NFPA 269 and ISO 13344. Levin and colleagues [50] have also developed a 7-gas model that considers the toxic effects of all the gases in the 6-gas model together with $NO_2$.


## 12.5    Health Hazards of Fibres

### 12.5.1    FIBRES IN SMOKE

Another health hazard of burning composites is the release of small fibre particles that can cause skin irritation, sore eyes and breathing problems. The size and amount of fibres released from burning carbon/epoxy composites has been extensively studied by fire testing [52-58], and Gandhi et al. [59] have reviewed the key findings of these tests. However, much less information is available on the release of other types of fibres, such as glass or boron.

Fibres released from a burning composite occur in various forms ranging from single filaments to small fragments that can contain up to several hundred fibres bound together by char or resin. Large variations in the amount of fibres released from carbon/epoxy composites have been reported, with values as low as 1% and as high as 23% of the original fibre content [53,56,58]. This variation occurs because the release of fibres depends on several factors, including the fibre volume content, fibre architecture (ie. unidirectional, woven, stitched), original fibre length (ie. continuous, chopped strand or milled), char yield, and ablative properties of the composite. Other factors can also contribute to the release of airborne fibres, including the fire temperature and wind speed.

Several studies have measured the fibre content within the smoke plume of burning carbon/epoxy composites [52,53,54,57,59]. In one study, a 45 kg aircraft composite component was burnt within a jet fuel pool fire for 20 minutes [53]. Filters suspended in the smoke plume were used to collect airborne carbon fibres, and the average fibre concentration was measured to be under 0.14 fibres/cm$^3$ of smoke. This fibre concentration is below the permissible fibre exposure level specified by health organisations in many countries. For example, the United States Navy specifies a time-weighed average exposure limit for their personal of 3.5 carbon fibres per cm$^3$ of air over a 10 hour workday and 40 hour working week.

Another safety issue with the release of fibres from burning composites is their size. Airborne fibres within a certain size range can be easily inhaled and deposited throughout the respiratory system. Hertzberg [58] performed a variety of combustion and fire tests on carbon fibres and carbon/polymer composites to determine the conditions needed to generate airborne fibres. It was found that the risk of producing carbon fibres that are sufficiently small to be inhaled increased with the temperature and oxygen content of the fire atmosphere because this caused fibre thinning by oxidation. However, a minimum temperature of 600-700°C is needed to cause oxidation, and even then the oxidation (ie. thinning) rate of the fibre is very slow. The generation of airborne fibres also increased with the speed of airflow over the decomposing material surface. Hertzberg concluded that the fires that pose the greatest risk in the generation of fibres that can be inhaled are flashover, extremely high temperature conflagrations (eg. liquid fuel fire) or cases where structural damage is part of the fire scenario.

Figure 12.6 and Table 12.3 show the ability of humans to inhale particles with different aerodynamic size ranges. The aerodynamic diameter of a fibre refers to the diameter of an equivalent spherical particle that has the same terminal velocity as the fibre. Airborne fibres longer than about 50 μm are usually not inhaled because they quickly fall to the ground under their own weight. Any particles greater than 50 μm that are inhaled are long enough to be trapped in the nose and throat, and then rejected by sneezing or coughing. The likelihood of inhaling fibres in the size range of ~7 to 50 μm is much greater than larger fibres because they remain suspended for a longer time in the smoke plume. When fibres in this size range are inhaled they are deposited in the nose, throat and upper regions of the respiratory system, although as with larger particles, they are often quickly rejected. The size range of greatest concern is between ~0.7 and 7 μm because the fibres can be easily inhaled into the respiratory system and deposited on pulmonary regions, including the alveoli of the lungs. The alveoli are small sacs along the inside lining of the lungs that have the primary function of transferring gases between the lungs and blood stream. The risk of developing acute lung injury, including hemorrhages, tumors and cancers, is highest when fibres reach the alveoli. For pulmonary regions, the number of deposited fibres increases dramatically when the aerodynamic diameter is 2 to 3 μm.

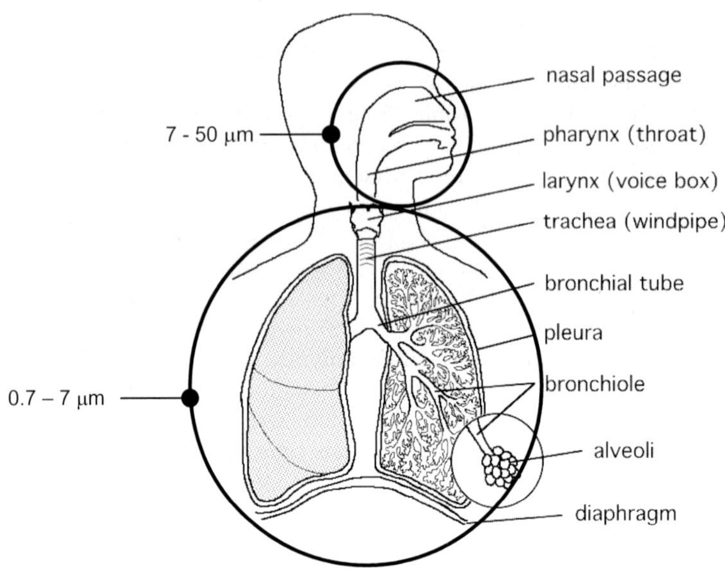


*Figure 12.6. Schematic of the human pulmonary system. The locations that fibre particles of different sizes can reach are indicated.*

*Table 12.3. Effect of fibre size on their ability to be inhaled.*

| Aerodynamic Diameter | Effect |
|---|---|
| >50 μm | Usually not in air long enough to be inhaled. |
| ~ 7-50 μm | Particles in this size range are often large enough to be caught by nose and throat, and are often ejected by coughing or sneezing. Usually filtered out by the nose, although can be deposited in cilia or airways. |
| 0.7 – 7 μm (respirable dust) | This particle size range presents the greatest hazard. They are small enough to reach the lungs when inhaled, yet large enough to remain in the lungs when we breathe out. Deposited in the lower bronchioles and alveoli. |
| <0.5 μm | Usually remain airborne and are exhaled. |

## 12.5.2   INHALATION TOXICITY OF CARBON FIBRES

The diameter of virgin carbon fibre is typically 7 μm, which is slightly too wide to be inhaled deep into the respiratory system. However, the diameter of carbon fibres can be reduced in a fire by oxidation and fibrillation. An analysis of carbon fibres collected in the smoke plume of burning carbon/epoxy composites has revealed that the mean fibre diameter can be under 7 μm, which is within the respiratory range. Figure 12.7 shows a frequency distribution plot giving the range of fibre sizes collected from the smoke of a carbon/epoxy laminate [50]. Included in the figure is the range of fibre diameters before fire testing, and this was between 5 and 7.5 μm. It is seen that the fibre diameter after fire testing is between 1.5 and 7.5 μm, with about 60% of the total fibre population being small enough to be inhaled deep into the lungs.

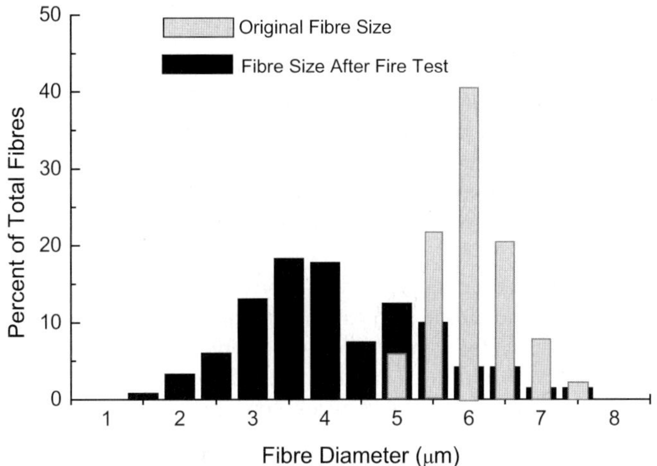


*Figure 12.7. Histogram showing the variation in fibre diameter before and after fire testing of a carbon/epoxy composite. Reproduced from Gandhi et al. [58].*

The inhalation toxicity of carbon fibres has been studied extensively over twenty years [59-70], and Gandhi et al. [59] give an excellent review of the topic. Most toxicological and epidemiological studies have focussed on the occupational health of factory workers who manufacture carbon fibres. In these studies, the health effects of inhaling virgin fibres have been evaluated. However, the toxicity of fibres released from a burning composite may be substantially different to virgin fibres because they are contaminated with various chemicals and combustion products, including char and other residuals of the polymer matrix. The health problems associated with inhaling contaminated carbon fibres are not as well understood as for virgin carbon fibres, and this is an issue that fails to attract much attention from the medical research community. Despite this, the toxicology studies performed on carbon dust and virgin fibres provide valuable insights into the potential health problems of inhaling carbon fibres from burning composite materials.

Numerous toxicology studies have been performed on laboratory animals to determine the health effects of inhaling increasing doses of virgin carbon fibres. Thomson et al. [64] exposed rats to airborne carbon fibres at concentrations of 40, 60 and 80 fibres/cm$^3$ for one hour each day over a nine-day period. The carbon fibres were 3.5 mm long and 3.5 $\mu$m in diameter, which is about the average diameter of airborne carbon fibres in a smoke plume. Post-mortem examination of lung tissue taken from the rats was performed at different times between 1 and 14 days after the exposure, and no signs were found of pulmonary infection or damage. Furthermore, no carbon fibres were observed in the lung tissue, presumably because the fibres were too long to reach the alveolar area located deep in the respiratory system. Warheit et al. [65] also studied the respiratory response of rats when exposed to increasing concentrations of carbon fibres suspended in air. In this study, rats were exposed to doses ranging from 50 to 90 fibres/cm$^3$ for between one and five days. The carbon fibres were 1 to 4 $\mu$m wide and 10 to 60 $\mu$m long. The exposure caused inflammation of the lung tissue, although the reaction was not permanent with the inflammation subsiding within ten days of the exposure. It is important to note that while the inhalation caused a short-term reaction, the toxicity test was performed at exposure levels well in excess of the fibre content measured in the smoke plume of a burning carbon/epoxy composite.

The effect of the exposure period to carbon fibres on the respiratory system has also been investigated. Holt and Horne [68] exposed guinea pigs to carbon fibre dust for 104 hours and did not observe any significant health problems. Owen [61] and Waritz et al. [66] performed inhalation studies on rats that were exposed to carbon fibres for 6 hours per day for 5 days each week over a 16 week period. No changes to the lung function response were detected at the end of the tests. Owen [61] killed the rats immediately after testing and then performed a detailed post-mortem examination of the brain, lungs, esophagas, stomach, duodenum, jejunum, ileum, colon, nasal turbinates, larynx, trachea and lymph nodes. No damage to any of these organs or tissues was observed. Histologic examination of the lung tissue revealed fibre particles that were coated with alvelor macrophages[1], which indicated the body's self-defence mechanism worked by dissolving the fibres by phagocytosis. A recent *in vivo* toxicity study by Zhang et al. [67] involved intratracheal injection of a liquid suspension of carbon fibres or carbon fibre/polymer dust directly into the lungs of laboratory rats. 93% of the fibres had a diameter under 5 $\mu$m and the medium length was 37.5 $\mu$m. Post-mortem examination of lavage cells extracted from the rat lungs at different times after the injection revealed

---

[1] Alvelor macrophages are a large white blood cell present within the lung (and mainly located within the alveoli) with the ability to scavenge for microbes and particulate matter. Macrophages contain chemicals and enzymes that serve the purpose of phagocytosing (ie. ingesting and destroying) microbes, antigens and other foreign substances, including certain types of foreign fibres.

that the fibres persisted for at least one month. However, at longer times no fibres were found. Zhang and colleagues did not find any adverse reactions to the carbon fibres and carbon/polymer dust, and concluded these materials are biologically inert at the dose level studied.

Martin et al. [62] also studied the inhalation toxicity of carbon/epoxy composite dust particles. Martin performed the study *in vitro* using rabbit alvelor macrophages as a measure of toxicity and *in vivo* using direct intratracheal injection in rat lungs. The toxicity of five types of carbon fibre composites were investigated: four types of carbon/epoxy and one type of carbon/PEEK. The carbon/epoxy composites differed in the type of fibre (ie. PAN or pitch-based fibres) and type of epoxy resin (ie. cured with an amine or aromatic amide reactive agent). The composites were pulversied into dust particles to an aerodynamic diameter of 1.1 to 1.9 µm before testing. It was found that the type of carbon fibre had no significant affect on toxicity, although the type of epoxy matrix had a major influence. It was discovered that the composites containing the epoxy matrix cured with the aromatic amine-curing agent had the highest cytoxicity. A significant increase in the amount of alveolar macrophages and neutrophils[2] were observed in samples of the lung tissue taken from the animals exposed to these composites. However, the infection was not lethal, and the carbon/epoxy dust was found to be much less toxic than known fibrous carcinogens such as quartz dust. Martin et al. [62] suggest that dust particles from carbon/epoxy composites cured with an aromatic amide agent have the potential to cause biologic effects in the human lung, although the types and extent of acute lung damage that would occur are not known.

Whitehead et al. [71] have published the only toxicology study into the health effects of inhaling carbon fibres in the smoke plume of a burning carbon/epoxy composite. In this study, the composite was ground into fine powder to ensure the release of fibres into the smoke when the material was burnt. Analysis of the smoke revealed the presence of various combustion gases, including $CO$, $CO_2$, $SO_2$ and nitrogen oxides, as well as carbon fibres with a median diameter of 1.6 µm. The condition of the smoke exposure was severe enough to cause several rats to die by asphyxia during the test. The lung function of the surviving rats was studied over a seven-day period following the smoke exposure. Inflammable of lung tissue was diagnosed; although Whitehead and colleagues believe that this was a typical reaction to severe smoke inhalation rather than a health problem unique to the inhalation of smoke from a composite material. No sign of more serious lung damage was found, such as hemorrhaging or distorted lung architecture, which is associated with acute lung injury. However, Whitehead et al. concluded that while the smoke from the burning carbon/epoxy composite did not cause serious damage to the respiratory system, under other fire conditions the smoke could cause acute lung injury.

---

[2] Neutrophils are leukocytes (white blood cells) that form a primary defence against bacterial infection. The body produces an increased number of neutrophils when suffering from a serious bacterial infection. Neutrophils perform their function partially through the process of phagocytosis.

In summary, the toxicology studies performed to date have found no major health problems associated with inhaling carbon fibres or carbon/polymer composite dust. The studies have been performed using fibre dose levels and exposure times far in excess of those expected in fire, suggesting that acute lung injury and other long-term respiratory problems should not be experienced from inhaling carbon fibres. However, no epidemiological data is available on the long-term health problems that may arise from inhaling smoke from composite materials containing carbon fibres that are contaminated with char and other solid degradation products from the polymer matrix. Lipscomb et al. [24] detected a variety of hazardous compounds on the surface of carbon fibres following fire, including nitrogenous aromatic compounds, phenols and polycyclic aromatic hydrocarbons (PAH), which can be carcinogenic. Therefore, contaminated fibres may possess different toxicological properties to virgin fibres, but as yet there has not been a thorough analysis of the health hazards.

## 12.5.3   INHALATION TOXICITY OF GLASS FIBRES

The health effects of inhaling glass fibres have also been studied, and as with carbon fibres, most of the toxicology studies have been performed on virgin fibres rather than glass fibres contaminated with combustion products released in fire. The diameter of virgin glass fibre is about 12 $\mu$m, which is above the critical size range of 0.7 to 7 $\mu$m that can be inhaled deep into the respiratory system. Unlike carbon fibre, the diameter of glass fibre is not reduced in a fire because oxidation or fibrillation does not occur. However, glass fibres will soften and melt when the temperature exceeds ~1100°C, although this usually causes the fibres to drip or flow together which reduces the risk of them becoming airborne. The only cases when significant amounts of glass fibre would be found in the smoke plume of a burning composite are when the material contains milled dust particles or the fibres have been pulverised due to a collision or other type of impact event.

A large number of toxicology and epidemiology studies have been performed on the hazards of inhaling virgin glass fibres, and all found no serious adverse health effects [62,72-78]. In one study, Hesterberg et al. [73] forced rats to inhale different doses (3, 16 or 30 mg/m$^3$) of milled glass fibres for 6 hours per day, 5 days per week over 24 months. This long-term exposure caused inflammation of the lung tissue, although this subsided when exposure to the fibres was stopped. Examination of the lung tissue did not reveal evidence of acute lung injury, and the study concluded that respirable glass fibres do not represent a significant hazard for lung disease in humans.

A key reason for glass fibres not being a serious health hazard is that they are rapidly dissolved by alveoli macrophages in the lung, particularly when under 10-20 $\mu$m [66,68,71,72]. For example, Eastes and Hadley [74] showed that a 1 $\mu$m diameter fibre of glass wool dissolves in a rat lung within about 0.14 years, whereas complete dissolution by macrophages of chrysotile and crocidolite asbestos fibres of the same size takes 7 and 52 years, respectively. Another important reason for the low health risk of glass fibres is they rarely penetrate further than the upper respiratory system because of their large diameter, whereas asbestos fibres are smaller and can therefore reach deeper into the lungs. In 1997, the World Health Organisation through the International Agency for Research on Cancer (IARC) concluded that glass fibres are not carcinogenic. However, no research has been reported on the health problems experienced with inhaling contaminated glass fibres from a burning composite material.

## 12.5.4   INHALATION TOXICITY OF ORGANIC FIBRES

The health problems associated with inhaling organic fibres used in composite materials, such as aramid or UHMW polyethylene, have not been extensively studied. Polyethylene is thermally degraded over the temperature range of 290 to 390°C, and does not yield a significant amount of char [44,45]. Therefore, it is unlikely that polyethylene fibres would survive within the hot smoke plume of a burning composite material. Aramid fibres decompose between 500 and 550°C with a significant amount of their original mass converted to char, and therefore it is possible to inhale charred or partially decomposed fibres [41-43]. The inhalation toxicity of charred aramid fibres has not been investigated.  Searl [76] investigated the effects of inhaling virgin aramid fibre on rats, and found that macrophages rapidly cleared these fibres from the lungs and therefore did not cause acute damage to the pulmonary system.

The inhalation toxicity of char is another aspect of the health hazards of burning composite materials that is poorly understood. It is well known that certain carbon-rich materials, particularly coal dust, cause long-term health problems when inhaled at high concentrations over a long period. However, many differences exist between coal dust and char particles in terms of chemical composition, particle size and shape, and therefore it is not possible to extrapolate the toxicological problems of coal to char. Carbon black has a similar composition to many types of char produced by aromatic polymers, however the particle size is much smaller (typically 10-800 nm). Studies have shown that exposure to carbon black dust, particularly at very high concentrations over a long period, causes pulmonary fibrosis, bronchitis and emphysema as well as skin irritation, although it is not considered to be carcinogenic [81].  Purser [9] believes the most serious hazard occurs when carbonaceous smoke particles contain adsorbed carcinogens, such as polyaromatic hydrocarbons, dibenzodioxins and dibenzofurans, or free radicals that are formed by the thermal decomposition of certain polymers. These compounds may condense or be absorbed by the soot particles and then inhaled along with combustion gases and fibre fragments.

## 12.5.5   SKIN & EYE IRRITATIONS

Apart from concerns about smoke inhalation, people have experienced skin and eye irritation when exposed to burning composite materials [1,3,4]. Broken fibres released from composites can have sharp tips that easily embed into the skin and eyes, as shown in Fig. 12.8. Such fibres can cause severe irritation, and in some cases skin rashes can develop. However, the effects are not permanent and any discomfort usually eases when the affected area is cleansed.

*Figure 12.8. Broken fibres can cause skin and eye irritations.*

## 12.6    Personal Protective Wear against Burning Composite Materials

Fire-fighters exposed to burning composite materials have suffered skin and eye irritations as well as breathing problems when not wearing appropriate protective clothing, eye-wear and respirators. Most fire-fighting agencies specify the full complement of safety wear when combating a burning composite. The United States Airforce, as an example, have developed guidelines for the minimum safety and health protection requirements for fire-fighters which specify self-contained breathing apparatus, sealed goggle-type eye protection, chemical protective clothing including aluminised proximity suits, puncture resistant gloves, neoprene overalls and hard soled boots [2,82,83]. No reports have been made of fire-fighters experiencing health problems when wearing the full complement of safety gear.

## 12.7    Concluding Remarks

An overview of the health hazards associated with the inhalation and exposure to the smoke released by burning composite materials has been presented. People have suffered a variety of health problems when exposed to burning composites, which vary in severity from skin and eye irritation through to severe coughing and respiratory problems that have required short-stay hospitalisation. The health problems suffered from smoke inhalation are usually quickly alleviated when the person is removed from the scene of the fire and exposed to fresh air.  As yet, no human deaths or long-term health problems have been attributed to smoke inhalation from composites, although much remains unknown about the toxic potency of the combustion gases, fibre fragments and soot particles present in the smoke.

A considerable amount of experimental (animal) research has been conducted into the toxicity of the primary combustion gases produced by polymers, such as CO, $CO_2$, HCN, HCl and $NO_2$, and the smoke toxicity can be predicted using the N-gas model. However, the polymers commonly used in composites also release a large variety of low molecular weight organic compounds, and the toxicity and toxic interactions of many of these gases are not well understood. A great deal more research is required into the toxic potency of combustion gases and the associated acute health problems. Fibres released from burning composite materials may also pose a health hazard. It is now widely accepted that carbon and glass fibres will not cause long-term health problems at the concentrations found in smoke. However, most studies into the toxicity of carbon and glass have been performed using virgin fibres. The fibres and soot particles within smoke are often coated with free radicals and organic compounds, some of which may be hazardous when inhaled. Again, further research into the toxic potency and health problems when exposed to contaminated fibres and soot particles is required.

## References


1.  C. Bickers. Danger: toxic aircraft. *Jane's Defence Weekly*, 1991; 711.
2.  M. Gaines. Composites menace crash teams. *Flight International*, 1991; 17.
3.  K. Tannen. Advanced composite materials. *Fire & Arson Investigator*, 1993; 1:50-51.
4.  F. Barthorpe. Danger: fibre on fire. *Professional Engineering*, 1995; 8:10-12.
5.  J. Anderson. Aircraft composite materials. In: *Proceedings of the 2nd California Symposium on Aviation Emergencies*, 1-3 October 2001, Santa Monica, CA.
6.  H.L. Kaplan, A.F. Grand and G.E. Hartzell. *Combustion Toxicology, Principles and Test Methods*, Lancaster PA: Technomic Publishing Co., 1983.
7.  G.E. Hartzell. Assessment of the toxicity of smoke. In: *Advances in Combustion Toxicology*, Lancaster PA: Technomic Publishing Co., 1988, pp. 1:8-18.
8.  B.C. Levin and R.G. Gann. Toxic potency of fire smoke: measurement and use. In: *Fire and Polymers: Hazards Identification and Prevention*, ed. G.L. Nelson, ACS Symposium Series 425, Washington DC: American Chemical Society, 1990, pp. 3-11.
9.  D.A. Purser. Toxicity assessment of combustion products. In: *SFPE Handbook of Fire Protection Engineering*, ed. P.J. DiNenno, Society of Fire Protection Engineers, pp. 85-146.
10. A. Vogt. Thermal analysis of epoxy-resins: identification of decomposition products. *Thermochemica Acta*, 1985; 85:407-410.
11. B.C. Levin. A summary of the NBS literature reviews on the chemical nature and toxicity of the pyrolysis and combustion products from seven plastics: acrylonitrile-butadiene-styrene (ABS), nylons, polyesters, polyethylenes, polystyrenes, poly(vinyl chlorides), and rigid polyurethane foams. *National Bureau of Standards Report NBSIR 85-3267*, 1986.
12. D.J. Caldwell and Y.C. Alarie. A method to determine the potential toxicity of smoke from buring polymers: III. Comparison of synthetic polymers to Douglas Fir using the Upitt II flaming combustion/toxicity of smoke apparatus. *Journal of Fire Sciences*, 1991; 9:470-518.
13. B.C. Levin. The development of a new small-scale smoke toxicity test method and its comparison with real-scale fire tests. *Toxicology Letters*, 1992; 64/65:257-264.
14. B.C. Levin. The development of a new small-scale smoke toxicity test method and its comparison with real-scale fire tests. In: *Toxicology from Discovery and Experimentation to the Human Perspective*, ed. P.L. Chambers, C.M. Chambers, H.M. Bolt and P. Preziosi, Amsterdam: Elsevier Science, 1992, pp. 257-264.

15. V. Babrauskas, R.H. Harris, E. Braun, B.C. Levin, M. Paabo and R.G. Gann. Large-scale validation of bench-scale fire toxicity tests. *Journal of Fire Sciences*, 1991; 9:125-148.

16. B.C. Levin, M. Paabo, J.L. Gurman and S.E. Harris. Effects of exposure to single or multiple combinations of the predominant toxic gases and low oxygen atmospheres produced in fires. *Fundamental & Applied Toxicity*, 1987; 9:236-250.

17. Y. Alarie and R.C. Anderson. Toxicologic and acute lethal hazard of thermal decomposition products of synthetic and natural polymers. *Journal of Toxicology & Applied Pharmacology*, 1979; 51:341-362.

18. B.C. Levin. Combustion toxicology. In: *Encyclopedia of Toxicology*, ed. P. Wexler and S.C. Gad, San Diego: Academic Press, 1998, pp. 360-374.

19. E. Braun, R.G. Gann, B.C. Levin and M. Paabo. Combustion product toxic potency measurements: comparison of a small scale test and 'real-world' fires. *Journal of Fire Sciences*, 1990; 8:63-79.

20. V. Babrauskas. The generation of CO in bench-scale fire tests and the prediction for real-scale fires. In: *Proceedings of the International Fire & Materials Conference*, London: Interscience Communications Ltd., 1992, pp. 155-177.

21. V. Babrauskas. Effective measurement techniques for heat, smoke, and toxic fire gases. *Fire Safety Journal*, 1991; 17:13-26.

22. B.C. Levin, M. Paabo, J.L. Gurman, S.E. Harris and E. Braun. Toxicological interactions between carbon monoxide and carbon dioxide. *Toxicology*, 1987; 47:135-164.

23. B.C. Levin, E. Braun, M. Navarro and M. Paabo. Further development of the N-gas mathematical model: An approach for predicting the toxic potency of complex combustion mixtures. In: *Fire and Polymers II: Materials and Tests for Hazard Prevention*, ACS Symposium Series No. 599, ed. G.L. Nelson, 21-26 August 1994, Washington DC: American Chemical Society, pp. 292-311.

24. J.C. Lipscomb, K.J. Kuhlmann, J.M. Cline, R.E. Larcomb, R.D. Peterson and D.L. Courson. Combustion products from advanced composite materials. *Drug Chemistry & Toxicology*, 1997; 20: 281-292.

25. E.C. Kimmel, J.E. Reboulet, D.L. Courson and K.R. Still. Airway reactivity response to aged carbon-graphite/epoxy composite material smoke. *Journal of Applied Toxicology*, 2002; 22:193-206.

26. A.N. Montestruc, M.A. Stubblefield, S.-S. Pang, V.A. Cundy and R.H. Lea. Smoke and toxicity tests of fiberglass-resin composite pipe samples. *Composites*, 1997; 28B:287-293.

27. J. Hunter and K.L. Forsdyke. Phenolic glass fiber-reinforced plastic and its recent applications. *Polymer Composites*, 1989; 2:169-185.

28. Tewarson and D.P. Macaione. Polymers and composites – an examination of fire spread and generation of heat and fire products. *Journal of Fire Sciences*, 1983; 11:421-441.

29. U. Sorathia, T. Dapp and J. Kerr. Flammability characteristics of composites for shipboard and submarine internal applications. In: *Proceedings of the 36th International SAMPE Symposium*, 15-18 April 1991, San Diego, CA, pp. 1868-1878.

30. J. Hume. Assessing the fire performance characteristics of GRP composites. In: *Proceedings of the International Conference on Materials and Design Against Fire*, London, 1992, pp. 11-15.

31. U. Sorathia. Flammability and fire safety of composite materials. In: *Proceedings of the 1st International Workshop on Composite Materials for Offshore Operations*, Houston, Texas, 26-28 October 1993, pp. 309-317.

32. J.R. Brown, P.D. Fawell and Z. Mathys. Fire-hazard assessment of extended-chain polyethylene and aramid composites by cone calorimetry. *Fire & Materials*, 1994; 18:167-172.

33. J.R. Brown and Z. Mathys. Reinforcement and matrix effects on the combustion properties of glass reinforced polymer composites. *Composites*, 1997; 28A:675-681.

34. S.B. Sastri, J.P. Armistead, T.M. Keller and U. Sorathia. Flammability characteristics of phthalonitrile composites. In: *Proceedings of the 42th International SAMPE Symposium*, 4-8 May 1997, pp. 1032-1038.

35. J.H. Koo, B. Muskopf, S. Venumbaka, R. Van Dine, B. Spencer and U. Sorathia. Flammability properties of polymer composites for marine applications. In: *Proceedings of the 32th International SAMPE Technical Conference*, 5-9 November 2000, Paper No. 136.

36. E. Braun and B.C. Levin. Polyesters: a review of the literature on products of combustion and toxicity. *Fire & Materials*, 1986; 10:107-123.

37. B.K. Kandola and A.R. Horrocks. Composites. In: *Fire Retardant Materials*, ed. A.R. Horrocks and D. Price, Cambridge: Woodhead Publishing Ltd, 2002, pp. 182-203.

38.  T.H. Dailey and J. Shuff. Phenolic resins enhance public safety by reducing smoke, fire and toxicity in composites. In: *Proceedings of the 46th Annual Conference of the Composites Institute*, 18-21 February 1991.

39.  D. Purser. Toxicity of fire retardants in relation to life safety and environmental hazards. In: *Fire Retardant Materials*, ed. A.R. Horrocks and D. Price, Cambridge: Woodhead Publishing Ltd., 2003, pp. 69-127.

40.  R.M. Morchat and J.A. Hiltz. Fire-safe composites for marine applications. In: *Proceedings of the 24th International SAMPE Technical Conference*, 20-22 October 1992, pp. T153-T163.

41.  J.R. Brown and D.K.C. Hodgeman. An e.s.r. study of the thermal degradation of Kevlar 49 aramid. *Polymer*, 1982; 23:365-368.

42.  S. Bourbigot, X. Flambard and F. Poutch. Study of the thermal degradation of high performance fibres – application to polybenzazole and p-aramid fibres. *Polymer Degradation & Stability*, 2001; 74:283-290.

43.  S. Bourbigot, X. Flambard, F. Poutch and S. Duquesne. Cone calorimeter study of high performance fibres – application to polybenzazole and p-aramid fibres. *Polymer Degradation & Stability*, 2001; 74:481-486.

44.  J.A. Coneas, A. Marcilla, R. Font and J.A. Caballero. Thermogravimetric studies on the thermal decomposition of polyethylene. *Journal of Analytical & Applied Pyrolysis*, 1996; 36:1-15.

45.  H. Bockburn, A. Hornung, U. Hornung and D. Schawaller. Kinetic study on the thermal degradation of polypropylene and polyethylene. *Journal of Analytical & Applied Pyrolysis*, 1999; 48:93-109.

46.  U. Sorathia, H. Telegadas and M. Bergen. Mechanical and flammability characteristics of phenolic composites for naval applications. In: *Proceedings of the 39th International SAMPE Symposium*, 11-14 April, 1994, pp. 2991-3002.

47.  D.E. Casliostro. Combustion toxicology of epoxy/carbon fiber composites. NASA Center for Aerospace Information, 1981.

48.  V. Babrauskas, B.C. Levin and R.G. Gann. A new approach to fire toxicity data for hazard evaluation', *ASTM Standardisation News*. 1986; 14:28-33.

49.  B.C. Levin, M. Paabo, J. Gurman, S.E. Harris and C.S. Bailey. Toxicity effects of the interactions of fire gases and their use in a hazard assessment computer model. *The Toxicologist*, 1987; 5:127.

50.  B.C. Levin. A new approach for predicting the toxic potency of complex combustion mixtures. *Proceedings of the American Chemical Society Division of Polymeric Materials, Science & Engineering*, 1994; 71:173-174.

51.  G.E. Hartzell, A.F. Grand and W.G. Switzer. Modeling of toxicological effects of fire gases: VII. Studies on evaluation of animal models in combustion toxicity. *Journal of Fire Sciences*, 1988; 6:411-431.

52.  V.L. Bell. Potential release of fibrers from burning carbon composites. NASA Report N80-29431, 1980.

53.  Anon. Carbon/Graphite Composite Materials Study. Third Annual Report, Office of Science & Technology Policy, Washington DC, 1980.

54.  A. Sussholz. Evaluation of micron size carbon fibers released from burning graphite composites. *NASA Report CR-159217*, 1980.

55.  J.F. Seibert. Composite fiber hazard. Air Force Occupational and Environmental Health Laboratory Report, *AFOEHL Report 90-EI00178MGA*, 1990.

56.  E. Glougherty, J. Gerren, J. Greene, D. Haagensen and R.G. Zalosh. Graphite fiber emissions from burning composite helicopter components. United States Coast Guard, 1997.

57.  S. Mahar. Particulate exposures from the investigation and remediation of a crash site of an aircraft containing carbon composites. *Journal of the American Industrial Hygiene Association*, 1990; 51: 459-465.

58.  T. Hertzberg. Dangers relating to fires in carbon-fibre based composite material. *Fire & Materials*, (in press).

59.  S. Gandhi, R. Lyon and L. Speitel. Potential health hazards from burning aircraft composites. *Journal of Fire Sciences*, 1999; 17:20-41.

60.  H.D. Jones, T.R. Jones, W.H. Lyle. Carbon fiber: results of a survey of process workers and their environment in a factory producing continuous filament. *Annals of Occupational Hygiene*, 1982; 26:861-868.

61.  P.E. Owen, J.R. Glazier, B. Ballantyne and J.J. Clary. Subchronic inhalation toxicology of carbon fibres. *Journal of Occupational Medicine*, 1986; 28:373-376.

62. T.R. Martin, S.W. Meyer and D.R. Luchtel. An evaluation of the toxicity of carbon fiber composites for lung cells *in vitro* and *in vivo*. *Environmental Research*, 1989; 49:246-261.

63. S.A. Thomson. Toxicology of carbon fibers. In: *Proceedings Occupational Health Aspects of Advanced Composite Technology in the Aerospace Industry, Health Effects and Exposure Considerations*, 1989, pp. 164-176.

64. S.A. Thomson, R.J. Hilaski, R. Wright and D. Mattie. Nonrespirability of carbon fibers in rats from repeated inhalation exposure. Chemical Research, Development and Engineering Center, Aberdeen Proving Ground, MD, *Report AD-A228-196/HDT*, 1990.

65. D.B. Warheit, J.F. Hansen, M.C. Carakostas and M.A. Hartsky. Acute inhalation toxicity studies in rats with a respirable-sized experimental carbon fibre: pulmonary biochemical and cellular effects. *American Occupational Hygiene*, 1994; 38:769-776.

66. R.S. Waritz, C.J. Collins, B. Ballantyne and J.J. Clary. Chronic inhalation of 3 μm diameter carbon fibres. *The Toxicologist*, 1990; 19:70-71.

67. Z. Zhang, X. Wang, L. Lin, S. Xing, Y. Wu, Y. Li, L. Wu and B. Gang. The effects of carbon fibre and carbon fibre composite dusts on bronchoalvelar lavage component of rates. *Journal of Occupational Health*, 2001; 43:75-79.

68. P.F. Holt and M. Horne. Dust from carbon fiber. *Environmental Research*, 1978; 17:276-283.

69. D.L. Luchtel. Carbon/graphite toxicology. In: *Fiber Toxicology*, ed. D.B. Warheit, New York: Academy Press, 1993, pp. 493-521.

70. D.J. Caldwell, K.J. Kuhlman and J.A. Roop. Smoke production in advanced composites. In *Fire & Polymers II*, ed. G.L. Nelson, Washington DC: ACS Symposium Series, American Chemical Society, 1995, pp. 366-375.

71. G.S. Whitehead, K.A. Grasman and E.C. Kimmel. Lung function and airway inflammation in rats following exposure to combustion products of carbon-graphite/epoxy composite material: comparison to a rodent model of acute lung function. *Toxicology*, 2003; 183:175-197.

72. L. Chiazze, D.K. Watkins and C. Fryar. A case-control study of malignant and nonmaglignant respiratory disease among employees of a fibreglass manufacturing facility. *British Journal of Industrial Medicine*, 1992; 49:326-331.

73. T.W. Hesterberg, W.C. Miller, E.E. McConnell, J. Chevalier, J.G. Hadley, D.M. Bernstein, P. Thevenaz and R. Anderson. Chronic inhalation toxicity of size-separated glass fibres in Fisher 344 rats. *Fundamental and Applied Toxicity*, 1993; 20:464-476.

74. W. Eastes and J.G. Hadley. Dissolution of fibers inhaled by rats. *Journal of Inhalation Toxicology*, 1995; 7:179-196.

75. P. Bofetta, R. Saracci, A. Andersen, P.A. Bertazzi, J. Chang-Claude, J. Cherrie, G. Ferro, R. Frentzel-Beyme, J. Hansen, J. Olsen, N. Plato, L. Teppo, P. Westerholm, P.D. Winter and C. Zocchetti. Cancer mortality among man-made vitreous fiber production workers. *Epidemiology*, 1997; 8:259-268.

76. A. Searl. A comparative study of the clearance of respirable para-aramid, chrysotile and glass fibres from rat lungs. *The Annals of Occupational Hygiene*, 1997; 41:217-233.

77. W.E. Fayerweather, J.R. Bender, J.G. Hadley and W. Eastes. Quantitative risk assessment for a glass fiber insulation product. *Regulatory Toxicology & Pharmacology*, 1997; 25:103-120.

78. European Glass Fibre Producers Association. Continuous filament glass fibre and human health. 30 March 2002.

79. D.B. Warheit. Contemporary issues in fiber toxicology. *Fundamental Applied Toxicology*, 1995; 25: 171-183.

80. S.M. Mattson. Glass fibers in simulated lung fluid: dissolution behaviour and analytical requirements. *Annals of Occupational Hygiene*, 1994; 38:857-877.

81. R. Risquez-Iribarren. Carbon black. In: *Encyclopedia of Occupational Health Safety*, ILO, Geneva, 1995, pp. 390.

82. J.M. Olson. Mishap risk control for advanced aerospace/composite materials. *Airforce Systems Command, Advanced Composite Program Office, Report AJ554083*, McClellan AFB, CA, 1994.

83. J.W.T. Andrews. Post crash management: the Royal Air Force approach. In: *Proceedings of Aircraft Fire Safety, Advisory Group for Aerospace Research and Development, AGARD-CP-587*, Germany, 1997.

# Subject Index

# Mechanics

# Mechanics

# Mechanics

***SOLID* MECHANICS AND ITS APPLICATIONS**

*Series Editor*: G.M.L. Gladwell


49.  J.R. Willis (ed.): *IUTAM Symposium on Nonlinear Analysis of Fracture.* Proceedings of the IUTAM Symposium held in Cambridge, U.K. 1997                ISBN 0-7923-4378-6
50.  A. Preumont: *Vibration Control of Active Structures.* An Introduction. 1997
     ISBN 0-7923-4392-1
51.  G.P. Cherepanov: *Methods of Fracture Mechanics: Solid Matter Physics.* 1997
     ISBN 0-7923-4408-1
52.  D.H. van Campen (ed.): *IUTAM Symposium on Interaction between Dynamics and Control in Advanced Mechanical Systems.* Proceedings of the IUTAM Symposium held in Eindhoven, The Netherlands. 1997                ISBN 0-7923-4429-4
53.  N.A. Fleck and A.C.F. Cocks (eds.): *IUTAM Symposium on Mechanics of Granular and Porous Materials.* Proceedings of the IUTAM Symposium held in Cambridge, U.K. 1997
     ISBN 0-7923-4553-3
54.  J. Roorda and N.K. Srivastava (eds.): *Trends in Structural Mechanics.* Theory, Practice, Education. 1997                ISBN 0-7923-4603-3
55.  Yu.A. Mitropolskii and N. Van Dao: *Applied Asymptotic Methods in Nonlinear Oscillations.* 1997                ISBN 0-7923-4605-X
56.  C. Guedes Soares (ed.): *Probabilistic Methods for Structural Design.* 1997
     ISBN 0-7923-4670-X
57.  D. François, A. Pineau and A. Zaoui: *Mechanical Behaviour of Materials.* Volume I: Elasticity and Plasticity. 1998                ISBN 0-7923-4894-X
58.  D. François, A. Pineau and A. Zaoui: *Mechanical Behaviour of Materials.* Volume II: Viscoplasticity, Damage, Fracture and Contact Mechanics. 1998                ISBN 0-7923-4895-8
59.  L.T. Tenek and J. Argyris: *Finite Element Analysis for Composite Structures.* 1998
     ISBN 0-7923-4899-0
60.  Y.A. Bahei-El-Din and G.J. Dvorak (eds.): *IUTAM Symposium on Transformation Problems in Composite and Active Materials.* Proceedings of the IUTAM Symposium held in Cairo, Egypt. 1998                ISBN 0-7923-5122-3
61.  I.G. Goryacheva: *Contact Mechanics in Tribology.* 1998                ISBN 0-7923-5257-2
62.  O.T. Bruhns and E. Stein (eds.): *IUTAM Symposium on Micro- and Macrostructural Aspects of Thermoplasticity.* Proceedings of the IUTAM Symposium held in Bochum, Germany. 1999
     ISBN 0-7923-5265-3
63.  F.C. Moon: *IUTAM Symposium on New Applications of Nonlinear and Chaotic Dynamics in Mechanics.* Proceedings of the IUTAM Symposium held in Ithaca, NY, USA. 1998
     ISBN 0-7923-5276-9
64.  R. Wang: *IUTAM Symposium on Rheology of Bodies with Defects.* Proceedings of the IUTAM Symposium held in Beijing, China. 1999                ISBN 0-7923-5297-1
65.  Yu.I. Dimitrienko: *Thermomechanics of Composites under High Temperatures.* 1999
     ISBN 0-7923-4899-0
66.  P. Argoul, M. Frémond and Q.S. Nguyen (eds.): *IUTAM Symposium on Variations of Domains and Free-Boundary Problems in Solid Mechanics.* Proceedings of the IUTAM Symposium held in Paris, France. 1999                ISBN 0-7923-5450-8
67.  F.J. Fahy and W.G. Price (eds.): *IUTAM Symposium on Statistical Energy Analysis.* Proceedings of the IUTAM Symposium held in Southampton, U.K. 1999                ISBN 0-7923-5457-5
68.  H.A. Mang and F.G. Rammerstorfer (eds.): *IUTAM Symposium on Discretization Methods in Structural Mechanics.* Proceedings of the IUTAM Symposium held in Vienna, Austria. 1999
     ISBN 0-7923-5591-1

# Mechanics

***SOLID* MECHANICS AND ITS APPLICATIONS**
*Series Editor*: G.M.L. Gladwell

69.  P. Pedersen and M.P. Bendsøe (eds.): *IUTAM Symposium on Synthesis in Bio Solid Mechanics.* Proceedings of the IUTAM Symposium held in Copenhagen, Denmark. 1999
ISBN 0-7923-5615-2

70.  S.K. Agrawal and B.C. Fabien: *Optimization of Dynamic Systems.* 1999
ISBN 0-7923-5681-0

71.  A. Carpinteri: *Nonlinear Crack Models for Nonmetallic Materials.* 1999
ISBN 0-7923-5750-7

72.  F. Pfeifer (ed.): *IUTAM Symposium on Unilateral Multibody Contacts.* Proceedings of the IUTAM Symposium held in Munich, Germany. 1999     ISBN 0-7923-6030-3

73.  E. Lavendelis and M. Zakrzhevsky (eds.): *IUTAM/IFToMM Symposium on Synthesis of Nonlinear Dynamical Systems.* Proceedings of the IUTAM/IFToMM Symposium held in Riga, Latvia. 2000     ISBN 0-7923-6106-7

74.  J.-P. Merlet: *Parallel Robots.* 2000     ISBN 0-7923-6308-6

75.  J.T. Pindera: *Techniques of Tomographic Isodyne Stress Analysis.* 2000  ISBN 0-7923-6388-4

76.  G.A. Maugin, R. Drouot and F. Sidoroff (eds.): *Continuum Thermomechanics.* The Art and Science of Modelling Material Behaviour. 2000     ISBN 0-7923-6407-4

77.  N. Van Dao and E.J. Kreuzer (eds.): *IUTAM Symposium on Recent Developments in Non-linear Oscillations of Mechanical Systems.* 2000     ISBN 0-7923-6470-8

78.  S.D. Akbarov and A.N. Guz: *Mechanics of Curved Composites.* 2000   ISBN 0-7923-6477-5

79.  M.B. Rubin: *Cosserat Theories: Shells, Rods and Points.* 2000     ISBN 0-7923-6489-9

80.  S. Pellegrino and S.D. Guest (eds.): *IUTAM-IASS Symposium on Deployable Structures: Theory and Applications.* Proceedings of the IUTAM-IASS Symposium held in Cambridge, U.K., 6–9 September 1998. 2000     ISBN 0-7923-6516-X

81.  A.D. Rosato and D.L. Blackmore (eds.): *IUTAM Symposium on Segregation in Granular Flows.* Proceedings of the IUTAM Symposium held in Cape May, NJ, U.S.A., June 5–10, 1999. 2000     ISBN 0-7923-6547-X

82.  A. Lagarde (ed.): *IUTAM Symposium on Advanced Optical Methods and Applications in Solid Mechanics.* Proceedings of the IUTAM Symposium held in Futuroscope, Poitiers, France, August 31–September 4, 1998. 2000     ISBN 0-7923-6604-2

83.  D. Weichert and G. Maier (eds.): *Inelastic Analysis of Structures under Variable Loads.* Theory and Engineering Applications. 2000     ISBN 0-7923-6645-X

84.  T.-J. Chuang and J.W. Rudnicki (eds.): *Multiscale Deformation and Fracture in Materials and Structures.* The James R. Rice 60th Anniversary Volume. 2001     ISBN 0-7923-6718-9

85.  S. Narayanan and R.N. Iyengar (eds.): *IUTAM Symposium on Nonlinearity and Stochastic Structural Dynamics.* Proceedings of the IUTAM Symposium held in Madras, Chennai, India, 4–8 January 1999     ISBN 0-7923-6733-2

86.  S. Murakami and N. Ohno (eds.): *IUTAM Symposium on Creep in Structures.* Proceedings of the IUTAM Symposium held in Nagoya, Japan, 3-7 April 2000. 2001    ISBN 0-7923-6737-5

87.  W. Ehlers (ed.): *IUTAM Symposium on Theoretical and Numerical Methods in Continuum Mechanics of Porous Materials.* Proceedings of the IUTAM Symposium held at the University of Stuttgart, Germany, September 5-10, 1999. 2001     ISBN 0-7923-6766-9

88.  D. Durban, D. Givoli and J.G. Simmonds (eds.): *Advances in the Mechanis of Plates and Shells* *The Avinoam Libai Anniversary Volume.* 2001     ISBN 0-7923-6785-5

89.  U. Gabbert and H.-S. Tzou (eds.): *IUTAM Symposium on Smart Structures and Structonic Systems.* Proceedings of the IUTAM Symposium held in Magdeburg, Germany, 26–29 September 2000. 2001     ISBN 0-7923-6968-8

# Mechanics

*SOLID* MECHANICS AND ITS APPLICATIONS

*Series Editor*: G.M.L. Gladwell

90. Y. Ivanov, V. Cheshkov and M. Natova: *Polymer Composite Materials – Interface Phenomena & Processes.* 2001    ISBN 0-7923-7008-2

91. R.C. McPhedran, L.C. Botten and N.A. Nicorovici (eds.): *IUTAM Symposium on Mechanical and Electromagnetic Waves in Structured Media.* Proceedings of the IUTAM Symposium held in Sydney, NSW, Australia, 18-22 Januari 1999. 2001    ISBN 0-7923-7038-4

92. D.A. Sotiropoulos (ed.): *IUTAM Symposium on Mechanical Waves for Composite Structures Characterization.* Proceedings of the IUTAM Symposium held in Chania, Crete, Greece, June 14-17, 2000. 2001    ISBN 0-7923-7164-X

93. V.M. Alexandrov and D.A. Pozharskii: *Three-Dimensional Contact Problems.* 2001    ISBN 0-7923-7165-8

94. J.P. Dempsey and H.H. Shen (eds.): *IUTAM Symposium on Scaling Laws in Ice Mechanics and Ice Dynamics.* Proceedings of the IUTAM Symposium held in Fairbanks, Alaska, U.S.A., 13-16 June 2000. 2001    ISBN 1-4020-0171-1

95. U. Kirsch: *Design-Oriented Analysis of Structures.* A Unified Approach. 2002    ISBN 1-4020-0443-5

96. A. Preumont: *Vibration Control of Active Structures.* An Introduction ($2^{nd}$ Edition). 2002    ISBN 1-4020-0496-6

97. B.L. Karihaloo (ed.): *IUTAM Symposium on Analytical and Computational Fracture Mechanics of Non-Homogeneous Materials.* Proceedings of the IUTAM Symposium held in Cardiff, U.K., 18-22 June 2001. 2002    ISBN 1-4020-0510-5

98. S.M. Han and H. Benaroya: *Nonlinear and Stochastic Dynamics of Compliant Offshore Structures.* 2002    ISBN 1-4020-0573-3

99. A.M. Linkov: *Boundary Integral Equations in Elasticity Theory.* 2002    ISBN 1-4020-0574-1

100. L.P. Lebedev, I.I. Vorovich and G.M.L. Gladwell: *Functional Analysis.* Applications in Mechanics and Inverse Problems ($2^{nd}$ Edition). 2002    ISBN 1-4020-0667-5; Pb: 1-4020-0756-6

101. Q.P. Sun (ed.): *IUTAM Symposium on Mechanics of Martensitic Phase Transformation in Solids.* Proceedings of the IUTAM Symposium held in Hong Kong, China, 11-15 June 2001. 2002    ISBN 1-4020-0741-8

102. M.L. Munjal (ed.): *IUTAM Symposium on Designing for Quietness.* Proceedings of the IUTAM Symposium held in Bangkok, India, 12-14 December 2000. 2002    ISBN 1-4020-0765-5

103. J.A.C. Martins and M.D.P. Monteiro Marques (eds.): *Contact Mechanics.* Proceedings of the $3^{rd}$ Contact Mechanics International Symposium, Praia da Consolação, Peniche, Portugal, 17-21 June 2001. 2002    ISBN 1-4020-0811-2

104. H.R. Drew and S. Pellegrino (eds.): *New Approaches to Structural Mechanics, Shells and Biological Structures.* 2002    ISBN 1-4020-0862-7

105. J.R. Vinson and R.L. Sierakowski: *The Behavior of Structures Composed of Composite Materials.* Second Edition. 2002    ISBN 1-4020-0904-6

106. Not yet published.

107. J.R. Barber: *Elasticity.* Second Edition. 2002    ISBN Hb 1-4020-0964-X; Pb 1-4020-0966-6

108. C. Miehe (ed.): *IUTAM Symposium on Computational Mechanics of Solid Materials at Large Strains.* Proceedings of the IUTAM Symposium held in Stuttgart, Germany, 20-24 August 2001. 2003    ISBN 1-4020-1170-9

# Mechanics

*SOLID* **MECHANICS AND ITS APPLICATIONS**
*Series Editor*: G.M.L. Gladwell

109.  P. Ståhle and K.G. Sundin (eds.): *IUTAM Symposium on Field Analyses for Determination of Material Parameters – Experimental and Numerical Aspects.* Proceedings of the IUTAM Symposium held in Abisko National Park, Kiruna, Sweden, July 31 – August 4, 2000. 2003
ISBN 1-4020-1283-7

110.  N. Sri Namachchivaya and Y.K. Lin (eds.): *IUTAM Symposium on Nonlinear Stochastic Dynamics.* Proceedings of the IUTAM Symposium held in Monticello, IL, USA, 26 – 30 August, 2000. 2003
ISBN 1-4020-1471-6

111.  H. Sobieckzky (ed.): *IUTAM Symposium Transsonicum IV.* Proceedings of the IUTAM Symposium held in Göttingen, Germany, 2–6 September 2002, 2003
ISBN 1-4020-1608-5

112.  J.-C. Samin and P. Fisette: *Symbolic Modeling of Multibody Systems.* 2003
ISBN 1-4020-1629-8

113.  A.B. Movchan (ed.): *IUTAM Symposium on Asymptotics, Singularities and Homogenisation in Problems of Mechanics.* Proceedings of the IUTAM Symposium held in Liverpool, United Kingdom, 8-11 July 2002. 2003
ISBN 1-4020-1780-4

114.  S. Ahzi, M. Cherkaoui, M.A. Khaleel, H.M. Zbib, M.A. Zikry and B. LaMatina (eds.): *IUTAM Symposium on Multiscale Modeling and Characterization of Elastic-Inelastic Behavior of Engineering Materials.* Proceedings of the IUTAM Symposium held in Marrakech, Morocco, 20-25 October 2002. 2004
ISBN 1-4020-1861-4

115.  H. Kitagawa and Y. Shibutani (eds.): *IUTAM Symposium on Mesoscopic Dynamics of Fracture Process and Materials Strength.* Proceedings of the IUTAM Symposium held in Osaka, Japan, 6-11 July 2003. Volume in celebration of Professor Kitagawa's retirement. 2004
ISBN 1-4020-2037-6

116.  E.H. Dowell, R.L. Clark, D. Cox, H.C. Curtiss, Jr., K.C. Hall, D.A. Peters, R.H. Scanlan, E. Simiu, F. Sisto and D. Tang: *A Modern Course in Aeroelasticity.* 4th Edition, 2004
ISBN 1-4020-2039-2

117.  T. Burczyński and A. Osyczka (eds.): *IUTAM Symposium on Evolutionary Methods in Mechanics.* Proceedings of the IUTAM Symposium held in Cracow, Poland, 24-27 September 2002. 2004
ISBN 1-4020-2266-2

118.  D. Ieşan: *Thermoelastic Models of Continua.* 2004        ISBN 1-4020-2309-X

119.  G.M.L. Gladwell: *Inverse Problems in Vibration.* Second Edition. 2004 ISBN 1-4020-2670-6

120.  J.R. Vinson: *Plate and Panel Structures of Isotropic, Composite and Piezoelectric Materials, Including Sandwich Construction.* 2005        ISBN 1-4020-3110-6

121.  *Forthcoming*

122.  G. Rega and F. Vestroni (eds.): *IUTAM Symposium on Chaotic Dynamics and Control of Systems and Processes in Mechanics.* Proceedings of the IUTAM Symposium held in Rome, Italy, 8–13 June 2003. 2005        ISBN 1-4020-3267-6

123.  E.E. Gdoutos: *Fracture Mechanics. An Introduction.* 2nd edition. 2005  ISBN 1-4020-3267-6

124.  M.D. Gilchrist (ed.): *IUTAM Symposium on Impact Biomechanics from Fundamental Insights to Applications.* 2005        ISBN 1-4020-3795-3

125.  J.M. Huyghe, P.A.C. Raats and S. C. Cowin (eds.): *IUTAM Symposium on Physicochemical and Electromechanical Interactions in Porous Media.* 2005        ISBN 1-4020-3864-X

126.  H. Ding, W. Chen and L. Zhang: *Elasticity of Transversely Isotropic Materials.* 2005
ISBN 1-4020-4033-4

127.  W. Yang (ed): *IUTAM Symposium on Mechanics and Reliability of Actuating Materials.* Proceedings of the IUTAM Symposium held in Beijing, China, 1–3 September 2004. 2005
ISBN 1-4020-4131-6

# Mechanics


***SOLID* MECHANICS AND ITS APPLICATIONS**

*Series Editor*: G.M.L. Gladwell

128. J.-P. Merlet: *Parallel Robots.* 2006  ISBN 1-4020-4132-2

129. G.E.A. Meier and K.R. Sreenivasan (eds.): *IUTAM Symposium on One Hundred Years of Boundary Layer Research.* Proceedings of the IUTAM Symposium held at DLR-Göttingen, Germany, August 12–14, 2004. 2006  ISBN 1-4020-4149-7

130. H. Ulbrich and W. Günthner (eds.): *IUTAM Symposium on Vibration Control of Nonlinear Mechanisms and Structures.* 2006  ISBN 1-4020-4160-8

131. L. Librescu and O. Song: *Thin-Walled Composite Beams.* Theory and Application. 2006  ISBN 1-4020-3457-1

132. G. Ben-Dor, A. Dubinsky and T. Elperin: *Applied High-Speed Plate Penetration Dynamics.* 2006  ISBN 1-4020-3452-0

133. X. Markenscoff and A. Gupta (eds.): *Collected Works of J. D. Eshelby.* Mechanics of Defects and Inhomogeneities. 2006  ISBN 1-4020-4416-X

134. R.W. Snidle and H.P. Evans (eds.): *IUTAM Symposium on Elastohydrodynamics and Microelastohydrodynamics.* Proceedings of the IUTAM Symposium held in Cardiff, UK, 1–3 September, 2004. 2006  ISBN 1-4020-4532-8

135. T. Sadowski (ed.): *IUTAM Symposium on Multiscale Modelling of Damage and Fracture Processes in Composite Materials.* Proceedings of the IUTAM Symposium held in Kazimierz Dolny, Poland, 23–27 May 2005. 2006  ISBN 1-4020-4565-4

136. A. Preumont: *Mechatronics.* Dynamics of Electromechanical and Piezoelectric Systems. 2006  ISBN 1-4020-4695-2

137. M.P. Bendsøe, N. Olhoff and O. Sigmund (eds.): *IUTAM Symposium on Topological Design Optimization of Structures, Machines and Materials.* Status and Perspectives. 2006  ISBN 1-4020-4729-0

138. A. Klarbring: *Models of Mechanics.* 2006  ISBN 1-4020-4834-3

139. H.D. Bui: *Fracture Mechanics.* Inverse Problems and Solutions. 2006  ISBN 1-4020-4836-X

140. M. Pandey, W.-C. Xie and L. Xu (eds.): *Advances in Engineering Structures, Mechanics and Construction.* Proceedings of an International Conference on Advances in Engineering Structures, Mechanics & Construction, held in Waterloo, Ontario, Canada, May 14–17, 2006. 2006  ISBN 1-4020-4890-4

141. G.Q. Zhang, W.D. van Driel and X. J. Fan: *Mechanics of Microelectronics.* 2006  ISBN 1-4020-4934-X

142. Q.P. Sun and P. Tong (eds.): *IUTAM Symposium on Size Effects on Material and Structural Behavior at Micron- and Nano-Scales.* Proceedings of the IUTAM Symposium held in Hong Kong, China, 31 May–4 June, 2004. 2006  ISBN 1-4020-4945-5

143. A.P. Mouritz and A.G. Gibson: *Fire Properties of Polymer Composite Materials.* 2006  ISBN 1-4020-5355-X